LONDON MATHEMATICAL SOCIETY LECTURE NOTE SERIES   T3-BOR-577

Managing Editor: Professor N.J. Hitchin, Mathematical Institute,
University of Oxford, 24–29 St Giles, Oxford OX1 3LB, United Kingdom

The titles below are available from booksellers, or, in case of difficulty, from Cambridge University Press.

London Mathematical Society Lecture Note Series. 277

Descriptive Set Theory and Dynamical Systems

Edited by

M. Foreman
University of California, Irvine

A. S. Kechris
California Institute of Technology

A. Louveau
CNRS, Université de Paris VI

B. Weiss
Hebrew University of Jerusalem

CAMBRIDGE
UNIVERSITY PRESS

PUBLISHED BY THE PRESS SYNDICATE OF THE UNIVERSITY OF CAMBRIDGE
The Pitt Building, Trumpington Street, Cambridge, United Kingdom

CAMBRIDGE UNIVERSITY PRESS
The Edinburgh Building, Cambridge, CB2 2RU, UK
40 West 20th Street, New York, NY 10011–4211, USA
10 Stamford Road, Oakleigh, Melbourne 3166, Australia
Ruiz de Alarcón 13, 28014 Madrid, Spain

http://www.cup.cam.ac.uk
http://www.cup.org

First published 2000

Printed in the United Kingdom at the University Press, Cambridge

A catalogue record for this book is available from the British Library

ISBN   0 521 78644 4 paperback

Contents

Contents

Preface

During July 1-5, 1996 an International Workshop on Descriptive Set Theory and Dynamical Systems took place at the Centre International de Rencontres Mathématiques (CIRM) of Marseille-Luminy, France. The aim of this meeting was to bring together mathematicians interested in these two areas and their interrelationships. The idea came out of this, very successful, conference to put together a collection of survey papers related to these fields, and their growing interactions, that would be accessible to a wide audience of students and researchers. The present volume is the realization of this idea.

The contributions provide introductions to a broad spectrum of topics and are meant to guide the reader to the most recent developments in many active areas of research. Here is a bird's eye view of the subjects covered here:

1) J. Aaronson: ergodic theory of non-singular transformations with particular emphasis on transformations admitting σ-finite infinite invariant measures.

2) V. Bergelson: recurrence theorems in ergodic theory and their applications in combinatorics.

3) S. Bezuglyi: structure of cocycles for non-singular group actions.

4) M. Foreman: descriptive aspects of ergodic theory.

5) E. Glasner: topological dynamics with particular emphasis on the structure theory of minimal dynamical systems and its applications.

6) V. Ya. Golodets, V. M. Kulagin, and S.D. Sinel'shchikov: cocycles in generic topological dynamics.

7) A.S. Kechris: the descriptive theory of Polish group actions.

8) A. B. Ramsay: aspects of the theory of groupoids.

9) B. Weiss: survey of topics in generic dynamics.

We would like to thank the CNRS, the Société Mathématique de France, and the Equipe d' Analyse, Université Paris VI, for their financial support of the workshop, and the Mathematics Department at Caltech for its help in putting together this volume.

November 1999   M. Foreman, A.S. Kechris, A. Louveau, and B. Weiss

An Overview of Infinite Ergodic Theory

Jon Aaronson
School of Mathematical Sciences
Tel Aviv University
Ramat Aviv, 69978 Tel Aviv, Israel

Abstract. We review the basic ergodic theory of non-singular transformations placing special emphasis on those transformations admitting σ-finite, infinite invariant measures. The topics to be discussed include invariant measures, recurrence, ergodic theorems, pointwise dual ergodicity, distributional limits, structure and intrinsic normalizing constants.

0   Introduction

Infinite ergodic theory is the study of measure preserving transformations of infinite measure spaces. It is part of the more general study of non-singular transformations (since a measure preserving transformation is also a non-singular transformation).

This paper is an attempt at an introductory overview of the subject, and is necessarily incomplete. More information on most topics discussed here can be found in [1]. Other references are also given in the text.

Before discussing the special properties of infinite measure preserving transformations, we need to review some basic non-singular ergodic theory first.

Let $(X, \mathcal{B}, m)$ be a standard σ-finite measure space. A *non-singular transformation* of X is only defined modulo nullsets, and is a map $T : X_0 \to X_0$ (where $X_0 \subset X$ has full measure), which is measurable and has the *non-singularity property* that for $A \in \mathcal{B}$, $m(T^{-1}A) = 0$ if and only if $m(A) = 0$. A *measure preserving transformation* of X is a non-singular transformation T with the additional property that $m(T^{-1}A) = m(A) \ \forall \ A \in \mathcal{B}$.

If T is a non-singular transformation of a σ-finite measure space $(X, \mathcal{B}, m)$, and p is another measure on $(X, \mathcal{B})$ *equivalent* to m (denoted $p \sim m$ and meaning that p and m have the same nullsets), then T is a non-singular transformation of $(X, \mathcal{B}, p)$.

Thus, a non-singular transformation of a σ-finite measure space is actually a non-singular transformation of a probability space.

Considering a non-singular transformation $(X, \mathcal{B}, m, T)$ of a probability space as a *dynamical system* (see [23], [29]), the measure space X represents the set of *"configurations"* of the system, and T represents the change under *"passage of time"*. The non-singularity of T reflects the assumed property of the system that configuration sets that are impossible sometimes are always impossible. A probability preserving transformation would describe a system in a "steady state", where configuration sets occur with the same likelihood at all times.

1 Invariant Measures

Given a particular non-singular transformation, one of the first tasks is to ascertain whether it could have been obtained by starting with a measure preserving transformation, and then "passing" to some equivalent measure.

If $T : X \to X$ is non-singular then $f \to f \circ T$ defines a linear isometry of $L^\infty(m)$. There is a predual called the *Frobenius-Perron* or *transfer* operator, $\widehat{T} : L^1(m) \to L^1(m)$, which is defined by

$$f \mapsto \nu_f(\cdot) = \int f dm \mapsto \widehat{T}f = \frac{d\nu_f \circ T^{-1}}{dm}$$

and satisfies

$$\int_X \widehat{T}f.gdm = \int_X f.g \circ Tdm \quad f \in L^1(m), \ g \in L^\infty(m).$$

Note that the domain of definition of $\widehat{T}$ can be extended to all non-negative measurable functions. This definition can be made when m is infinite, but σ-finite.

Evidently the density $h \geq 0$ of an absolutely continuous invariant measure μ satisfies $\widehat{T}h = h$, since for any $g \geq 0$ measurable,

$$\int_X \widehat{T}hgdm = \int_X hg \circ Tdm = \int_X g \circ Td\mu = \int_X gd\mu = \int_X hgdm.$$

Clearly, if T is invertible, then

$$\widehat{T}f = \frac{dm \circ T^{-1}}{dm} f \circ T^{-1},$$

and if the non-singular transformation T of X is *locally invertible* in the sense that there are disjoint measurable sets $\{A_j : j \in J\}$ (J finite or countable) such that $m(X \setminus \bigcup_{j \in J} A_j) = 0$, and T is invertible on each A_j, then

$$\widehat{T}f = \sum_{j \in J} 1_{TA_j} \frac{dm \circ v_j}{dm} f \circ v_j$$

where $v_j : TA_j \to A_j$ is measurable and satisfies $T \circ v_j \equiv \mathrm{Id}$.

Boole's transformations I

For some locally invertible non-singular transformations T and measurable functions f, $\widehat{T}f$ can be computed explicitly. For example, consider the transformations $T : \mathbb{R} \to \mathbb{R}$ defined by

$$Tx = \alpha x + \beta + \sum_{k=1}^{n} \frac{p_k}{t_k - x} \tag{1.1}$$

where $n \geq 1$, α, $p_1, \ldots, p_n \geq 0$ (not all zero) and $\beta, t_1, \ldots, t_n \in \mathbb{R}$.

These transformations (called *Boole's transformations*) were considered by G. Boole in [15]. They are non-singular transformations of $\mathbb{R}$ equipped with Lebesgue measure $m_\mathbb{R}$, and for $h : \mathbb{R} \to \mathbb{R}$ a non-negative measurable function,

$$\widehat{T}h(x) = \sum_{y \in \mathbb{R}, \; Ty=x} \frac{h(y)}{T'(y)}.$$

Note that the $1 - 1$ Boole's transformations are the real Möbius transformations.

1.1 Boole's Formula [15] *For T as above, $x \in \mathbb{R}$ and $\omega \in \mathbb{C}$, $T\omega \neq x$,*

$$\sum_{y \in \mathbb{R}, \; Ty=x} \frac{1}{(y - \omega)T'(y)} = \frac{1}{x - T(\omega)}.$$

If $\omega \in \mathbb{R}^{2+}$, the upper half plane, and $\omega = a + ib$, $a, b \in \mathbb{R}$, $b > 0$ then

$$\text{Im} \frac{1}{x - \omega} = \frac{b}{(x - a)^2 + b^2} = \pi \varphi_\omega(x)$$

where φ_ω is the well known *Cauchy density* and Boole's formula has the immediate corollary:

$$\widehat{T}\varphi_\omega = \varphi_{T(\omega)}, \text{ or } P_\omega \circ T^{-1} = P_{T\omega} \tag{1.2}$$

where $dP_\omega := \varphi_\omega dm_\mathbb{R}$.

The original proof of Boole's formula in [15] uses the

1.2 Proposition [15] *Suppose that $F : \mathbb{C} \to \overline{\mathbb{C}}$ is rational, and $E : \mathbb{C} \to \mathbb{C}$ is a polynomial. Then*

$$\sum_{x \in \mathbb{C}: \; E(x)=0} F(x) = - \sum_{a \; a \; pole \; of \; F} \text{Res}(F(\log E)'; a) + \text{Res}(F(\log E)'; 0).$$

Modern proofs of (1.2) use the fact that Boole's transformations are $\mathbb{R}$-restrictions of analytic endomorphisms of $\mathbb{R}^{2+}$ (*inner functions*). Many of the results given here for Boole's transformations remain valid for arbitrary inner functions ([4], [38]).

Bounded analytic functions on $\mathbb{R}^{2+}$ are Cauchy integrals of their restrictions to $\mathbb{R}$ and writing $dP_\omega := \varphi_\omega dm_\mathbb{R}$, one sees ([36]) that for $t > 0$,

$$\widehat{P_\omega \circ T^{-1}}(t) = \int_\mathbb{R} e^{itT(x)} dP_\omega(x) = e^{itT(\omega)} = \widehat{P_{T(\omega)}}(t),$$

whence (1.2).

As a consequence of (1.2), we see that the Boole transformation T has an absolutely continuous invariant probability if $\exists\ \omega \in \mathbb{R}^{2+}$ with $T(\omega) = \omega$ (in which case P_ω is T-invariant). It turns out that this is the only way it can admit an absolutely continuous invariant probability ([38]).

If $\alpha > 0$ in (1.1), then ([36]) using the fact that $\pi b\varphi_{a+ib} \to 1$ as $b \to \infty$, $\frac{a}{b} \to 0$ we see that $m_\mathbb{R} \circ T^{-1} = \alpha^{-1}m_\mathbb{R}$, whence T preserves $m_\mathbb{R}$ if $\alpha = 1$ in (1.1). By considering other analogous limits, one also sees that:

$Tx = \tan x$ preserves the measure $d\mu_0(x) := \frac{dx}{x^2}$ (see [4]);

and

if $f(\omega) = \int_\mathbb{R} \frac{d\nu(t)}{t-\omega}$ (the complex Hilbert transform of ν) where $\nu \perp m_\mathbb{R}$, and $T(x) := \lim_{y\downarrow 0} f(x + iy)$ a.e. (the Hilbert transform of ν), then $m_\mathbb{R} \circ T^{-1} = \nu(\mathbb{R})\mu_0$ (see [22]).

CONDITIONS FOR EXISTENCE OF INVARIANT MEASURES

Unfortunately, one cannot expect always to be able to identify absolutely continuous invariant measures by explicit computation. To help remedy this situation, there are many conditions for existence of such (see [35]). Some conditions for existence of absolutely continuous invariant probabilities depend on the following:

1.3 Proposition *Let T be a non-singular transformation of $(X, \mathcal{B}, m)$. If $\exists\ f \in L^1(m)$, $f \geq 0$, $\int_X f dm > 0$ such that $\{\frac{1}{n}\sum_{k=1}^n \hat{T}^k f : n \geq 1\}$ is a uniformly integrable family, then $\exists$ a T-invariant probability $P << m$.*

The invariant probability's density $h \in L^1(m)$, $h \geq 0$ is found as a weak limit point of $\{\frac{1}{n}\sum_{k=1}^n \hat{T}^k f : n \geq 1\}$ which is weakly sequentially precompact in $L^1(m)$ owing to the assumed uniform integrability.

Expanding interval maps I

Let $I = [0, 1]$, m_I be Lebesgue measure on I, and α be a collection of disjoint open subintervals of I such that

$$m(I \setminus U_\alpha) = 0 \text{ where } U_\alpha = \bigcup_{a\in\alpha} a.$$

A piecewise onto, C^2 *interval map* with *basic partition* α is a map $T : I \to I$ is such that

For each $a \in \alpha$, $T|_a$ extends to a C^2 diffeomorphism $T : \bar{a} \to I$.   (1.3)

Note that if $a_1, ..., a_n \in \alpha$ then $a = \bigcap_{k=1}^n T^{-(k-1)}a_k$ is an open interval, and $T^n : \bar{a} \to I$ is a C^2 diffeomorphism. Hence, if T is a piecewise onto, C^2 interval map with basic partition α then, for $n \geq 1$, T^n is an interval map with basic partition

$$\alpha_0^{n-1} := \bigvee_{k=0}^{n-1} T^{-k}\alpha = \{\bigcap_{k=1}^{n} T^{-(k-1)}a_k \,:\, a_1, ..., a_n \in \alpha\}.$$

The piecewise onto, C^2 interval map T is called *expanding* if

$$\exists \lambda > 1 \ni |T'x| \geq \lambda \,\forall\, x \in I. \tag{1.4}$$

The following condition limits the multiplicative variation (or *distortion*) of v_a' ($a \in \alpha$). It is known as *Adler's condition* or *bounded distortion*. It follows from (1.3) and (1.4) in case α is finite.

$$\exists M > 1 \ni \frac{|T''x|}{|T'x|^2} \leq M \,\forall\, x \in I.$$

Given a piecewise onto, C^2 interval map T with basic partition α, and $a \in \alpha_0^{n-1}$; denote by v_a the C^2 diffeomorphism $v_a : I \to \overline{a}$ satisfying $T^n \circ v_a(x) = x$.

The basic result concerning piecewise onto, C^2 expanding interval maps satisfying Adler's condition is that Adler's condition holds uniformly for all powers of T ([12]):

$$\frac{|T^{n''}x|}{|T^{n'}x|^2} \leq K = \frac{\lambda M}{\lambda - 1} \,\forall\, x \in I,\ n \geq 1,$$

whence Renyi's distortion property (see [41]):

$$|v_a'(x)| = e^{\pm K} m_I(a) \,\forall\, x \in I, \quad a \in \bigcup_{n=1}^{\infty} \alpha_0^{n-1}. \tag{1.5}$$

As a consequence of Renyi's distortion property, we have that

$$\widehat{T}^n 1 = \sum_{a \in \alpha_0^{n-1}} |v_a'| = e^{\pm K},$$

and so by proposition 1.3 (a uniformly bounded sequence being uniformly integrable) $\exists$ an absolutely continuous, invariant probability with density h satisfying $h = \widehat{T}h = e^{\pm K}$.

It is shown in [27] that h is Lipschitz continuous. To see this using [21] note that for f Lipschitz continuous (hence differentiable a.e.),

$$(\widehat{T}^n f)' = \sum_{a \in \alpha_0^{n-1}} v_a'' f \circ v_a + \sum_{a \in \alpha_0^{n-1}} v_a'^2 f' \circ v_a$$

whence

$$\|(\widehat{T^n}f)'\|_\infty \le \frac{e^K}{\lambda^n}\|f'\|_\infty + e^{2K}\|f\|_\infty$$

and it follows from [21] that $\exists\, h_0$ Lipschitz continuous and $M > 0$, $r \in (0,1)$ such that $\widehat{T}h_0 = h_0$ and $\|\widehat{T^n}f - h_0\int_I f\,dm\|_L \le Mr^n\|f\|_L\ \forall\, f$ Lipschitz continuous where $\|f\|_L := \|f\|_\infty + \|f'\|_\infty$. In particular, $h = h_0$ mod m.

The assumption that the interval map is expanding is not crucial. Gauss' continued fraction map $Tx = \{\frac{1}{x}\}$ is not expanding, but $(T^2)' \ge 4$ and the above applies.

However the piecewise onto, C^2 interval map $Tx = \{\frac{1}{1-x}\}$ satisfies Adler's condition, and no power is expanding, as $T(0) = 0$, $T'(0) = 1$. In fact T admits no absolutely continuous, invariant probability, the infinite measure $d\nu(x) := \frac{dx}{x}$ being T-invariant.

Conditions for the existence of absolutely continuous, infinite, invariant measures depend on recurrence properties.

2   Recurrence and Conservativity

There are non-singular transformations T of $(X, \mathcal{B}, m)$ which are *recurrent* in the sense that

$$\liminf_{n\to\infty} |h \circ T^n - h| = 0 \text{ a.e. } \forall\, h : X \to \mathbb{R} \text{ measurable.}$$

One extreme form of non-recurrent (or *transient*) behaviour is exhibited by *wandering sets*.

Let T be a non-singular transformation of the standard measure space $(X, \mathcal{B}, m)$.

A set $W \subset X$ is called a *wandering* set (for T) if the sets $\{T^{-n}W\}_{n=0}^\infty$ are disjoint. Let $\mathcal{W} = \mathcal{W}(T)$ denote the collection of measurable wandering sets.

Evidently, the collection of measurable wandering sets is a hereditary collection (any subset of a wandering set is also wandering), and T-invariant (W a wandering set $\implies T^{-1}W$ a wandering set).

Using a standard exhaustion argument it can be shown that $\exists$ a countable union of wandering sets $\mathfrak{D}(T) \in \mathcal{B}$ with the property that any wandering set $W \in \mathcal{B}$ is contained in $\mathfrak{D}(T)$ mod m (i.e. $m(W \setminus \mathfrak{D}(T)) = 0$). Evidently $\mathfrak{D}(T)$ is unique mod m and $T^{-1}\mathfrak{D} \subseteq \mathfrak{D}$ mod m. It is called the *dissipative part* of the non-singular transformation T. In case T is invertible, it can be shown that $\exists$ a wandering set $W \in \mathcal{B}$ such that $\mathfrak{D}(T) = \bigcup_{n\in\mathbb{Z}} T^n W$, whence $T^{-1}\mathfrak{D} = \mathfrak{D}$.

The *conservative part* of T is defined to be $\mathfrak{C}(T) := X \setminus \mathfrak{D}(T)$ and the partition $\{\mathfrak{C}(T), \mathfrak{D}(T)\}$ is called the *Hopf decomposition* of T.

The non-singular transformation T is called *conservative* if $\mathfrak{C}(T) = X$ mod m, and (totally) *dissipative* if $\mathfrak{D}(T) = X$ mod m.

Using

Halmos' recurrence theorem [26] *If T is conservative, then*

$$\sum_{n=1}^{\infty} 1_B \circ T^n = \infty \ a.e. \ on \ B, \ \ \forall B \in \mathcal{B}.$$

one can show that a non-singular transformation is recurrent if and only if it is conservative.

CONDITIONS FOR CONSERVATIVITY

If there exists a finite, T-invariant measure $q << m$, then clearly there can be no wandering sets with positive q-measure, whence $q(\mathfrak{D}) = 0$ and $[\frac{dq}{dm} > 0] \subseteq \mathfrak{C}$ mod m. In particular ([40]) any probability preserving transformation is conservative.

A measure preserving transformation of a σ-finite, infinite measure space is not necessarily conservative. For example $x \mapsto x+1$ is a measure preserving transformation of $\mathbb{R}$ equipped with Borel sets, and Lebesgue measure, which is totally dissipative.

2.1 Proposition

1) If $T : X \to X$ is non-singular, then

$$\mathfrak{C}(T) = [\sum_{n=1}^{\infty} \widehat{T}^k f = \infty] \quad mod \ m, \ \forall \ f \in L^1(m), f > 0.$$

2) If $T : X \to X$ is a measure preserving transformation, then $T^{-1}\mathfrak{C}(T) = \mathfrak{C}(T)$ mod m. Indeed

$$\mathfrak{C}(T) = \left[\sum_{n=1}^{\infty} f \circ T^n = \infty\right] \quad mod \ m, \ \ \forall \ f \in L^1(m), f > 0 \ a.e..$$

Boole's transformations II

If

$$Tx = \alpha x + \beta + \sum_{k=1}^{n} \frac{p_k}{t_k - x}$$

where $n \geq 1$, $\alpha > 0$, $p_1, \ldots, p_n \geq 0$ and β, $t_1, \ldots, t_n \in \mathbb{R}$ then (as deduced from Boole's formula) $m_{\mathbb{R}} \circ T^{-1} = \alpha^{-1} m_{\mathbb{R}}$.

When $\alpha > 1$,

$$\sum_{n=1}^{\infty} \widehat{T}^n f < \infty \ a.e. \ \forall \ f \in L^1(m) \cap L^{\infty}(m), \ f > 0 \ a.e.,$$

whence by proposition 2.1(1), T is totally dissipative.

Now let

$$Tx = x + \beta + \sum_{k=1}^{n} \frac{p_k}{t_k - x} \qquad (2.1)$$

where $n \geq 1$, $p_1, \ldots, p_n \geq 0$ (not all zero) and β, $t_1, \ldots, t_n \in \mathbb{R}$. As above, $m_{\mathbb{R}} \circ T^{-1} = m_{\mathbb{R}}$.

To see when T is conservative, we use proposition 2.1. For T defined by (2.1) and $\omega \in \mathbb{R}^{2+}$, $T^n(\omega) \to \infty$. Write $T^n(\omega) := u_n + iv_n \to \infty$, whence $\pi \widehat{T}^n \varphi_\omega(x) = \text{Im} \frac{1}{x - T^n(\omega)} = \frac{v_n}{(x-u_n)^2 + v_n^2} \sim \frac{v_n}{u_n^2 + v_n^2}$ and T is conservative iff

$$\sum_{n=1}^{\infty} \frac{v_n}{u_n^2 + v_n^2} = \infty.$$

We see (as in [4], [5]) that when $\beta \neq 0$,

$$v_n \uparrow v_\infty < \infty, \quad u_n = \beta n - \frac{\nu}{\beta} \log n + O(1) \text{ as } n \to \infty; \qquad (2.2)$$

and when $\beta = 0$,

$$\sup_{n \geq 1} |u_n| < \infty, \quad v_n \sim \sqrt{2\nu n} \text{ as } n \to \infty \qquad (2.3)$$

where $\nu := \sum_{k=1}^{n} p_k$.

It follows that T is conservative when $\beta = 0$ ($\sum_{n=1}^{\infty} \frac{v_n}{u_n^2 + v_n^2} = \infty$); and totally dissipative when $\beta \neq 0$ ($\sum_{n=1}^{\infty} \frac{v_n}{u_n^2 + v_n^2} < \infty$).

INDUCED TRANSFORMATIONS, CONSERVATIVITY AND INVARIANT MEASURES

Suppose T is conservative and non–singular, and let $A \in \mathcal{B}_+$, then m–a.e. point of A returns infinitely often to A under iterations of T, and in particular the *return time* function to A, defined for $x \in A$ by $\varphi_A(x) := \min\{n \geq 1 : T^n x \in A\}$ is finite m–a.e. on A.

The *induced transformation* ([33]) on A is defined by $T_A x = T^{\varphi_A(x)} x$, and can be defined whenever the return time function is finite m–a.e. on A (whether T is conservative, or not).

The first key observation is that $m|_A \circ T_A^{-1} << m|_A$. This is because

$$T_A^{-1} B = \bigcup_{n=1}^{\infty} [\varphi = n] \cap T^{-n} B.$$

It follows that $\varphi_A \circ T_A$ is defined a.e. on A and an induction now shows that all powers $\{T_A^k\}_{k \in \mathbb{N}}$ are defined a.e. on A, and satisfy

$$T_A^k x = T^{(\varphi_A)_k(x)} x \text{ where } (\varphi_A)_1 = \varphi_A, \quad (\varphi_A)_k = \sum_{j=0}^{k-1} \varphi_A \circ T_A^j.$$

2.2 Proposition (c.f. [33]) *Let T be a non–singular transformation of $(X, \mathcal{B}, m)$, and suppose that $A \in \mathcal{B}$, $m(A) > 0$ satisfies $\varphi_A < \infty$ a.e. on A.*

1) If T is conservative, then the induced transformation T_A is a conservative, non–singular transformation of $(A, \mathcal{B} \cap A, m|_A)$.

2) If T is a measure preserving transformation, then T_A is a measure preserving transformation of $(A, \mathcal{B} \cap A, m|_A)$, and in case $\bigcup_{n=0}^{\infty} T^{-n} A = X$ mod m, T_A a conservative iff T is conservative.

2.3 Proposition (c.f. [32]) *Let T be a non–singular transformation of $(X, \mathcal{B}, m)$, and suppose that the return time function to $A \in \mathcal{B}$, $m(A) > 0$ is finite m–a.e. on A.*

Let $A \in \mathcal{B}_+$, and suppose that $q << m|_A$ is a T_A–invariant measure. Set, for $B \in \mathcal{B}$,

$$\mu(B) = \sum_{k=0}^{\infty} q(A \cap T^{-k} B \setminus \bigcup_{j=1}^{k} T^{-j} A).$$

Then $\mu << m$ is a T–invariant measure.

Non-expanding interval maps I

Let T be a piecewise onto, C^2 interval map with basic partition $\alpha = \{(0, u)\} \cup \alpha_0$ satisfying Adler's condition. Suppose that

$$Tx = x + cx^{1+p} + o(x^{1+p}) \quad \text{as } x \to 0$$

where $c > 0$, $p \geq 1$, $T(u) = 1$, $T'' \geq 0$ on $[0, u]$ and $\exists \, \kappa > 1$ such that

$$T'x \geq \kappa \quad \forall \, x \in \, a \in \alpha_0$$

(e.g. $Tx = \{\frac{1}{1-x}\}$ where $p = c = 1$).

Evidently the return time function to $[u, 1]$ is finite on $[u, 1]$, and $T_{[u,1]}$ is an expanding, piecewise onto, C^2 interval map of $[u, 1]$.

It turns out that $T_{[u,1]}$ also satisfies Adler's condition. We sketch a way to see this (the full proof is in [45]). Let $x \in [u, 1]$, then $T_{[u,1]} x = v_0^{-n} \circ (x)$ for some $n \geq 0$, where $v_0 : I \to [0, u]$, $T \circ v_0 = \text{Id}$. It follows that

$$\frac{|T_{[u,1]}'' x|}{(T_{[u,1]}' x)^2} \leq \frac{|v_0^{-n''}(Tx)|}{v_0^{-n'}(Tx)^2} + 1.$$

Adler's condition for $T_{[u,1]}$ will now follow from

$$\sup_{y \in (u,1), \, n \geq 1} \frac{|v_0^{n''}(y)|}{v_0^{n'}(y)} < \infty.$$

To show this, calculate first that

$$v_0^{n''}(y) = v_0^{n'}(y) \sum_{k=0}^{n-1} \frac{v_0''(v_0^k y)}{v_0'(v_0^k y)} v_0^{k'}(y),$$

whence

$$\frac{|v_0^{n''}(y)|}{v_0^{n'}(y)} \leq M \sum_{k=0}^{n-1} v_0^{k'}(y) \leq M \sum_{k=0}^{\infty} \frac{v_0^k(y) - v_0^{k+1}(y)}{y - v_0(y)}$$
$$= M \frac{y}{y - v_0(y)} \leq M' < \infty \ \forall \ y \in (u, 1).$$

As in section 1, there is a $T_{[u,1]}$-invariant probability on $[u, 1]$ with Lipschitz continuous density bounded away from 0.

By proposition 2.3, there is an absolutely continuous T-invariant σ-finite measure and μ on I, and by proposition 2.2 T is a conservative measure preserving transformation of $(I, \mathcal{B}, \mu)$.

It can be shown (see [46]) that the density h of μ has the form

$$h(x) = \frac{\varkappa(x)}{x^p}$$

where $\varkappa$ is Lipschitz continuous, and bounded away from zero. Therefore μ is an infinite measure.

This method for proving conservativity and existence of invariant measure is difficult to apply when no suitable set to induce on presents itself. More widely applicable developments of the method use a "jump transformation" to replace induction (see [43], [45]).

3 Ergodicity and Exactness

Ergodicity is an irreducibility property. A non-singular transformation T of the measure space $(X, \mathcal{B}, m)$ is called *ergodic* if

$$A \in \mathcal{B}, \ T^{-1}A = A \ \mod m \ \Rightarrow \ m(A) = 0, \ \text{or} \ m(A^c) = 0.$$

An invertible ergodic non-singular transformation T of a non-atomic measure space is necessarily conservative.

3.1 Proposition *Let T be a non-singular transformation. The following are equivalent:*

T *is conservative and ergodic;*

$$\sum_{n=1}^{\infty} 1_A \circ T^n = \infty \ \ a.e. \ \ \forall A \in \mathcal{B}_+;$$

$$\sum_{n=0}^{\infty} \widehat{T}^n f = \infty \ \ a.e. \ \forall f \in L^1(m), \ \ f \geq 0 \ a.e. \ , \ \int_X f dm > 0.$$

3.2 Proposition *Suppose that T is conservative, and $A \in \mathcal{B}_+$. Then*

$$T \text{ ergodic} \Rightarrow T_A \text{ ergodic},$$

$$T_A \text{ is ergodic, and } \bigcup_{n=1}^{\infty} T^{-n} A = X \text{ mod } m \Rightarrow T \text{ is ergodic}.$$

3.3 Theorem (Unicity of invariant measure) *Let T be a conservative, ergodic, non-singular transformation of $(X, \mathcal{B}, m)$. Then, up to multiplication by constants, there is at most one m-absolutely continuous, σ-finite T-invariant measure μ and (if there is one) $\mu \sim m$.*

EXACTNESS

There are many ways of proving ergodicity. Sometimes, it's easier to prove a stronger property.

A non-singular transformation T of the measure space $(X, \mathcal{B}, m)$ is called *exact* if

$$\bigcap_{n=1}^{\infty} T^{-n} \mathcal{B} = \{\emptyset, X\} \mod m.$$

As above, the σ-algebra $\mathfrak{T}(T) := \bigcap_{n=1}^{\infty} T^{-n} \mathcal{B}$ is called the *tail* of T.

Evidently T-invariant measurable sets are in the tail and so exactness implies ergodicity. The converse is false due to the existence of invertible, ergodic transformations.

We show first that expanding maps of the unit interval are exact. Let T be a piecewise onto, C^2 interval map satisfying Adler's condition and with basic partition α.

It follows from (1.5) that for $n \geq 1$, $a \in \alpha_0^{n-1}$ and $B \in \mathcal{B}$,

$$m(a \cap T^{-n} B) = \int_B \widehat{T}^n 1_a dm = e^{\pm K} m(B) m(a).$$

Now suppose that $B \in \mathfrak{T}$, and $a \in \alpha_0^{n-1}$; then $\exists B_n \in \mathcal{B}$ such that $B = T^{-n} B_n$. Thus

$$m(a \cap B) = m(a \cap T^{-n} B_n) = e^{\pm K} m(B_n) m(a) = e^{\pm 2K} m(B) m(a).$$

This remains true for $a \in \mathcal{B}$, whence $0 = m(B^c \cap B) = e^{\pm 2K} m(B) m(B^c)$ and $m(B) = 0, 1$.

Non-expanding interval maps II

As a consequence of this, we obtain from proposition 3.2 that the non-expanding maps of the unit interval are also ergodic. It now follows proposition 3.3 that (having an absolutely continuous invariant infinite measure) they admit no absolutely continuous invariant probability. In fact ([45]) the non-expanding maps of the unit interval considered here are themselves exact.

The following is a useful criterion for exactness.

3.4 Theorem [37] *Let T be a non-singular transformation of $(X, \mathcal{B}, m)$. Then*

$$T \text{ is exact } \Leftrightarrow \|\widehat{T}^n f\|_1 \to_{n \to \infty} 0 \ \forall f \in L^1, \int_X f dm = 0.$$

Boole's transformations III

1) Suppose that T is a non-Möbius Boole's transformation and suppose that $\omega \in \mathbb{R}^{2+}$ and $T(\omega) = \omega$. As shown above, the measure P_ω is T-invariant. We'll show using theorem 3.4 that T is exact.

Since T is not Möbius, $|T'(\omega)| < 1$ and $T^n(z) \to \omega \ \forall \ z \in \mathbb{R}^{2+}$.

We'll show exactness via theorem 3.4 using $f * \varphi_{ib} \to f$ in L^1 as $b \to 0$, $\forall \ f \in L^1$ where $f * g(x) := \int_\mathbb{R} f(x-y)g(y)dy$.

Fix $f \in L^1$. For $b > 0$, $f * \varphi_{ib} = \int_\mathbb{R} f(y)\varphi_{y+ib}dy$ whence $\widehat{T}^n(f * \varphi_{ib}) = \int_\mathbb{R} f(y)\varphi_{T^n(y+ib)}dy$ and

$$\|\widehat{T}^n(f * \varphi_{ib}) - \int_\mathbb{R} f dm_\mathbb{R} \varphi_\omega\|_1 \leq \int_\mathbb{R} |f(y)| \|\varphi_{T^n(y+ib)} - \varphi_\omega\|_1 \to 0,$$

$$\limsup_{n \to \infty} \|\widehat{T}^n f - \int_\mathbb{R} f dm_\mathbb{R} \varphi_\omega\|_1 \leq \limsup_{n \to \infty} \|\widehat{T}^n f - \widehat{T}^n(f * \varphi_{ib})\|_1$$
$$\leq \|f - f * \varphi_{ib}\|_1 \to 0$$

as $b \to 0$. Exactness follows from theorem 3.4.

2) Now consider

$$Tx = x + \sum_{k=1}^{n} \frac{p_k}{t_k - x}$$

where $n \geq 1$, $p_1, \ldots, p_n \geq 0$ (not all zero) and $t_1, \ldots, t_n \in \mathbb{R}$.

Recall from above that T is a conservative, measure preserving transformation of $\mathbb{R}$ equipped with Lebesgue measure.

We claim that T is exact. This can be shown using theorem 3.4 in the following steps ([4]):

- By (2.3), $\|\varphi_{T^n(\omega)} - \varphi_{T^n(\omega')}\|_1 \to 0$ as $n \to \infty \ \forall \ \omega, \ \omega' \in \mathbb{R}^{2+}$.
- For $f \in L^1$, $\int_X f dm = 0$ fixed, if $b > 0$ then $f * \varphi_{ib} = \int_\mathbb{R} f(y)\varphi_{y+ib}dy$ whence

$$|\widehat{T}^n(f * \varphi_{ib})| = |\int_\mathbb{R} f(y)\varphi_{T^n(y+ib)}dy| \leq \int_\mathbb{R} |f(y)| |\varphi_{T^n(y+ib)} - \varphi_{T^n(ib)}|dy,$$

and

$$\|\widehat{T}^n(f * \varphi_{ib})\|_1 \leq \int_\mathbb{R} |f(y)| \|\varphi_{T^n(y+ib)} - \varphi_{T^n(ib)}\|_1 dy \to 0 \text{ as } n \to \infty.$$

• As above,

$$\limsup_{n\to\infty} \|\widehat{T}^n f\|_1$$

$$\leq \limsup_{n\to\infty} \|\widehat{T}^n (f * \varphi_{ib})\|_1 + \limsup_{n\to\infty} \|\widehat{T}^n (f * \varphi_{ib} - f)\|_1$$

$$\leq \|f * \varphi_{ib} - f\|_1 \to 0 \text{ as } b \to 0.$$

4   Ergodic Theorems

The original "proof of the ergodic theorem" was

4.1 Birkhoff's ergodic theorem [14] *Suppose that T is a probability preserving transformation of $(X, \mathcal{B}, m)$. Then*

$$\frac{1}{n} \sum_{k=1}^{n} f(T^k x) \to E(f|\mathfrak{I})(x) \text{ as } n \to \infty \text{ for a.e. } x \in X, \ \forall f \in L^1(m),$$

where $\mathfrak{I}$ is the σ-algebra of T-invariant sets in $\mathcal{B}$.

Generalizations of Birkhoff's theorem were given by E.Hopf and Stepanov ([30], [44]) for infinite measure preserving transformations, W.Hurewicz [31] (for non-singular transformations), and R.Chacon, D.Ornstein [18] (for Markov operators).

4.2 Hopf-Stepanov ergodic Theorem [30], [44] *Suppose that T is a conservative, measure preserving transformation of $(X, \mathcal{B}, m)$. Then*

$$\frac{\sum_{k=1}^{n} f \circ T^k(x)}{\sum_{k=1}^{n} p \circ T^k(x)} \to E_{m_p}(\frac{f}{p}|\mathfrak{I})(x),$$

as $n \to \infty$ for a.e. $x \in X$, $\forall f, p \in L^1(m)$, $p > 0$, where $dm_p = p dm$.

4.3 Hurewicz's Ergodic Theorem [31] *Suppose that T is a conservative non-singular transformation of $(X, \mathcal{B}, m)$. Then*

$$\frac{\sum_{k=1}^{n} \widehat{T}^k f(x)}{\sum_{k=1}^{n} \widehat{T}^k p(x)} \to E_{m_p}(\frac{f}{p}|\mathfrak{I})(x)$$

as $n \to \infty$ for a.e. $x \in X$, $\forall f, p \in L^1(m)$, $p > 0$.

Note that, when T is ergodic, $E_{m_p}(\frac{f}{p}|\mathfrak{I})(x) = \frac{\int_X f dm}{\int_X p dm}$.

5   Absolutely Normalized Convergence

Given a conservative, ergodic measure preserving transformation T, it is natural to ask the rate at which $\sum_{k=1}^{n} p \circ T^k \to \infty$ for $p \in L^1$, $p \geq 0$. For probability preserving T, $\sum_{k=1}^{n} p \circ T^k \sim n \int_X p\,dm$ a.e., however for a conservative, ergodic infinite measure preserving transformation T one only obtains (from the Hopf-Stepanov ergodic theorem) that $\sum_{k=1}^{n} p \circ T^k = o(n)$ a.e. as $n \to \infty$ for $p \in L^1$, $p \geq 0$.

We are led to ask for "absolutely normalized ergodic convergence" i.e. for constants $a_n > 0$ such that $\frac{1}{a_n} \sum_{k=1}^{n} p \circ T^k \to \int_X p\,dm$ in some sense.

The Hopf-Stepanov ergodic theorem does not provide this absolutely normalized convergence for ergodic, infinite measure preserving transformations, and the next result shows that absolutely normalized pointwise convergence is impossible for ergodic, infinite measure preserving transformations.

5.1 Theorem [1] *Suppose that T is a conservative, ergodic measure preserving transformation of the σ-finite, infinite measure space $(X, \mathcal{B}, m)$, and let $a_n > 0$ $(n \geq 1)$. Then*

$$either \ \liminf_{n \to \infty} \frac{S_n(f)}{a_n} = 0 \ a.e., \ \forall f \in L^1(m)_+, \tag{1}$$

$$or \ \exists n_k \uparrow \infty \ such \ that \ \frac{S_{n_k}(f)}{a_{n_k}} \to_{n \to \infty} \infty \ a.e., \ \forall f \in L^1(m)_+. \tag{2}$$

POINTWISE DUAL ERGODICITY

The situation is different for the duals of some non-invertible transformations.

Boole's transformations IV

Suppose that $T : \mathbb{R} \to \mathbb{R}$ is defined by $Tx = x + \sum_{k=1}^{n} \frac{p_k}{t_k - x}$ where $n \geq 1$, $p_1, \ldots, p_n \geq 0$ and $t_1, \ldots, t_n \in \mathbb{R}$. Then

$$\frac{1}{a_n} \sum_{k=1}^{n} \widehat{T}^k f \ \to \ \int_{\mathbb{R}} f\,dm_{\mathbb{R}} \ a.e. \ as \ n \to \infty, \ \ \forall f \in L^1(m_{\mathbb{R}}),$$

where $a_n := \sqrt{\frac{2n}{\pi^2 \nu}}$ and $\nu := \sum_{k=1}^{n} p_k$.

Proof ([4])

For $\omega \in \mathbb{R}^{2+}$, $T^n(\omega) = u_n + iv_n$ where by (2.3) $\sup_{n \geq 1} |u_n| < \infty$ and $v_n \sim \sqrt{2\nu n}$ as $n \to \infty$. It follows that

$$\pi \widehat{T}^n \varphi_\omega(x) = \text{Im} \frac{1}{x - T^n(\omega)} = \frac{v_n}{(x - u_n)^2 + v_n^2} \sim \frac{1}{\sqrt{2\nu n}}$$

whence

$$\sum_{k=1}^{n} \widehat{T}^k \varphi_\omega \sim \sum_{k=1}^{n} \frac{1}{\pi\sqrt{2\nu n}} = a_n$$

and for $f \in L^1(m_{\mathbb{R}})$, using Hurewicz's theorem,

$$\frac{\sum_{k=1}^{n} \widehat{T}^k f(x)}{a_n} \approx \frac{\sum_{k=1}^{n} \widehat{T}^k f(x)}{\sum_{k=1}^{n} \widehat{T}^k \varphi_\omega(x)} \to \int_X f \, dm \text{ a.e. as } n \to \infty.$$

$\square$

A conservative, ergodic, measure preserving transformation is called *pointwise dual ergodic* if there are constants a_n such that

$$\frac{1}{a_n} \sum_{k=0}^{n-1} \widehat{T}^k f \to \int_X f \, dm_T \text{ a.e. as } n \to \infty \ \forall \ f \in L^1(X_T).$$

The sequence a_n is called a *return sequence* of T and denoted $a_n(T)$. Its asymptotic proportionality class $\mathcal{A}(T) := \{(a'_n)_{n\in\mathbb{N}} : \exists \lim_{n\to\infty} \frac{a'_n}{a_n(T)} \in \mathbb{R}_+\}$ is called the *asymptotic type* of T.

Non-expanding interval maps III

Let T be a piecewise onto, C^2 interval map with basic partition $\alpha = \{(0, u)\} \cup \alpha_0$ satisfying Adler's condition. Suppose that

$$Tx = x + cx^{1+p} + o(x^{1+p}) \quad \text{as } x \to 0$$

where $c > 0$, $p \geq 1$, $T(u) = 1$, $T'' \geq 0$ on $[0, u]$ and $\exists \ \kappa > 1$ such that

$$T'x \geq \kappa \ \forall \ x \in \ a \in \alpha_0,$$

then ([7], [47]) T is pointwise dual ergodic and

$$a_n(T) \propto \begin{cases} n^{\frac{1}{p}} & p > 1, \\ \frac{n}{\log n} & p = 1. \end{cases}$$

Examples of piecewise onto, C^2 interval maps with more general return sequences can be found in [1], §4.8.

Markov shifts

Let S be a countable set, $T : S^{\mathbb{N}} \to S^{\mathbb{N}}$ be the *shift* map, defined by $T(x_1, x_2, ...) = (x_2, x_3, ...)$.

Given a *stochastic matrix* P on S ($p : S \times S \to \mathbb{R}$, $p_{s,t} \geq 0$, $\sum_{t\in S} p_{s,t} = 1 \ \forall s \in S$) and a probability π on S such that $\pi_s > 0 \ \forall s \in S$, we specify the probability $m = m_\pi$ on $S^{\mathbb{N}}$ such that

$$m([s_1, \ldots, s_n]) = \pi_{s_1} p_{s_1,s_2} \cdots p_{s_{n-1},s_n} \ \forall s_1, \ldots, s_n \in S^n, \ n \in \mathbb{N}$$

where $[s_1, \ldots, s_n] := \{x = (x_1, x_2, \ldots) \in S^{\mathbb{N}} : x_1 = s_1, \ldots, x_n = s_n\}$.

The *Markov shift* of P (with initial distribution π) is the quadruple

$$(S^{\mathbb{N}}, \mathcal{B}, m, T),$$

where $\mathcal{B} = \sigma(\{[s_1, \ldots, s_n] : s_1, \ldots, s_n \in S^n, \, n \in \mathbb{N}\}$.

Since $T[s] = \bigcup_{t \in S, p_{s,t} > 0}[t]$ mod m, we see T is a non-singular transformation of $(S^{\mathbb{N}}, \mathcal{B}, m)$ iff

$$\forall t \in S, \, \exists s \in S, \, p_{s,t} > 0$$

and in this situation,

$$\widehat{T}^k 1_{[t]}(x) = \frac{\pi_t}{\pi_{x_1}} p_{t,x_1}^{(k)} \; \forall \, k \in \mathbb{N}, \, t \in S.$$

A state $s \in S$ is called *recurrent* if $\sum_{n=0}^{\infty} p_{s,s}^{(n)} = \infty$.

We denote the collection of recurrent states by S_r and call P *recurrent* if $S_r = S$.

Evidently, if P is recurrent, then T is non-singular.

As shown in [28], if T is non-singular, then $\mathfrak{C}(T) = \bigcup_{s \in S_r}[s]$ (this also follows from proposition 2.1). In particular, if P is recurrent, then T is non-singular and conservative.

A stochastic matrix is called *irreducible* if

$$\forall s, t \in S, \, \exists n \in \mathbb{N} \ni p_{s,t}^{(n)} > 0.$$

Evidently again, if a stochastic matrix is irreducible, then its shift is non-singular.

We'll sketch how to show that the shift of an irreducible, recurrent stochastic matrix is ergodic ([28]), and has a σ-finite invariant measure which comes from an invariant distribution on states ([19]).

First, fix $s \in S_r$, and let m_s be that constant multiple of $m|_{[s]}$ with $m_s([s]) = 1$, then $T_{[s]}$ is an exact measure preserving transformation of $([s], \mathcal{B} \cap [s], m_s)$. This is because if $\alpha = \{[s, t_1, \ldots, t_{n-1}, s] : n \geq 1, \, t_1, \ldots, t_{n-1} \neq s\}$ then $\{T_{[s]}^{-n}\alpha\}$ are statistically independent and $\sigma(\{T_{[s]}^{-n}\alpha\}) = \mathcal{B} \cap [s]$ whence the tail is trivial by Kolmogorov's zero-one law ([34]).

Next, suppose that $T^{-1}A = A \in \mathcal{B}_+$ then $T_{[s]}^{-1}(A \cap [s]) = A \cap [s] \; \forall \, s \in S$, whence $m(A \cap [s]) > 0 \Rightarrow A \supset [s]$ mod m. Since $m(A) > 0$, $\exists \, t \in S$ such that $A \supset [t]$ mod m.

Let $s \in S$, then, by irreducibility, there exists $n \geq 1$ such that $p_{s,t}^{(n)} > 0$, whence

$$m([s] \cap A) = m([s] \cap T^{-n}A) \geq m([s] \cap T^{-n}[t]) > 0 \Rightarrow A \supset [s] \mod m.$$

This shows that T is ergodic.

For $s \in S$, let

$$\mu_s(B) = \int_{[s]} \left(\sum_{k=0}^{\varphi_{[s]}-1} 1_B \circ T^k \right) dm_s,$$

then, by proposition 2.5, $\mu_s \circ T^{-1} = \mu_s << m$.

By unicity of invariant measure (theorem 3.3), $\mu_t \equiv \mu_s([t])\mu_s \sim m \ \forall \, t \in S$, and, necessarily, $0 < \mu_s([t]) < \infty$ for $s, t \in S$.

Since $\mu_s|_{[s]} = m_s|_{[s]} \ \forall \, s \in S$, it follows that

$$\mu_s([t_1, ..., t_n]) = \mu_s([t_1]) \prod_{k=1}^{n-1} p_{t_k, t_{k+1}},$$

whence

$$\sum_{t \in S} \mu_s([t]) p_{t,u} = \mu_s(T^{-1}[u]) = \mu_s([u]).$$

A calculation shows ([19]) that

$$\mu_s([t]) = \sum_{n=1}^{\infty} {}_s p_{s,t}^{(n)}$$

where

$$_s p_{s,t}^{(1)} = p_{s,t}, \quad _s p_{s,t}^{(n+1)} = \sum_{u \neq s} {}_s p_{s,u}^{(n)} p_{u,t} \quad (n \geq 1).$$

Consider now T as a measure preserving transformation of $(X, \mathcal{B}, \mu)$ where $\mu = \mu_s$. Write $\mu_s([t]) = c_t$. It follows that

$$\widehat{T}^k 1_{[t]}(x) = \frac{c_t}{c_{x_1}} p_{t,x_1}^{(k)} \ \forall \, k \in \mathbb{N}, \ t \in S,$$

whence T is pointwise dual ergodic with $a_n(T) \sim \sum_{k=0}^{n-1} p_{s,s}^{(k)}$.

An irreducible stochastic matrix is said to be *aperiodic* if for some (and hence all) $s \in S$, g.c.d. $\{n \geq 1 : p_{s,s}^{(n)} > 0\} = 1$. It is shown in [16] that the Markov shift of an aperiodic, recurrent stochastic matrix is is exact (see also theorem 3.2 of [10]).

RATIONAL ERGODICITY

No invertible, conservative, ergodic, measure preserving transformation of an infinite measure space can be pointwise dual ergodic because $\widehat{T}f = f \circ T^{-1}$ and pointwise dual ergodicity would violate theorem 5.1. There are invertible rationally ergodic transformations.

A conservative, ergodic, measure preserving transformation T of $(X, \mathcal{B}, m)$ is called *rationally ergodic* if there is a set $A \in \mathcal{B}$, $0 < m(A) < \infty$ satisfying a *Renyi inequality*: $\exists \, M > 0$ such that

$$\int_A (S_n(1_A))^2 \, dm \leq M \left(\int_A S_n(1_A) dm \right)^2 \quad \forall \, n \geq 1.$$

5.2 Theorem [2], [3] *Suppose that T is rationally ergodic. Then there is a sequence of constants $a_n \uparrow \infty$, unique up to asymptotic equality, such that whenever $A \in \mathcal{B}$ satisfies a Renyi inequality,*

$$\frac{1}{a_n} \sum_{k=0}^{n-1} m(B \cap T^{-k}C) \to m(B)m(C) \text{ as } n \to \infty \, \forall \, B, C \in \mathcal{B} \cap A$$

$$\varlimsup_{n \to \infty} \frac{1}{a_n} \sum_{k=0}^{n-1} m(B \cap T^{-k}C) \geq m(B)m(C) \, \forall \, B, C \in \mathcal{B},$$

and

$$\forall \, m_\ell \uparrow \infty \, \exists \, n_k = m_{\ell_k} \uparrow \infty$$

such that

$$\frac{1}{N} \sum_{k=1}^{N} \frac{1}{a_{n_k}} \sum_{j=0}^{n_k-1} f \circ T^j \to \int_X f \, dm \text{ a.e. as } n \to \infty \, \forall \, f \in L^1(m). \tag{3}$$

A conservative, ergodic, measure preserving transformation satisfying (3) with respect to some sequence of constants $\{a_n\}$ is called *weakly homogeneous*.

5.3 Proposition [4] *Suppose that T is pointwise dual ergodic. Then T is rationally ergodic and satisfies (3) with respect to its return sequence $a_n(T)$.*

The sequence of constants $\{a_n\}$ appearing in (3) may therefore be also called the *return sequence* of T without ambiguity in this choice of name.

ASYMPTOTIC DISTRIBUTIONAL BEHAVIOUR

5.4 Proposition [6] *If T is a conservative, ergodic measure preserving transformation of $(X, \mathcal{B}, m)$, and $n_k \to \infty$, and $d_k > 0$, then $\exists \, m_\ell := n_{k_\ell} \to \infty$, and a random variable Y on $[0, \infty]$ such that*

$$g\left(\frac{1}{d_{k_\ell}} \sum_{j=0}^{m_\ell-1} f \circ T^j \right) \to E(g(\mu(f)Y)), \tag{$*$}$$

weak $*$ in L^∞, $\forall \, f \in L^1(m)$, $f \geq 0$, $g \in C([0, \infty])$, where $\mu(f) := \int_X f \, dm$.

In the situation of $(*)$, we'll write

$$\frac{S_{m_\ell}^T}{d_{k_\ell}} \xrightarrow{\partial} Y,$$

and call Y a *distributional limit of T along $\{m_\ell\}$*.

Clearly $Y = 0$ and $Y = \infty$ are distributional limits of any conservative, ergodic, measure preserving transformation. We'll consider a random variable Y on $[0, \infty]$ *trivial* if it is supported on $\{0, \infty\}$.

Mittag-Leffler distribution. Let $\alpha \in [0, 1]$. The random variable Y_α on $\mathbb{R}_+$ has the *normalized Mittag-Leffler distribution of order α* if

$$E(e^{zY_\alpha}) = \sum_{p=0}^{\infty} \frac{\Gamma(1+\alpha)^p z^p}{\Gamma(1+p\alpha)}$$

(the normalization being $E(Y_\alpha) = 1$).

As can be checked, Y_0 is exponentially distributed, $Y_{\frac{1}{2}}$ is distributed as the absolute value of a centered Gaussian random variable, and $Y_1 = 1$.

5.5 Darling- Kac Theorem ([6], c.f. [20]) *Suppose that T is pointwise dual ergodic, and that $a_n(T)$ is regularly varying with index $\alpha \in [0, 1]$ as $n \to \infty$. Then*

$$\frac{S_n^T}{a_n(T)} \xrightarrow{\partial} Y_\alpha,$$

where Y_α has the normalized Mittag-Leffler distribution of order α.

Examples

Boole's transformation $T : \mathbb{R} \to \mathbb{R}$ is defined by $Tx = x + \sum_{k=1}^n \frac{p_k}{t_k - x}$, where $n \geq 1$, $p_1, \ldots, p_n \geq 0$ and $t_1, \ldots, t_n \in \mathbb{R}$, satisfy $\frac{S_n^T}{a_n(T)} \xrightarrow{\partial} Y_{\frac{1}{2}}$, where $a_n := \sqrt{\frac{2n}{\pi^2 \nu}}$ and $\nu := \sum_{k=1}^n p_k$.

If $T : [0, 1] \to [0, 1]$ is defined by $Tx = \{\frac{1}{1-x}\}$, then $\frac{S_n^T}{a_n(T)} \xrightarrow{\partial} Y_1$ where $a_n := \frac{n}{\log n}$; equivalently $\frac{1}{a_n(T)} \sum_{k=0}^{n-1} f \circ T^k \to \int_I f(x) \frac{dx}{x}$ in measure $\forall f \in L^1(\frac{dx}{x})$.

The Mittag-Leffler distributions are related to the positive stable distributions, indeed $Y_\alpha^{-\alpha}$ has positive stable distribution of order α:

$$E(e^{-tY_\alpha^{-\frac{1}{\alpha}}}) = e^{-c_\alpha t^\alpha}. \tag{4}$$

See [17] and [6].

For distributional convergence to stable laws as a consequence of the Darling-Kac theorem, see [6].

6 Structure

For $c \in (0, \infty]$, a *c-factor map* from a measure preserving transformation $(X, \mathcal{B}, \mu, S)$ onto a measure preserving transformation $(Y, \mathcal{C}, \nu, T)$ is a measurable map $\pi : X' \to Y'$ (where $X' = S^{-1}X' \in \mathcal{B}$ and $Y' = T^{-1}Y' \in \mathcal{BC}$ are sets of full measure) satisfying

$$\mu \circ \pi^{-1}(A) = c\nu(A) \ \forall \ A \in \mathcal{B}_T, \text{ and } \pi \circ S = T \circ \pi.$$

We shall denote this situation by $\pi : S \xrightarrow{c} T$.

A *c-isomorphism* is an invertible *c*-factor map π (denoted $\pi : S \xleftrightarrow{c} T$).

It is necessary to consider *c*-factor maps with $c \neq 1$ as our measure spaces are not normalized (being infinite).

If, for some $c \in \mathbb{R}_+$ there exists $\pi : S \xrightarrow{c} T$, we shall call T a *factor* of S, and S an *extension* of T, denoting this by $S \to T$.

Clearly if $\pi : S \xrightarrow{c} T$, then $\pi^{-1}\mathcal{B}_T$ is a σ-finite, sub-σ-algebra of $\mathcal{B}_S$ which is *T-invariant* in the sense that $S^{-1}(\pi^{-1}\mathcal{B}_T) \subset \pi^{-1}\mathcal{B}_T$. In case T is invertible, $\pi^{-1}\mathcal{B}_T$ is *strictly T-invariant* : $S^{-1}(\pi^{-1}\mathcal{B}_T) = \pi^{-1}\mathcal{B}_T$.

6.1 Factor Proposition *Suppose that T is a measure preserving transformation of the σ-finite, standard measure space $(X, \mathcal{B}, m)$.*

For every T-invariant, sub-σ-algebra of $\mathcal{F} \subset \mathcal{B}$, there is a factor U, and a factor map $\pi : T \xrightarrow{1} U$ with $\mathcal{F} = \pi^{-1}\mathcal{B}_U$.

The factor is invertible iff the sub-σ-algebra is strictly T-invariant.

Two measure preserving transformations are called *strongly disjoint* if they have no common extension, and *similar* otherwise.

No two probability preserving transformations are strongly disjoint. Indeed, if S, T are probability preserving transformations, the Cartesian product transformation $R = S \times T$ is a probability preserving transformation, and

$$S \leftarrow R \to T.$$

Recall from [23] that the probability transformations S, T are called *disjoint* if any common extension R has the Cartesian product as factor (i.e. $R \to S \times T$).

When S, and T are infinite measure preserving transformations, the Cartesian product is not an extension of either transformation and moreover, strong disjointness is not uncommon among infinite measure preserving transformations.

Notwithstanding, similarity is an equivalence relation.

6.2 Proposition *Suppose that S and T are similar measure preserving transformations. Then S is conservative if and only if T is conservative.*

This follows from proposition 2.1.

6.3 Proposition [3] *Suppose that S and T are similar measure preserving transformations, both conservative and ergodic.*

If S is weakly homogeneous, then so is T and

$$\exists \lim_{n \to \infty} \frac{a_n(S)}{a_n(T)} \in \mathbb{R}_+.$$

Thus, Boole's transformations $Tx = x + \sum_{k=1}^{n} \frac{p_k}{t_k - x}$ (with return sequence $a_n(T) \propto \sqrt{n}$) are all strongly disjoint from $Sx = \{\frac{1}{1-x}\}$ (return sequence: $a_n(S) \propto \frac{n}{\log n}$).

A one parameter family of pairwise strongly disjoint pointwise dual ergodic transformations is given by the non-expanding interval maps considered above. For $p \geq 1$ let T_p be a piecewise onto, C^2 interval map with basic partition $\alpha = \{(0, \frac{1}{2}), (\frac{1}{2}, 1)\}$ with

$$T_p x = \begin{cases} x + c_p x^{1+p} + o(x^{1+p}) & x \in (0, \frac{1}{2}), \\ 2x - 1 & x \in (\frac{1}{2}, 1) \end{cases}$$

where $c_p > 0$ is fixed so that $T_p(x) \to 1$ as $t \to \frac{1}{2}-$.

As before, each T_p is pointwise dual ergodic and

$$a_n(T_p) \propto \begin{cases} n^{\frac{1}{p}} & p > 1, \\ \frac{n}{\log n} & p = 1, \end{cases}$$

whence T_p and $T_{p'}$ are strongly disjoint for $p \neq p'$.

NATURAL EXTENSION

Let $(X, \mathcal{B}, m, T)$ be a measure preserving transformation of a σ-finite, standard measure space. As in [42], a *natural extension* of T is an invertible extension $(X', \mathcal{B}', m', T')$ of T which is minimal in the sense that and $T'^n \pi^{-1} \mathcal{B} \uparrow \mathcal{B}'$ mod m' where $\pi : T' \overset{1}{\to} T$ is the extension map.

6.4 Theorem: Existence and uniqueness of natural extensions [42]

Any measure preserving transformation of a standard, σ-finite measure space has a natural extension (also on a standard space).

All natural extensions of the same measure preserving transformation are isomorphic.

6.5 Theorem [39] *The natural extension of a conservative, ergodic measure preserving transformation is conservative, and ergodic.*

7 Intrinsic Normalizing Constants and Laws of Large Numbers

Although infinite measure spaces are not canonically normalized, it may be that a measure preserving transformation T is intrinsically normalized, for example, in the sense that $T \overset{c}{\leftrightarrow} T$ for each $c \neq 1$. To this end, we consider (for T a measure preserving transformation)

$$\Delta_0(T) := \{c \in (0, \infty) : T \overset{c}{\leftrightarrow} T\}$$

first introduced in [24]. Clearly, this is a multiplicative subgroup of $\mathbb{R}_+$, and if S and T are isomorphic, then $\Delta_0(S) = \Delta_0(T)$.

7.1 Proposition *If T is a conservative, ergodic measure preserving transformation of $(X, \mathcal{B}, m)$, then $\Delta_0(T)$ is a Borel subset of $\mathbb{R}$.*

For example, let T be a totally dissipative measure preserving transformation of the standard σ-finite, non-atomic measure space $(X, \mathcal{B}, m)$. There is a wandering set $W \in \mathcal{B}$ such that $X = \bigcup_{n \in \mathbb{Z}} T^n W$, and it is not hard to see that

$$\Delta_0(T) = \begin{cases} \{1\} & m(W) < \infty, \\ \mathbb{R}_+ & m(W) = \infty. \end{cases}$$

The first conservative, ergodic example with $\Delta_0(T) = \{1\}$ was given in [25].

Unfortunately, it is not the case that $\Delta_0(S) = \Delta_0(T)$ for similar conservative, ergodic measure preserving transformations (see [8]).

Accordingly, we consider the *intrinsic normalizing constants* of a conservative, ergodic, measure preserving transformation T, namely the collection

$$\Delta_\infty(T) := \{c \in (0, \infty] : \exists \text{ a conservative, ergodic m.p.t. } R \overset{1}{\to} T, \ R \overset{c}{\to} T\}.$$

It is shown in [8] that $\Delta_\infty(T) \cap \mathbb{R}_+$ is an analytic subset of $\mathbb{R}_+$.

7.2 Proposition [8] *If S and T are similar conservative ergodic measure preserving transformations, then*

$$\Delta_\infty(S) = \Delta_\infty(T).$$

7.3 Proposition [8] *If T is a conservative, ergodic, measure preserving transformation, then $\Delta(T) := \Delta_\infty(T) \cap \mathbb{R}_+$ is a multiplicative subgroup of $\mathbb{R}_+$.*

7.4 Proposition [6] *If T is weakly homogeneous, then $\Delta_\infty(T) = \{1\}$.*

This will follow from proposition 7.8 below.

Examples where $\Delta_\infty(T) \neq \{1\}$

Let $\Omega = \{x = (x_1, x_2, \dots) : x_n = 0, 1\}$ be the group of dyadic integers, and let $(\tau x)_n := x_n + \epsilon_n \mod 2$ where $\epsilon_1 = 1$ and $\epsilon_{n+1} = 1$ if $x_n + \epsilon_n \geq 2$ and $\epsilon_{n+1} = 0$ otherwise.

For $p \in (0, 1)$, define a probability μ_p on Ω by

$$\mu_p([\epsilon_1, \dots, \epsilon_n]) = \prod_{k=1}^n p(\epsilon_k)$$

where $p(0) = 1 - p$ and $p(1) = p$.

Recall that $\mu_{\frac{1}{2}} = m$ is Haar measure on Ω, whence $\mu_{\frac{1}{2}} \circ \tau = \mu_{\frac{1}{2}}$. It is no longer true that τ preserves μ_p if $p \neq 1/2$, however $\mu_p \circ \tau \sim \mu_p$ and

$$\frac{d\mu_p \circ \tau}{d\mu_p} = \left(\frac{1-p}{p}\right)^\phi$$

where

$$\phi(x) = \min\{n \in \mathbb{N} : x_n = 0\} - 2.$$

7.5 Proposition τ *is an invertible, conservative, ergodic non-singular transformation of* $(\Omega, \mathcal{B}, \mu_p)$.

Proof sketch

A calculation shows that τ is non-singular (and indeed measure preserving when $p = \frac{1}{2}$).

All τ-invariant sets are in the *tail* σ-algebra $\mathfrak{T} := \bigcap_{n=1}^\infty \sigma(x_n, x_{n+1}, \dots)$, which is trivial by Kolmogorov's zero-one law ([34]). □

In fact, as shown in [13], there is no τ-invariant σ-finite, μ_p-absolutely continuous measure on Ω when $p \neq 1/2$. The next result is a strengthening of this.

For $0 < p < 1$, we consider the measure preserving transformations $T_p := (X, \mathcal{B}, m_p, T)$ where

$$X = \Omega \times \mathbb{Z}, \quad \mathcal{B} := \mathcal{B}(\Omega) \otimes 2^\mathbb{Z},$$

$$m_p(A \times \{n\}) = \mu_p(A)\left(\frac{1-p}{p}\right)^n, \text{ and } T(x, n) = (\tau x, n - \phi(x)).$$

7.6 Theorem [24] $(X, \mathcal{B}, m_p, T)$ *is a conservative, ergodic measure preserving transformation.*

The result of [13] follows from this since the existence of a τ-invariant σ-finite, μ_p-absolutely continuous measure on Ω when $p \neq 1/2$ entails $\phi = f - f \circ \tau$ for some $f : \Omega \to \mathbb{Z}$ measurable contradicting ergodicity of $(X, \mathcal{B}, m_p, T)$.

7.7 Proposition [8] *For $p \neq \frac{1}{2}$,*

$$\Delta_0(T_p) = \Delta_\infty(T_p) = \left\{ \left(\frac{1-p}{p} \right)^n : n \in \mathbb{Z} \right\}.$$

When $p = \frac{1}{2}$ the situation is different: $T_{\frac{1}{2}}$ is rationally ergodic with $a_n(T_{\frac{1}{2}}) \asymp \frac{n}{\sqrt{\log n}}$ (whence by proposition 7.4, $\Delta_\infty(T_{\frac{1}{2}}) = \{1\}$). For this, and further results on the transformations T_p, see [11].

In [8], examples of conservative, ergodic, measure preserving transformations T with $\Delta(T) = \Delta_0(T)$ of arbitrary Hausdorff dimension were given.

For a conservative, ergodic, measure preserving transformation T with $\Delta_\infty(T) = \{1, \infty\}$, see proposition 2.1 of [9].

Definition: Law of large numbers A *law of large numbers* for a conservative, ergodic, measure preserving transformation T of $(X, \mathcal{B}, m)$ is a function $L : \{0, 1\}^{\mathbb{N}} \to [0, \infty]$ such that $\forall A \in \mathcal{B}$, for a.e. $x \in X$,

$$L(1_A(x), 1_A(Tx), \dots) = m(A).$$

For example, if $(X, \mathcal{B}, m, T)$ is weakly homogeneous, then $\exists \, n_k \to \infty$ such that

$$\frac{1}{N} \sum_{k=1}^{N} \frac{1}{a_{n_k}(T)} \sum_{j=0}^{n_k - 1} f \circ T^j \to \int_X f \, dm \text{ a.e. as } n \to \infty \; \forall \, f \in L^1(m)$$

and a law of large numbers for T is given by

$$L(\epsilon_1, \epsilon_2, \dots) := \limsup_{N \to \infty} \frac{1}{N} \sum_{k=1}^{N} \frac{1}{a_{n_k}(T)} \sum_{j=0}^{n_k - 1} \epsilon_j.$$

7.8 Proposition [1] *If $(X, \mathcal{B}, m, T)$ is conservative and ergodic, and every conservative, ergodic measure preserving transformation similar to T has a law of large numbers, then $\Delta_\infty(T) = \{1\}$.*

Proof

Let $c \in \Delta_\infty(T)$. Suppose that $(Y \mathcal{C}, \mu, U)$ is conservative and ergodic, and let $\varphi : U \xrightarrow{1} T$, and $\psi : U \xrightarrow{c} T$ be 1-, and c-factor maps (respectively). Let L be a law of large numbers for U, and fix $A \in \mathcal{B}$, $m(A) = 1$ then

$$L(1_A(\varphi x), 1_A(T\varphi x), \dots) = L(1_{\varphi^{-1}A}(x), 1_{\varphi^{-1}A}(Ux), \dots) = \nu(\varphi^{-1}A) = 1,$$

ν-a.e., whence $L(1_A(x), 1_A(Tx), \dots) = 1$ m-a.e., but also

$$L(1_A(\psi x), 1_A(T\psi x), \dots) = L(1_{\psi^{-1}A}(x), 1_{\psi^{-1}A}(Ux), \dots) = \nu(\psi^{-1}A) = c$$

ν-a.e., whence $L(1_A(x), 1_A(Tx), \dots) = c$ m-a.e. with the conclusion that $c = 1$. $\square$

It turns out that the existence of laws of large numbers is close to characterising absence of intrinsic normalising constants.

7.9 Theorem [1] *If* $\Delta_\infty(T) = \{1\}$, *then* T *has a law of large numbers.*

This result can be used to prove the converse of proposition 7.8.

As another application, let $(X, \mathcal{B}, m)$ be a compact, metric group equipped with normalised Haar measure and let T be an ergodic translation of X. Suppose that $d \in \mathbb{N}$ and that $\phi : X \to \mathbb{Z}^d$ is measurable. Define the skew product transformation $T_\phi : X \times \mathbb{Z}^d \to X \times \mathbb{Z}^d$ by $T_\phi(x, y) = (Tx, y + \phi(x))$. This preserves the product measure $m \times$ counting measure. It is shown in [1] (by considering self joinings of T_ϕ) that if T_ϕ is ergodic, then $\Delta_\infty(T_\phi) = \{1\}$ and thus, by theorem 7.9, T_ϕ has a law of large numbers.

References

[1] J. Aaronson, *An Introduction to Infinite Ergodic Theory*, Mathematical Surveys and Monographs 50, American Mathematical Society, Providence RI, USA, 1997.

[2] J. Aaronson, Rational ergodicity and a metric invariant for Markov shifts, *Israel Journal of Math.*, **27**, (1977), 93-123.

[3] J. Aaronson, On the pointwise ergodic theory of transformations preserving infinite measures, *Israel Journal of Math.*, **32**, (1978), 67-82.

[4] J. Aaronson, Ergodic theory for inner functions of the upper half plane, *Ann. Inst. H. Poincaré (B)* , **XIV**, (1978), 233-253.

[5] J. Aaronson, A remark on the exactness of inner functions, *Jour. L.M.S.*, **23**, (1981), 469-474.

[6] J. Aaronson, The asymptotic distributional behavior of transformations preserving infinite measures, *J. D'Analyse Math.*, **39**, (1981), 203-234.

[7] J. Aaronson, Random f-expansions, *Ann. Probab.*, **14**, (1986), 1037-1057.

[8] J. Aaronson, The intrinsic normalising constants of transformations preserving infinite measures, *J. D'Analyse Math.*, **49**, (1987), 239-270.

[9] J. Aaronson, M.Lemańczyk, D. Volný, A cut salad of cocycles , *Fund. Math.*, **157**, (1998), 99-119.

[10] J. Aaronson, M. Denker, M. Urbański, Ergodic theory for Markov fibred systems and parabolic rational maps. *Trans. Amer. Math. Soc.*, **337**, (1993), 495-548.

[11] J. Aaronson and B. Weiss, On the asymptotics of a 1-parameter family of infinite measure preserving transformations, *Bol. Soc. Brasil. Mat. (N.S.)*, **29**, (1998), 181–193.

[12] R. Adler, *F*-expansions revisited, *Recent Advances in Topological Dynamics*, (S.L.N.Math. 318), Springer, Berlin, Heidelberg, New York, (1973), 1-5.

[13] L. K. Arnold, On σ-finite invariant measures, *Z. Wahrsch. u.v. Geb.*, **9**, (1968), 85-97.

[14] G. D. Birkhoff , Proof of the ergodic theorem, *Proc. Nat. Acad. Sci. USA*, **17**, (1931), 656-660.

[15] G. Boole, On the comparison of transcendents with certain applications to the theory of definite integrals, *Philos. Transac. R. Soc. London*, **147**, (1857), 745-803.

[16] L. Breiman, *Probability*, Addison-Wesley, Reading, Mass., U.S. 1968.

[17] P.J. Brockwell, B.M. Brown, Expansions for positive stable laws, *Z. Wahrsch. u.v. Geb.*, **45**, (1978), 213-224.

[18] R. Chacon, D. Ornstein, A general ergodic theorem, *Illinois J. Math.*, **4**, (1960), 153-160.

[19] K.L. Chung, *Markov Chains with Stationary Transition Probabilities*, Springer, Heidelberg, 1960.

[20] D.A. Darling, M. Kac, On occupation times for Markov processes, *Trans. Amer. Math. Soc.*, **84**, (1957), 444-458.

[21] W. Doeblin, R. Fortet, Sur des chaines a liaison complètes, *Bull. Soc. Math. de France*, **65**, (1937), 132-148.

[22] E. M. Dynkin, Methods of the theory of singular integrals: Hilbert transform and Calderón-Zygmund theory, *Commutative Harmonic Analysis I*, Encyclopaedia of Mathematical Sciences, 15, Springer, Berlin, Heidelberg, New York, 1987.

[23] H. Furstenberg, *Recurrence in Ergodic Theory and Combinatorial Number theory* , Princeton University Press , Princeton, N.J., 1981.

[24] A. Hajian, Y. Ito, S. Kakutani, Invariant measures and orbits of dissipative transformations, *Adv. Math* , **9**, (1972), 52-65.

[25] A. Hajian, S. Kakutani, An example of an ergodic measure preserving transformation defined on an infinite measure space, *Contributions to ergodic theory and probability* (S.L.N. Math. 160), (1970), 45-52.

[26] P. Halmos, *Lectures on Ergodic Theory*, Chelsea, New York, 1956.

[27] M. Halfant, Analytic properties of Renyi's invariant density, *Israel Journal of Math.*, **27**, (1977), 1-20.

[28] T. E. Harris, H. Robbins, Ergodic theory of Markov chains admitting an infinite invariant measure, *Proc. Nat. Acad. Sci. USA*, **39**, (1953), 860-864.

[29] B. Hasselblatt, A. Katok, *Introduction to the Modern Theory of Dynamical Systems* , Cambridge University Press, Cambridge, U.K, 1995

[30] E. Hopf, *Ergodentheorie*, Ergeb. Mat., 5, Springer, Berlin, 1937.

[31] W. Hurewicz, Ergodic theorem without invariant measure, *Ann. Math.*, **45**, (1944), 192-206.

[32] M. Kac, On the notion of recurrence in discrete stochastic processes, *Bull. Amer. Math. Soc.*, **53**, (1947), 1002-1010.

[33] S. Kakutani, Induced measure preserving transformations, *Proc. Imp. Acad. Sci. Tokyo*, **19**, (1943), 635-641.

[34] A.N. Kolmogorov, *Foundations of the Theory of Probability*, Chelsea, New York, 1956.

[35] U. Krengel, *Ergodic Theorems*, de Gruyter, Berlin, 1985

[36] G. Letac, Which functions preserve Cauchy laws? , *Proc. Amer. Math. Soc.*, **67**, (1977), 277-286.

[37] M. Lin, Mixing for Markov operators, *Z. Wahrsch. u.v. Geb.*, **19**, (1971), 231-243.

[38] J. H. Neuwirth, Ergodicity of some mappings of the circle and the line, *Israel Journal of Math.*, **31**, (1978), 359-367.

[39] W. Parry, Ergodic and spectral analysis of certain infinite measure preserving transformat ions, *Proc. Am. Math. Soc.*, **16**, (1965), 960-966.

[40] H. Poincaré, *Les Méthodes Nouvelles de la Mécanique Céleste*, 3, Gauthier-Villars, Paris, 1899

[41] A. Rényi, Representations for real numbers and their ergodic properties, *Acta. Math. Acad. Sci. Hung.*, **8**, (1957), 477-493.

[42] V.A. Rokhlin, Selected topics from the metric theory of dynamical systems, *Uspehi Mat. Nauk.*, **4**, (1949), 57-125; *A.M.S.Transl. Ser. 2*, **49**, (1966)

[43] F. Schweiger, *Ergodic Theory of Fibred Systems and Metric Number Theory*, Clarendon Press, Oxford, 1995

[44] V. V. Stepanov, Sur une extension du theoreme ergodique, *Compositio Math.*, **3**, (1936), 239-253.

[45] M. Thaler, Estimates of the invariant densities of endomorphisms with indifferent fixed points, *Israel Journal of Math.*, **37**, (1980), 303-314.

[46] M. Thaler, Transformations on [0, 1] with infinite invariant measures, *Israel Journal of Math.*, **46**, (1983), 67-96.

[47] M. Thaler, A limit theorem for the Perron-Frobenius operator of transformations on [0,1] with indifferent fixed points, *Israel Journal of Math.*, **91**, (1995), 111-127.

The Multifarious Poincaré Recurrence Theorem

Vitaly Bergelson
Department of Mathematics
The Ohio State University
Columbus, OH 43210

Acknowledgement. I wish to express my sincere appreciation to Steve Cook and Sasha Leibman for their help in preparing these notes. I also would like to thank A. Dynin for useful discussions on the philosophical aspects of Poincaré recurrence theorem. This work was supported by NSF grant DMS-9706057.

1   Some history and background

The Poincaré recurrence theorem (PRT), which one can find in virtually any book on ergodic theory, is usually stated as follows:

PRT *If T is a measure-preserving transformation of a probability space $(\Omega, \mathcal{B}, \mu)$ and $A \in \mathcal{B}$ with $\mu(A) > 0$, then there exists a measurable subset $A_0 \subseteq A$ with $\mu(A_0) = \mu(A)$ such that for any $x \in A_0$ there exists an infinite sequence $(n_i)_{i=1}^{\infty}$ such that $T^{n_i}x \in A_0$ for all i.*

This theorem is considered so basic that some books do not give a reference for it, and those which do either quote [P3] (cf. [Ha] or [Ho]), or refer to the three-volume, 1000-plus-page *Les méthodes nouvelles de la mécanique céleste* [P2], usually without giving the reader any more specific directions. (For the reader's information: the version of PRT appearing in [P2] is to be found in Ch. 26, Vol. 3). Yet prior to 1899, the year of publication of the third volume of *Méthodes nouvelles*, this theorem was at the center of quite stormy discussions related to Zermelo's *Wiederkehreinwand* [1]. (See [Z1], [Z2], [Bo1], [Bo2], [Bo3].) We shall return to Zermelo's argument involving PRT later, but first we want to formulate PRT as it appeared in Poincaré's King Oscar Prize-winning memoir [P1]. (This memoir was itself a source of some controversy. See, for example, [B] or [Gor], Section 1.3.) While not resolving the (still open) problem of the stability of the solar system, this work of Poincaré was, according to Weierstrass, "of such importance that its publication will open a new era in the history of celestial mechanics." The object of our interest in this essay, PRT, is referred to by Poincaré in the introduction to [P1] in the following way:

> J'ai étudié plus spécialement un cas particulier du problème des trois corps, celui où l'une des masses est nulle et où le mouvement des deux autres est circulaire; j'ai reconnu que dans ce cas les trois corps repasseront une infinité de fois aussi près que l'on veut de leur position initiale, à moins que les conditions initiales du mouvement ne soient exceptionnelles.

---

[1] According to M. Moravcsik, translator of [EE], this cumbersome word means something like "objection or counter-argument based on reasoning involving a return to the same state."

Comme on le voit, ces résultats ne nous apprennent que peu de chose sur le cas général du problème; mais ce qui peut leur donner quelque prix, c'est qu'ils sont établis avec rigueur, tandis que le problème des trois corps ne paraissait jusqu'ici abordable que par des méthodes d'approximation successive où l'on faisait bon marché de cette rigueur absolue qui est exigée dans les autres parties des mathématiques.

Here then is Poincaré's original formulation of PRT, *Théorème* I from [P1]. There are only three statements in this 270-page memoir, all of them in Section 8, which Poincaré calls *Théorème*.

Théorème I *Supposons que le point P reste à distance finie, et que le volume $\int dx_1\, dx_2\, dx_3$ soit un invariant intégral; si l'on considère une région r_0 quelconque, quelque petite que soit cette région, il y aura des trajectoires qui la traverseront une infinité de fois.*

After formulating the recurrence theorem, Poincaré first establishes a combinatorial principle (See Principle P below), on the basis of which he proceeds to discuss two different approaches to the question of recurrence or, as he calls it at the beginning of Section 8 in [P1] and in the introduction to Ch. 26 in [P2], *stability in the sense of Poisson*. These two approaches are, essentially, a topological one and a probabilistic, or rather, measure preserving one; and while the modern reader may find it hard to agree with Poincaré's claim of *"rigueur absolue"*, he will undoubtedly recognize in the discussions of Section 8 of [P1] and Ch. 26 of [P2], the familiar elements of modern versions of recurrence theorems. The reader is referred to [C] for the first modern rendition of PRT. See also [Hi2] and [O] for a discussion of a category statement which, according to J. Oxtoby, "has to be read between the lines of Poincaré's discussion."

The combinatorial principle mentioned above is nothing but a "crossbreeding" between the pigeon-hole principle and the *stationarity* assumption. Enhanced further by the possibility of repeated *iterations*, this principle not only leads to PRT and some of its numerous refinements, but, as we shall try to demonstrate, provides a simplified and unified approach to many number-theoretical and combinatorial results. Here is the relevant passage from [P1]:

En effet le point P restant à distance finie, ne sortira jamais d'une région limitée R. J'appelle V le volume de cette région R.

Imaginons maintenant une région très petite r_0, j'appelle v le volume de cette région. Par chacun des points de r_0 passe une trajectoire que l'on peut regarder comme parcourue par un point mobile suivant la loi définie par nos équations différentielles. Considérons donc une infinité de points mobiles remplissant au temps 0 la région r_0 et se mouvant ensuite conformément à cette loi. Au temps τ ils rempliront une certaine région r_1, au temps 2τ une région r_2, etc. au temps $n\tau$ une région r_n. Je puis supposer que τ est assez grand et r_0 assez petit pour que r_0 et r_1 n'aient aucun point commun.

Le volume étant un invariant intégral, ces diverse régions $r_0, r_1, \cdots$, r_n auront même volume v. Si ces régions n'avaient aucun point commun, le volume total serait plus grand que nv; mais d'autre part toute ces régions sont intérieures à R, le volume total est donc plus petit que V. Si donc on a:

$$n > \frac{V}{v},$$

il faut que deux au moins de nos régions aient une partie commune. Soient r_p et r_q ces deux régions $(q > p)$. Si r_p et r_q ont une partie commune, il est clair que r_0 et r_{q-p} devront avoir une partie commune.

Here now is a formulation in modern terms:

Principle P *Let μ be a finitely additive probability measure defined on an algebra $\mathcal{B}$ of subsets of a set X. Assume further that the sets $A_n \in \mathcal{B}$, $n = 0, 1, 2, \cdots$ satisfy, for any $n \geq m \geq 0$, the* stationarity *condition*

$$\mu(A_n \cap A_m) = \mu(A_0 \cap A_{n-m})$$

and that $\mu(A_0) = a > 0$. Then there exists a positive integer $k \leq \left[\frac{1}{a}\right] + 1$ such that $\mu(A_0 \cap A_k) > 0$.

A natural question is: what is the best $\delta = \delta(a)$ such that for some $k > 0$ one has $\mu(A_0 \cap A_k) > \delta$? Taking any sequence of pairwise independent sets (say, $A_n = \bigcup_{i=1}^{2^{n-1}} \left[\frac{2i-2}{2^n}, \frac{2i-1}{2^n}\right] \subseteq [0,1]$) shows that $\delta(a) \leq a^2$. The following useful statement, which we shall call EPP (Enhanced Principle P), supplies quite a satisfactory answer to the question above.

EPP *Under the assumptions of Principle P, for any $0 < \lambda < 1$ there exists $c = c(a, \lambda)$ such that for some $0 < k < c$ one has $\mu(A_0 \cap A_k) > \lambda a^2$.*

Proof. We are going to utilize an idea due to Gillis ([G]). (He worked with σ-additive measures, but, as we shall see, it does not really matter.) Given a simple function $f = \sum_{i=1}^n \alpha_i 1_{A_i}$, write $\int f\, d\mu = \sum_{i=1}^n \alpha_i \mu(A_i)$. It is trivial to see that $\int f\, d\mu$ does not depend on the representation of f. What is important to us is that this limited notion of integral obeys the Cauchy-Schwartz inequality:

$$\left(\int fg\, d\mu\right)^2 \leq \int f^2\, d\mu \int g^2\, d\mu.$$

Remembering that $\mu(X) = 1$, we get for $g \equiv 1$

$$\left(\int f \, d\mu \right)^2 \leq \int f^2 \, d\mu.$$

That is all one needs to conclude the proof of EPP, since if no $c = c(a, \lambda)$ with the desired property exists, the following inequality would be contradictory for large enough n:

$$n^2 a^2 = \left(\int \sum_{i=1}^{n} 1_{A_i} \, d\mu \right)^2 \leq \int \left(\sum_{i=1}^{n} 1_{A_i} \right)^2 d\mu$$

$$= \sum_{i=1}^{n} \mu(A_i) + 2 \sum_{1 \leq i < j \leq n} \mu(A_i \cap A_j).$$

∎

Remark 1.1 The reader may wonder why we bothered to formulate Principle P and its enhanced version in terms of finitely additive rather than countably additive measures. The answer is that in many situations outside the realm of ergodic theory and dynamical systems one often does not have the luxury of countable additivity. We shall see, however, that finite additivity is quite sufficient for many applications of Principle P.

To see that Principle P is all that one needs to prove PRT as stated at the beginning of this section, let us note first that when applied to the sequence $A_n = T^{-n}A$, $n = 0, 1, 2, \cdots$, where $\mu(A) > 0$, Principle P implies the existence of $k \in \mathbb{N}$ such that $\mu(A \cap T^{-k}A) > 0$. For $n \in \mathbb{N}$, let $B_n \subseteq A$ be the set of those $x \in A$ which do not return to A under $S = T^n$. Formally, $B_n = A \cap \left(\bigcap_{i=1}^{\infty} S^{-i}(A^c) \right)$. (In particular, B_n is measurable.) We claim that $\mu(B_n) = 0$. Indeed, if $\mu(B_n) > 0$ then for some $k \in \mathbb{N}$, $\mu(B_n \cap S^{-k}B_n) > 0$. But then for any $x \in B_n \cap S^{-k}B_n$ one has $S^k x = T^{kn}x \in B_n \subseteq A$ which contradicts the definition of B_n. It follows that for any $n \in \mathbb{N}$ the measurable subset $C_n \subseteq A$ defined by $C_n = \{x \in A : \exists m > n : T^m x \in A\}$ satisfies $\mu(C_n) = \mu(A)$. We are done since all the points of the set $A_0 = \bigcap_{n=1}^{\infty} C_n$ return to A infinitely many times and $\mu(A_0) = \mu(\bigcap_{n=1}^{\infty} C_n) = \mu(A)$.

It is hard not to agree with Marc Kac who, after remarking that "there are many proofs of this theorem [i.e. PRT] all of which are almost trivial," added a footnote:

We have here another example of an important and even profound fact whose purely mathematical content is very much on the surface. ([Kac], p. 63)

The examples which we bring in the next section will provide further support to the validity of Kac's remark.

We want to conclude this introductory section with an excerpt from [Z1] in the translation of S.G. Brush ([Br], pp. 208-209). For Boltzmann's response and for the ensuing discussion, the reader is referred to [Br], and for a neat analysis of the *Wiederkehreinwand*, to [EE].

In the second chapter of Poincaré's prize essay on the three-body problem, there is proved a theorem from which it follows that the usual description of the thermal motion of molecules, on which is based for example the kinetic theory of gases, requires an important modification in order that it be consistent with the thermodynamic law of increase of entropy. Poincaré's theorem says that *in a system of mass-points under the influence of forces that depend only on position in space, in general any state of motion (characterized by configurations and velocities) must recur arbitrarily often, at least to any arbitrary degree of approximation even if not exactly, provided that the coordinates and velocities cannot increase to infinity*. Hence, in such a system *irreversible processes are impossible* since (aside from singular initial states) no single-valued continuous function of the state variables, such as entropy, can continually increase; if there is a finite increase, then there must be a corresponding decrease when the initial state recurs. Poincaré, in the essay cited, used his theorem for astronomical discussions on the stability of sun systems; he does not seem to have noticed its applicability to systems of molecules or atoms and thus to the mechanical theory of heat...

2 Combinatorial richness of large sets in countable amenable groups

Recall that a discrete group G is called *amenable* if there exists a finitely additive measure μ on $\mathcal{P}(G)$ such that $\mu(G) = 1$ and for any $g \in G$ and $A \subseteq G$, $\mu(gA) = \mu(A)$ (i.e., μ is left-invariant). The notion of amenability may be defined in many equivalent ways. (One of them, via the Følner condition, is especially useful and will be given below.) The class of amenable groups includes solvable (in particular, abelian) groups as well as the profinite groups, such as, say, the group S_∞ of finite permutations of $\mathbb{N}$. On the other hand, the non-abelian free group $F_2 = \langle a, b \rangle$ is a classical example of a non-amenable group. We cannot resist the temptation to show that all one needs for the proof of this fact is to apply Principle P. The proof is by contradiction. Assume that μ is a left-invariant, finitely-additive probability measure on $\mathcal{P}(F_2)$. Consider the partition $F_2 = A^+ \cup A^- \cup B^+ \cup B^- \cup \{e\}$, where e is the unit of F_2 ($\equiv$ the "empty word") and the sets A^+, A^-, B^+, and

B^- consist of the (reduced) words starting, respectively, with a, a^{-1}, b, and b^{-1}. Since μ is finitely additive, one of these five sets has to have positive measure. Clearly, $\mu(\{e\}) = 0$. (If $\mu(\{e\}) = c > 0$ then, by shift-invariance, $\mu(\{a^n\}) = c$ for all $n \in \mathbb{Z}$ and one gets a contradiction by taking $N \geq \frac{1}{c}$ and considering the set $B = \{a^i,\ i = 0, 1, 2, \cdots, N\}$ which has to satisfy $\mu(B) = \sum_{i=0}^N \mu(\{a^i\}) = (N + 1)c > 1$.) So assume $\mu(A^+) = c > 0$. (The same proof will work for any set of the partition which happens to have positive measure.) Let $A_n = b^n A^+$, $n = 0, 1, 2, \cdots$. By shift-invariance we have, for any $n \geq m \geq 0$:

$$\mu(A_n \cap A_m) = \mu(b^m A^+ \cap b^n A^+) = \mu(A^+ \cap b^{n-m} A^+) = \mu(A_0 \cap A_{n-m}).$$

By Principle P there exists $k \in \mathbb{N}$ such that $\mu(A_0 \cap b^k A_0) = \mu(A^+ \cap b^k A^+) > 0$. But this is impossible since, obviously, $A^+ \cap b^k A^+ = \emptyset$. (No reduced word in F_2 can start with both a and b!)

From now on we shall, for the sake of convenience, deal exclusively with countable groups. It should be remarked, though, that many of the examples and results which we bring in this paper extend to general discrete and topological amenable semigroups.

A sequence of finite sets $\{F_n\}_{n=1}^\infty$ is called a *left Følner sequence*, if for any $g \in G$ one has

$$\lim_{n \to \infty} \frac{|gF_n \triangle F_n|}{|F_n|} = 0.$$

For example, in $\mathbb{Z}$, any sequence of intervals $[a_n, b_n]$ with $|b_n - a_n| \to \infty$ is a Følner sequence. Here is one more useful example. Let G be the direct sum $\mathbb{Z}_p^\infty$ of countably many copies of $\mathbb{Z}_p$. (It is convenient to envision $\mathbb{Z}_p^\infty$ as the set of infinite sequences $(a_1, a_2, \cdots)$ where $a_i \in \mathbb{Z}$, and all but finitely many $a_i = 0$, and addition is defined component-wise modulo p.) Let $F_n = \{(a_1, a_2, \cdots) \in \mathbb{Z}_p^\infty : a_i = 0\ \forall i \geq n + 1\}$. One can easily check that $\{F_n\}$ is a Følner sequence. We shall return to this example later.

The existence of a Følner sequence in a group is tightly related to amenability: a (countable) group is amenable if and only if it has a left Følner sequence. Følner sequences are also helpful in all kinds of Ramsey-theoretical questions since they allow one to define a notion of largeness in a natural way.

Definition 2.1 Given a left Følner sequence $\{F_n\}$ and a set $E \subseteq G$, the *upper density* of E with respect to $\{F_n\}$ is defined by

$$\bar{d}_{\{F_n\}}(E) = \limsup_{n \to \infty} \frac{|E \cap F_n|}{|F_n|}$$

We shall say that a set $E \subseteq G$ is *left-large* if for some left Følner sequence $\bar{d}_{\{F_n\}}(E) > 0$, and *left-conull* if $\bar{d}_{\{F_n\}}(E) = 1$.

According to the principles of Ramsey theory (see [GRS] and [Be3]), large sets, and especially conull sets, ought to be combinatorially rich. We are going to present a few results which substantiate this claim. Before doing so, we collect some useful facts about sets of positive upper density in the following Proposition.

Proposition 2.2 *Let* $\{F_n\}$ *be a left Følner sequence in* G*. For* $E \subseteq G$ *one has:*

(i) $\forall g \in G,\ \bar{d}_{\{F_n\}}(gE) = \bar{d}_{\{F_n\}}(E)$

(ii) *If* $\bar{d}_{\{F_n\}}(E) = c > 0$*, then for any* $0 < \lambda < 1$ *there exists* g *such that* $\bar{d}_{\{F_n\}}(E \cap gE) > \lambda c^2$*. The element* g *can be chosen to lie outside of any prescribed finite set* $F \subseteq G$*.*

Proof. (i) trivially follows from the definition of a Følner sequence. (ii) is just an application of the Enhanced Principle P (See Remark 1.1). ∎

Definition 2.3 Given a subset $\Gamma = \{g_i\}_{i \in I} \subseteq G$ (where I is a finite or countable subset of $\mathbb{N}$), an *FP-set*, generated by Γ, is the set of all finite products of distinct elements of Γ with ascending indices. More formally, writing $\mathcal{F}(I)$ for the set of finite, non-empty subsets of I, we have

$$\mathrm{FP}(\Gamma) = \big\{ g_{i_1} g_{i_2} \cdots g_{i_k},\ i_1 < i_2 < \cdots < i_k,\ \{i_1, \cdots, i_k\} \in \mathcal{F}(\Gamma) \big\}.$$

Remarks 2.4 1. There exists, of course, a dual notion of FP-sets which corresponds to taking products with descending indices. Our choice was dictated by the fact that we are working with *left* Følner sequences.
2. When G is an additive abelian group, one replaces products with sums and speaks of FS-sets. FS-sets which are formed with a countable set of indices I are called, in ergodic theory and topological dynamics, *IP-sets* (a term coined by Furstenberg and Weiss in [FW]). It turns out that many familiar ergodic and dynamical results involving group actions can be extended and refined to actions of IP-sets, which brings, in particular, some strong applications to combinatorics and number theory. See, for example, [FW], [F2], [FK], and [BeM2].

As the following Proposition shows, every large set contains translates of sets of the form $\mathrm{FP}(g_i)_{i=1}^n$ with arbitrarily large n (and pairwise distinct g_i's).

Proposition 2.5 *Let* $E \subseteq G$ *be left-large* $(\bar{d}_{\{F_n\}}(E) > 0)$*. Then for any* n *there are pairwise distinct* $g_0 = e, g_1, \cdots, g_n$ *such that* $\bar{d}_{\{F_n\}}\left(\bigcap_{g \in \mathrm{FP}(g_i)_{i=1}^n} g^{-1}E \right) > 0.$

Remark 2.6 Clearly, any $x \in \displaystyle\bigcap_{g \in \mathrm{FP}(g_i)_{i=1}^n} g^{-1}E$ satisfies $x\mathrm{FP}(g_i)_{i=1}^n \subseteq E.$

Proof. The proposition follows by iteration of property (ii) from Proposition 2.2 above. Let $g_1 \neq e$ be such that $\bar{d}_{\{F_n\}}(E \cap g_1^{-1}E) > 0$. Denoting $E_1 = E \cap g_1 E$, let $g_2 \notin \{e, g_1\}$ be such that $\bar{d}_{\{F_n\}}(E_1 \cap g_2^{-1}E_1) = \bar{d}_{\{F_n\}}(E \cap g_1^{-1} \cap g_2^{-1}E \cap g_2^{-1}g_1^{-1}E) > 0$. Continuing in this fashion one arrives at a sequence $g_0 = e, g_1, \cdots, g_n$ with the desired property. ∎

The simple Proposition which we have just proved contains, as a quite special case, the following result, whose proof occupies more than two pages in [Hi1]. (Hilbert needed it to show that if the polynomial $p(x, y) \in \mathbb{Z}[x, y]$ is irreducible then, for some $c \in \mathbb{N}$, $p(x, c) \in \mathbb{Z}[x]$ is irreducible.)

Proposition 2.7 (Hilbert, [Hi1], pp. 104-107) *For any $k, r \in \mathbb{N}$, if $\mathbb{N} = \bigcup_{i=1}^{r} C_i$, then one of the C_i contains infinitely many translates of a set of the form $\mathrm{FS}(n_i)_{i=1}^{k}$ (where the n_i's are pairwise distinct).*

Proof. Fix, in $\mathbb{N}$, any sequence of intervals $I_n = [a_n, b_n]$ with $|I_n| = |b_n - a_n| \to \infty$ and observe that one of the C_i satisfies $\bar{d}_{\{I_n\}}(C_i) > 0$. Apply Proposition 2.5. ∎

Hilbert's result is a forerunner of a much deeper modern theorem due to Hindman ([Hi3]), which claims that for any finite partition of $\mathbb{N}$, one of the cells of the partition contains an IP-set. We shall give a proof of Hindman's theorem below, but first we prove a related result about conull sets.

Proposition 2.8 *Any left-conull set $E \subseteq G$ contains an IP-set.*

Proof. The proof uses the same iterational idea as Proposition 2.5 above, but since this time we are dealing with a conull set, the iterations can be arranged in a more controlled fashion.

Let us fix a left Følner sequence with respect to which E is left-conull, and let us denote the corresponding upper density by $\bar{d}$. Choose $g_1 \in E$ arbitrarily. Clearly, $\bar{d}(E \cap g_1^{-1}E) = 1$. Pick $g_2 \in E_1 = E \cap g_1^{-1}E$ so that $g_2 \neq g_1$. Observe that $1 = \bar{d}(E_1 \cap g_2^{-1}E_1) = \bar{d}(E \cap g_1^{-1}E \cap g_2^{-1}E \cap g_2^{-1}g_1^{-1}E)$. Notice that any $g_3 \in E_2 = E_1 \cap g_2^{-1}E$ has the property that $\mathrm{FP}(g_i)_{i=1}^{3} \subseteq E$. Continuing in this fashion (and taking care to choose each successive g_k so that $g_k \notin \{g_1, g_2, \cdots, g_{k-1}\}$), we shall arrive at an infinite sequence $(g_i)_{i=1}^{\infty}$ such that for any $n \in \mathbb{N}$, $\mathrm{FP}(g_i)_{i=1}^{n} \subseteq E$. We are done. ∎

Extending the notion of an FP-set, let us, given a set $\Gamma = \{g_i\}_{i \in I} \subseteq G$ and $l \in \mathbb{N}$, define an $\mathrm{FP}^{(l)}$-*set generated by Γ as*

$$\mathrm{FP}^{(l)}(\Gamma) = \left\{ g_{i_1}^{c_1} g_{i_2}^{c_2} \cdots g_{i_k}^{c_k} : i_1 < \cdots < i_k, \{i_1, \cdots, i_k\} \subseteq \mathcal{F}(I), 1 \leq c_i \leq l \right\}.$$

In other words, in $\mathrm{FP}^{(l)}$-sets bounded repetitions of generators are allowed. Of course, $\mathrm{FP}^{(1)}$-sets are just the FP-sets defined above. Note that if G is an abelian group with uniformly bounded torsion, then for large enough l, any $FP^{(l)}$-set in G is a subgroup. As before, when the notation is additive, we shall talk about $FS^{(l)}$-sets, and (when the set of indices I is infinite) $IP^{(l)}$-sets.

While not every conull set in, say, $\mathbb{Z}$ or $\mathbb{Z}_p^\infty$ contains an $FS^{(l)}$-set for $l > 1$, the following Proposition gives a convenient criterion for a conull set to contain $IP^{(l)}$-sets for any l. We remark that (ii) below was proved by M. Karpovsky and V. Milman in [KaM] (by a different method).

Proposition 2.9 (i) *If the intervals* $I_n = [a_n, b_n] \subseteq \mathbb{Z}$, *with* $b_n - a_n \to \infty$, *satisfy* $[a_n, b_n] \subseteq [a_{n+1}, b_{n+1}]$ *for all* $n \in \mathbb{N}$, *then any* $E \subseteq \mathbb{Z}$ *such that* $\bar{d}_{\{I_n\}}(E) = 1$ *contains an* $IP^{(l)}$-*set for any* $l \in \mathbb{N}$.
(ii) *Let* $G = \mathbb{Z}_p^\infty$ *and* $F_n = \{(a_1, a_2, \cdots) \in \mathbb{Z}_p^\infty : a_k = 0 \,\forall k > h\}$. *If* $\bar{d}_{\{F_n\}}(E) = 1$, *then* $E \cup \{0\}$ *contains an infinite subgroup (which is isomorphic to* $\mathbb{Z}_p^\infty$).

Proof. We shall prove (ii) only, the proof of (i) being similar. For $k \in \mathbb{Z} \setminus \{0\}$ let

$$E/k = \{g \in \mathbb{Z}_p^\infty : kg \in E\}.$$

(This definition, of course, makes sense for any abelian group G with additive notation.) Writing $\bar{d}$ for the upper density defined by the sequence of subgroups $\{F_n\}$, observe that if $\bar{d}(E) = 1$, then for any $1 \leq k \leq p - 1$, $\bar{d}(E/k) = 1$. This observation will allow us to prove the desired fact by a simple iterative process. We start with the set $E_1 = \bigcap_{k=1}^{p-1} E/k$. Note that, $\bar{d}(E_1) = 1$. If $x_1 \in E_1, x_1 \neq 0$, then

$$S(x_1) = \{ix_1, i = 0, 1, \cdots, p - 1\} \subseteq E \cup \{0\},$$

and clearly, $S(x_1)$ is isomorphic to $\mathbb{Z}_p$. Now let $E_2 = \bigcap_{i_2=1}^{p-1} \bigcap_{i_1=0}^{p-1} (E - i_1 x_1)/i_2$. Then $\bar{d}(E_2) = 1$ and any $x_2 \in E_2$ such that $x_2 \notin S(x_1)$ has the property that

$$S(x_1, x_2) = \{i_1 x_1 + i_2 x_2 : i_1, i_2 \in \{0, 1, \cdots, p - 1\}\} \subseteq E \cup \{0\}$$

and, in addition, $S(x_1, x_2)$ is isomorphic to $\mathbb{Z}_p \oplus \mathbb{Z}_p$. At the next step one considers the set $E_3 = \bigcap_{i_3}^{p-1} \bigcap_{i_1,i_2=0}^{p-1} (E - i_1 x_1 - i_2 x_2)/i_3$, picks $x_3 \in E_3$ subject to the condition $x_3 \notin S(x_1, x_2)$, and so on. It is clear that, continuing in this fashion, one arrives at an infinite sequence $(x_i)_{i=1}^\infty$ such that for any $n \geq 1$,

$$x_{n+1} \notin S(x_1, x_2, \cdots, x_n) = \{i_1 x_1 + i_2 x_2 + \cdots + i_n x_n :$$
$$\cdot i_1, \cdots, i_n \in \{0, 1, \cdots, p - 1\}\} \subseteq E \cup \{0\}.$$

Since, for every n, $S(x_1, \cdots, x_n)$ is isomorphic to the direct sum of n copies of $\mathbb{Z}_p$, we are done. ∎

3   Principle P and ultrafilters

First, we are going to introduce an important family of finitely additive probability measures, the so-called *ultrafilters*. As we shall see, Hindman's theorem, alluded to in Section 2, follows in a natural way by repeated application of Principle P to a sequence of sets which are large with respect to a conveniently chosen ultrafilter.

We give only the minimal amount of background information on ultrafilters. The interested reader will find the missing details and much more discussion in [CN], [HiS], and [Be3].

Recall that a *filter p* on $\mathbb{N}$ is a set of subsets of $\mathbb{N}$ satisfying the following conditions:

(i) $\emptyset \notin p$.

(ii) $A \in p$ and $A \subseteq B$ imply $B \in p$.

(iii) $A \in p$ and $B \in p$ imply $A \cap B \in p$.

Now, an ultrafilter is a filter which, additionally, has the property

(iv) If $r \in \mathbb{N}$ and $\mathbb{N} = A_1 \cup A_2 \cup \cdots \cup A_r$, then $A_i \in p$ for some i, $1 \le i \le r$.

A rather dull class of examples of ultrafilters is provided by the so-called *principal* ones, which are defined, for any $n \in \mathbb{N}$, by $p_n = \{A \subseteq \mathbb{N} : n \in A\}$. To construct less trivial examples, one has to resort to Zorn's lemma. (One can show that this is unavoidable: see, for example, [CN], pp. 161-162.)

Assume that a family $\mathcal{C}$ of subsets of $\mathbb{N}$ satisfies conditions (i), (ii), and (iii). We claim there is an ultrafilter p such that for any $C \in \mathcal{C}$, one has $C \in p$. To see this, let

$$\tilde{\mathcal{C}} = \{\mathcal{D} \subseteq \mathcal{P}(\mathbb{N}) : \mathcal{D} \text{ satisfies (i), (ii) and (iii), and } \mathcal{C} \subseteq \mathcal{D}\}.$$

Since $\mathcal{C} \in \tilde{\mathcal{C}}$, $\tilde{\mathcal{C}} \ne \emptyset$. Also, the union of any chain in $\tilde{\mathcal{C}}$ is a member of $\tilde{\mathcal{C}}$. By Zorn's lemma, there is a maximal member p of $\tilde{\mathcal{C}}$ which, being maximal, has to satisfy (iv).

Here is a useful example. Let $I_n = [a_n, b_n]$ be a sequence of intervals in $\mathbb{N}$ satisfying $b_n - a_n \to \infty$, and let $\mathcal{C} = \{A \subseteq \mathbb{N} : \bar{d}_{\{I_n\}}(A) = 1\}$. $\mathcal{C}$ satisfies conditions (i), (ii), and (iii), and hence there exists an ultrafilter p such that $\mathcal{C} \subseteq p$. Note that if $B \in p$, then $\bar{d}_{\{I_n\}}(B) > 0$. (Otherwise, $B^c \in \mathcal{C}$.)

The set of ultrafilters on $\mathbb{N}$ is naturally identified with $\beta\mathbb{N}$, the Stone-Čech compactification of $\mathbb{N}$. The sets $\bar{A} = \{p \in \beta\mathbb{N}, A \in p\}$, where $A \subseteq \mathbb{N}$, form a basis for the open sets in $\beta\mathbb{N}$ (and a basis for the closed sets). With this topology $\beta\mathbb{N}$ is a compact Hausdorff space which is, in some respects, rather an odd object. In particular, it has a dense countable subset (namely, the set of principal ultrafilters), but has the cardinality of $\mathcal{P}(\mathcal{P}(\mathbb{N}))$ (and hence is non-metrizable).

Each ultrafilter $p \in \beta\mathbb{N}$ can be naturally identified with a finitely additive, zero-one valued probability measure μ_p on the power set $\mathcal{P}(\mathbb{N})$. Indeed, let $\mu_p(A) = 1$ if and only if $A \in p$. From now on we are going to view ultrafilters as measures, but we will prefer to write $A \in p$ instead of $\mu_p(A) = 1$.

In addition to being a compact Hausdorff space, $\beta\mathbb{N}$ has two natural algebraic structures which are induced by $(\mathbb{N}, +)$ and $(\mathbb{N}, \cdot)$. With respect to each of these, $\beta\mathbb{N}$ is a compact left-topological semigroup. We shall concentrate on the operation which comes from $(\mathbb{N}, +)$.

Definition 3.1 Given $p, q \in \beta\mathbb{N}$, define

$$p + q = \{A \subseteq \mathbb{N} : \{n \in \mathbb{N} : (A - n) \in p\} \in q\},$$

where $A - n$ is the set of all m for which $m + n \in A$.

Remarks 3.2 1. The operation just introduced is nothing but convolution of measures! The reader should find it instructive to compare it with familiar formulas for the convolution of measures μ, ν on a locally compact group G:

$$\mu * \nu(A) = \int_G \nu(x^{-1}A) \, d\mu(x) = \int_G \mu(Ay^{-1}) \, d\nu(y).$$

2. One easily checks that for principal ultrafilters, the operation $+$ corresponds to addition on $\mathbb{N}$.

It is not hard to verify that $p + q \in \beta\mathbb{N}$ and that the operation $+$ is associative. However, a word of warning is in place here: $+$ is, generally speaking, not commutative. One can actually show that the center of the semigroup $(\beta\mathbb{N}, +)$ contains only the principal ultrafilters. One can also show that for any fixed p, the function $f_p(q) = p + q$ is continuous. In other words, the operation $+$ is left-continuous.

By a theorem due to Ellis ([El]), any compact semigroup with a left-continuous operation has an idempotent. It is the idempotent ultrafilters which are the key to understanding Hindman's theorem, which we are going to prove now. Before embarking on the proof, let us look more closely at the notion of an idempotent ultrafilter. Given an ultrafilter p, call a set $C \subseteq \mathbb{N}$ p-*big* if $C \in p$. Assume now that $p \in \beta\mathbb{N}$ satisfies $p + p = p$. By the definition of $+$, this means that

$$A \in p \Longleftrightarrow A \in p + p \Longleftrightarrow \{n \in \mathbb{N} : (A - n) \in p\} \in p.$$

In other words, if p is an idempotent, then A is p-big if and only if for p-many $n \in \mathbb{N}$, the shifted set $(A - n)$ is p-big. This explains why idempotent ultrafilters are often called "almost shift invariant". Indeed, A is p-big if p-almost all shifts of A are also p-big. We are now in a position to state and prove Hindman's theorem.

Theorem 3.3 (Hindman, [Hi3]) *If, for some $r \in \mathbb{N}$, $\mathbb{N} = \bigcup_{i=1}^{r} C_i$ then one of the C_i contains an* IP*-set.*

Proof. We shall prove a stronger fact: if $p \in \beta\mathbb{N}$ is idempotent, then any $C \in p$ contains $\mathrm{FS}(n_j)_{j=1}^{\infty}$ for some increasing sequence $(n_j)_{j=1}^{\infty}$. We start by observing that if a set C is a member of the idempotent ultrafilter p, then the basic conclusion of Principle P is satisfied: for some n (and actually for p-almost all n) one has $C \cap (C - n) \in p$. Since $C \cap \{n : (C - n) \in p\} \in p$ as well, we see that one can always pick $n_1 \in C$ so that $C \cap (C - n_1) \in p$. The rest of the proof is virtually identical to that of Proposition 2.8 and follows by iteration. Let $C_1 = C \cap (C - n_1)$. Pick $n_2 \in C_1$ so that $C_2 = C_1 \cap (C_1 - n_2) \in p$. Since p is an idempotent, it is a non-principal ultrafilter, and hence its members are infinite sets. This allows us always to assume that any new element chosen from a member of p lies outside any given finite set. In particular, we can assume that $n_2 > n_1$. Now, $n_2 \in C_1 = C \cap (C - n_1)$ implies that $\mathrm{FS}(n_i)_{i=1}^{2} \subseteq C$. Continuing in the same fashion, let $n_3 \in C_2$ be such that $C_2 \cap (C_2 - n_3) \in p$, and $n_3 > n_1 + n_2$. Notice that $n_3 \in C_2 = C_1 \cap (C_1 - n_2) = C \cap (C - n_1) \cap (C - n_2) \cap (C - (n_1 + n_2))$ implies $\mathrm{FS}(n_i)_{i=1}^{3} \subseteq C$. And so on! The sequence $(n_i)_{i=1}^{\infty}$ created this way will have the property that for any $k \in \mathbb{N}$, $\mathrm{FS}(n_i)_{i=1}^{k} \subseteq C$. We are done. ∎

4 The law of return of large sets

As we saw in Sections 1 and 2, a typical application of Principle P is to ensure that large sets return to themselves under transformations which preserve the notion of largeness. For example, to prove PRT, one first establishes the fact that if $\mu(A) > 0$, then for some $n \in \mathbb{N}$, $\mu(A \cap T^{-n}A) > 0$. Hilbert's theorem (Proposition 2.7) hinges on a similar statement: if $\bar{d}(E) > 0$, then for some $n \neq 0$, $\bar{d}(E \cap (E - n)) > 0$. Finally, Hindman's theorem also starts with the same kind of statement: if $p \in \beta\mathbb{N}$ is idempotent and $C \in p$, then for p-many n one has $(C - n) \in p$ and hence $C \cap (C - n) \in p$.

In this Section we shall have a closer look at the phenomenon of the return of large sets and discuss some of its applications and refinements.

Given an abelian group $(G, +)$ and a set $S \subseteq G$, let $S - S = \{x - y : x, y \in S\}$. One often encounters results which can be expressed as follows: if S is large, then $S - S$ is very large. On many occasions, such statements are just simple corollaries of the law of return of large sets. Perhaps the best known result of this kind is the following useful theorem due to Steinhaus ([St]).

Theorem 4.1 *If A is a Lebesgue measurable subset of $\mathbb{R}$ with $\mu(A) > 0$, then $A - A$ contains an open interval around 0.*

Proof. All that one needs to demonstrate is a form of "local" Poincaré recurrence: if $\mu(A) > 0$, then for all small enough x, $\mu(A \cap (A - x)) > 0$.

This, in turn, follows directly from Lebesgue's theorem about points of density (which says that for almost every $x \in A$ one has $\lim_{\varepsilon \downarrow 0} \frac{\mu(A \cap (x-\varepsilon, x+\varepsilon))}{2\varepsilon} = 1$), but can also be shown to follow almost immediately from the mere definition of Lebesgue measure. The following is essentially Steinhaus' original argument. Assuming without loss of generality that $\mu(A) < \infty$, let (I_n) be a sequence of open intervals which cover A and satisfy $\sum \mu(I_n) < \frac{4}{3}\mu(A)$. It is easy to see that one of the I_n, call it I, satisfies $\mu(I \cap A) > \frac{3}{4}\mu(I)$ (otherwise, $\mu(A) \leq \sum \mu(A \cap I_n) \leq \frac{3}{4} \sum \mu(I_n) < \mu(A)$, a contradiction). But then for any x satisfying $|x| < \frac{1}{2}\mu(I)$ one has $\mu(A \cap (A-x)) \geq \mu((A \cap I) \cap ((A \cap I)-x)) > 0$ (otherwise, $\mu((A \cap I) \cup ((A \cap I) - x)) = 2\mu(A \cap I) > \frac{3}{2}\mu(I)$, which would contradict $\mu(I \cup (I - x)) < \frac{3}{2}\mu(I)$). We are done. ∎

The argument used in the above proof may be iterated to show that if $\mu(A) > 0$, then for sufficiently small $|x_i|$, $i = 1, 2, \cdots, k$, $\mu(A \cap (A - x_1) \cap \cdots \cap (A - x_k)) > 0$. This gives us the following refinement of Theorem 4.1.

Theorem 4.2 *If $A \subseteq \mathbb{R}$ is a Lebesgue measurable set of positive measure, then A contains an affine image of any finite subset of $\mathbb{R}$.*

Proof. Given $F = \{x_1, \cdots, x_k\} \subseteq \mathbb{R}$, observe that for small enough $t > 0$, one has

$$\mu(A \cap (A - tx_1) \cap (A - tx_2) \cap \cdots \cap (A - tx_k)) > 0.$$

This clearly implies that for some $a \in A$

$$a + tF = \{a + tx_i : i = 1, 2, \cdots, k\} \subseteq A.$$

∎

Here is a more recent result which deals with different notions of large and very large, but surely fits the pattern we are interested in.

Theorem 4.3 ([BeS]) *Let F be an infinite field and Γ a multiplicative subgroup of finite index in F^*. (F^* denotes the multiplicative group of the field F.) Then*

$$\Gamma - \Gamma = \{x - y : x, y \in \Gamma\} = F.$$

Theorem 4.3 has a "finitistic" version which says that if $n \in \mathbb{N}$ is fixed and a finite field F is large enough, then $\{x^n - y^n : x, y \in F\} = F$. (Note that $\{x^n, x \in F^*\}$ is a multiplicative subgroup.) As a corollary, one obtains an old result of Dickson ([D]): for fixed n and large enough prime p, the equation $x^n - y^n \equiv z^n \pmod{p}$ has non-trivial solutions (i.e. solutions with $x, y, z \neq 0$).

It was Schur who, in 1916 ([Sch]), gave a simple proof of Dickson's result. Schur's proof uses the following lemma, which is a (very) special case of Hindman's theorem.

Proposition 4.4 ([Sch]) *If* $r \in \mathbb{N}$ *and* $\mathbb{N} = \bigcup_{i=1}^{r} C_i$, *then one of the* C_i *contains* x, y, z *such that* $x + y = z$.

Schur's lemma, in turn, follows from the following result on returns of large sets.

Proposition 4.5 *Let* $I_n = [a_n, b_n]$, $n \in \mathbb{N}$, *be a sequence of increasing intervals in* $\mathbb{N}$. *If* $\mathbb{N} = \bigcup_{i=1}^{r} C_i$, *then one of the* C_i, *call it* C, *has the property that for some* $x \in C$, $\bar{d}_{\{I_n\}}(C \cap (C - x)) > 0$.

To see that Proposition 4.5 implies Proposition 4.4, notice that if $y \in C \cap (C - x)$, then x, y, and $z = x + y$ all lie in C. For a short proof and further discussion of Proposition 4.5 the reader is referred to [Be1]. See also [BeM1] for a treatment of Schur-type theorems in general amenable groups.

Motivated by Theorem 4.3, one may ask whether any set $A \subseteq \mathbb{N}$ with $\bar{d}(A) = \limsup_{n \to \infty} \frac{|A \cap \{1, \cdots, n\}|}{n} > 0$ contains an affine image of any finite set. The answer is yes, but the corresponding result is nontrivial and is, actually, equivalent to the following deep theorem due to Szemerédi.

Theorem 4.6 ([Sz]) *If* $A \subseteq \mathbb{N}$ *satisfies* $\bar{d}(A) > 0$, *then* A *contains arbitrarily long arithmetic progressions.*

Corollary 4.7 *If* $A \subseteq \mathbb{N}$, $\bar{d}(A) > 0$, *and* $F = \{x_1, \cdots, x_k\} \subseteq \mathbb{N}$, *then for some* $a \in A$ *and* $d \in \mathbb{N}$, *one has* $a + dF = \{a + dx_i : i = 1, 2, \cdots, k\} \subseteq A$.

Proof. Let $m \geq \max F$. By Szemerédi's theorem, there are a and d such that $\{a, a + d, a + 2d, \cdots, a + md\} \subseteq A$. Clearly $a + dF = \{a + dx_i : i = 1, 2, \cdots, k\} \subseteq A$. ■

Remark 4.8 One also immediately observes that Corollary 4.7, in its turn, implies Szemerédi's theorem. (Just take F to be of the form $\{1, 2, \cdots, m\}$.)

The proof of Szemerédi's theorem in [Sz] is elementary but very involved. A completely different, ergodic theoretic approach to Szemerédi's theorem and, indeed, to a variety of problems belonging to Ramsey Theory, was initiated by Furstenberg ([F1]), who derived Szemerédi's theorem from a far-reaching extension of Poincaré's recurrence theorem, which corresponds to $k = 1$ in the following.

Theorem 4.9 (Furstenberg, [F1]) *Let* $(X, \mathcal{B}, \mu, T)$ *be a probability measure-preserving system. For any* $k \in \mathbb{N}$, *and for any* $A \in \mathcal{B}$ *with* $\mu(A) > 0$, *there exists* $n \in \mathbb{N}$ *such that* $\mu(A \cap T^{-n}A \cap T^{-2n}A \cap \cdots \cap T^{-kn}A) > 0$.

In order to derive Szemerédi's theorem from Theorem 4.9, Furstenberg introduced a correspondence principle which allows one to translate recurrence results in ergodic theory into statements about returns of large sets.

Theorem 4.10 (Furstenberg's correspondence principle) *Given a set* $E \subseteq \mathbb{Z}$ *with*

$$d^*(E) = \lim_{N-M \to \infty} \frac{|E \cap \{M, M+1, \cdots, N\}|}{N-M+1} > 0,$$

there exists a probability measure-preserving system $(X, \mathcal{B}, \mu, T)$ *and a set* $A \in \mathcal{B}$, $\mu(A) = d^*(E)$, *such that for any* $k \in \mathbb{N}$ *and any* $n_1, n_2, \cdots, n_k \in \mathbb{Z}$ *one has:*

$$d^*\big(E \cap (E - n_1) \cap \cdots \cap (E - n_k)\big) \geq \mu(A \cap T^{-n_1}A \cap \cdots \cap T^{-n_k}A).$$

Remark 4.11 The quantity $d^*(E)$ featured in the formulation of Furstenberg's correspondence principle is called *upper Banach density*. Clearly, if $d^*(E) > 0$, then for some sequence of intervals $I_n = [a_n, b_n]$ with $|b_n - a_n| \to \infty$ one has $d^*(E) = \bar{d}_{\{I_n\}}(E) > 0$.

Furstenberg's seminal paper started a whole new area, Ergodic Ramsey Theory. See [F2] and [Be3] for further information. See also the recent work of Gowers ([Go1], [Go2]) for an approach to Szemerédi's theorem, which provides a strong estimate for the number $N(k, \delta)$, defined as the minimal natural number such that every subset of $\{1, 2, \cdots, n\}$ containing more than δn elements must contain a length k arithmetic progression whenever $n \geq N(k, \delta)$.

5 A generalization of Khintchine's recurrence theorem

Recall that a set S in a countable abelian group G is called *syndetic* (or, sometimes, relatively dense) if there exists a finite set $F \subseteq G$ such that $S + F = \{x + y : x \in S, y \in F\} = G$. The following result, originally proved by Khintchine ([Kh]) for measure-preserving $\mathbb{R}$-actions, is usually called Khintchine's recurrence theorem.

Theorem 5.1 *For any invertible probability measure-preserving system* $(X, \mathcal{B}, \mu, T)$, *any* $0 < \lambda < 1$, *and any* $A \in \mathcal{B}$ *with* $\mu(A) > 0$, *the set*

$$\{n \in \mathbb{Z} : \mu(A \cap T^n A) > \lambda \mu(A)^2\}$$

is syndetic.

Khintchine's recurrence theorem immediately follows from the following corollary to the classical von Neumann theorem.

Theorem 5.2 *For any probability measure-preserving system $(X, \mathcal{B}, \mu, T)$ and any $A \in \mathcal{B}$ one has:*

$$\lim_{N-M\to\infty} \frac{1}{N-M} \sum_{n=M}^{N-1} \mu(A \cap T^{-n}A) \geq \mu(A)^2.$$

In this section we are going to show that similar results hold for the second iteration – that is, for the analogous expression for $\mu((A \cap T^n A) \cap T^m(A \cap T^n A))$ – as well. Namely, we are going to establish the following theorems.

Theorem 5.3 *Let $(X, \mathcal{B}, \mu, T)$ be an invertible probability measure-preserving system. Then:*

(i) *For any $f_1, f_2, f_3 \in L^\infty(X, \mathcal{B}, \mu)$*

$$\lim_{N-M\to\infty} \frac{1}{(N-M)^2} \sum_{n,m=M}^{N-1} f_1(T^n x) f_2(T^m x) f_3(T^{n+m} x)$$

exists in L^2,

(ii) *For any $A \in \mathcal{B}$ with $\mu(A) > 0$*

$$\lim_{N-M\to\infty} \frac{1}{(N-M)^2} \sum_{n,m=M}^{N-1} \mu(A \cap T^n A \cap T^m A \cap T^{n+m} A) \geq \mu(A)^4.$$

Corollary 5.4 *For any invertible measure-preserving system $(X, \mathcal{B}, \mu, T)$, any $0 < \lambda < 1$, and any $A \in \mathcal{B}$, $\mu(A) > 0$, the set*

$$\{(n, m) \in \mathbb{Z}^2 : \mu((A \cap T^n A) \cap T^m(A \cap T^n A)) > \lambda\mu(A)^4\}$$

is syndetic.

Remark 5.5 Although the original paper [Kh], as well as numerous books on ergodic theory (see, for example, [Ho],[Pa],[Pe]), derive Khintchine's recurrence theorem from a much stronger Theorem 5.2, one can give a very simple proof based on the enhanced Principle P, which, moreover, works for measure-preserving actions of arbitrary (not necessarily amenable) semi-groups. (See [Be3], Section 5 for details.) Unfortunately, this simple approach does not seem to be easily modifiable to enable one to prove the two-parameter version, Corollary 5.4.

Before embarking on the proof of Theorem 5.3 we want to make some remarks and review some facts that we are going to use.

First of all, since Theorem 5.3 is trivial when μ is atomic (and since one can treat the atomic part of μ separately), we are going to assume that the measure μ is non-atomic. Having made this assumption, we can further assume that the measure space $(X, \mathcal{B}, \mu)$ is a Lebesgue space. To see this, note that given the measure-preserving transformation T and functions f_1, f_2, f_3 featured in the formulation of Theorem 5.3 (in part (ii) one takes $f_1 = f_2 = f_3 = 1_A$), we may restrict ourselves to a T-invariant separable sub-σ-algebra of $\mathcal{B}$, with respect to which all the functions $T^n f_i$, $n \in \mathbb{Z}$, $i = 1, 2, 3$ are measurable. Now, by Carathéodory's theorem (see [R], Ch. 15, Theorem 4), any separable atomless σ-algebra of subsets of a probability space is isomorphic to the σ-algebra $\mathcal{L}$ induced by Lebesgue measure on the unit interval. This isomorphism carries T into a Lebesgue measure-preserving isomorphism of $\mathcal{L}$ which, by a theorem due to von Neumann ([R], Ch. 15, Theorem 20), admits a realization as a point mapping. Having assumed that $(X, \mathcal{B}, \mu)$ is Lebesgue, we can (and will) further assume that T is ergodic with respect to μ. Indeed, if the measure-preserving system $(X, \mathcal{B}, \mu, T)$ is not ergodic, one considers the ergodic decomposition of μ, defined by

$$\mu(A) = \int_Y \mu_\omega(A) \, d\nu(\omega), \quad A \in \mathcal{B},$$

where μ_ω are T-invariant ergodic measures on $(X, \mathcal{B})$, indexed by elements of a Lebesgue space $(Y, \mathcal{D}, \nu)$ (where the measure ν may have atoms). It is not hard to see that the validity of Theorem 5.3 for ergodic measure-preserving systems $(X, \mathcal{B}, \mu_\omega, T)$ implies its validity for $(X, \mathcal{B}, \mu, T)$. To see that Corollary 5.4 is also implied by its validity in the ergodic case, one argues as follows. Assume that for any $\omega \in Y$ one has

$$\lim_{N-M \to \infty} \frac{1}{(N-M)^2} \sum_{n,m=M}^{N-1} \mu_\omega(A \cap T^n A \cap T^m A \cap T^{n+m} A) \geq \mu_\omega(A)^4.$$

Then we have

$$\lim_{N-M \to \infty} \frac{1}{(N-M)^2} \sum_{n,m=M}^{N-1} \mu(A \cap T^n A \cap T^m A \cap T^{n+m} A)$$

$$= \lim_{N-M \to \infty} \frac{1}{(N-M)^2} \sum_{n,m=M}^{N-1} \int_Y \mu_\omega(A \cap T^n A \cap T^m A \cap T^{n+m} A) \, d\nu(\omega)$$

$$= \int_Y \left(\lim_{N-M \to \infty} \frac{1}{(N-M)^2} \sum_{n,m=M}^{N-1} \mu_\omega(A \cap T^n A \cap T^m A \cap T^{n+m} A) \right) d\nu(\omega)$$

$$\geq \int_Y \mu_\omega(A)^4 \, d\nu(\omega) \geq \left(\int_Y \mu_\omega(A) \, d\nu(\omega) \right)^4 = \mu(A)^4.$$

We shall also need the following simple fact.

Proposition 5.6 *If the invertible measure preserving system* $(X, \mathcal{B}, \mu, T)$ *is ergodic, then for any* $h, f, g \in L^\infty(X, \mathcal{B}, \mu)$ *one has*

$$\lim_{N-M\to\infty} \frac{1}{(N-M)^2} \sum_{n,m=M}^{N-1} \int h(T^n x) f(T^m x) g(T^{n+m} x)\, d\mu(x)$$

$$= \int h\, d\mu \int f\, d\mu \int g\, d\mu.$$

Proof. We show first that

$$\frac{1}{(N-M)^2} \sum_{n,m=M}^{N-1} f(T^{m-n} x) g(T^m x) \to \int f\, d\mu \int g\, d\mu$$

in L^2 as $N - M \to \infty$. To verify this assertion, observe first that without loss of generality one can assume that $\int f\, d\mu = 0$. It follows from von Neumann's ergodic theorem that for every $\varepsilon > 0$ there exists $C > 0$ such that if $N - M > C$, then

$$\left\| \frac{1}{N-M} \sum_{n=M}^{N-1} f(T^{m-n} x) \right\|_2 < \varepsilon$$

uniformly in m. Hence, for $N - M > C$ one has

$$\left\| \frac{1}{(N-M)^2} \sum_{n,m=M}^{N-1} f(T^{m-n} x) g(T^m x) \right\|_2$$

$$= \left\| \frac{1}{N-M} \sum_{m=M}^{N-1} g(T^m x) \left(\frac{1}{N-M} \sum_{n=M}^{N-1} f(T^{m-n} x) \right) \right\|_2$$

$$\leq \frac{1}{N-M} \sum_{m=M}^{N-1} \left\| g(T^m x) \left(\frac{1}{N-M} \sum_{n=M}^{N-1} f(T^{m-n} x) \right) \right\|_2$$

$$\leq \varepsilon \|g\|_\infty.$$

Since ε was arbitrary, it follows that, under our assumptions,

$$\lim_{N-M\to\infty} \frac{1}{(N-M)^2} \sum_{n,m=M}^{N-1} f(T^{m-n} x) g(T^m x) = 0,$$

which implies that, for general $f \in L^\infty(X, \mathcal{B}, \mu)$,

$$\lim_{N-M\to\infty} \frac{1}{(N-M)^2} \sum_{n,m=M}^{N-1} f(T^{m-n} x) g(T^m x) = \int f\, d\mu \int g\, d\mu.$$

We have now:

$$\lim_{N-M\to\infty} \frac{1}{(N-M)^2} \sum_{n,m=M}^{N-1} \int h(T^n x) f(T^m x) g(T^{n+m} x) \, d\mu(x)$$

$$= \lim_{N-M\to\infty} \frac{1}{(N-M)^2} \sum_{n,m=M}^{N-1} \int h(x) f(T^{m-n} x) g(T^m x) \, d\mu(x)$$

$$= \int h(x) \Big(\lim_{N-M\to\infty} \frac{1}{(N-M)^2} \sum_{n,m=M}^{N-1} f(T^{m-n} x) g(T^m x) \Big) \, d\mu(x)$$

$$= \int h \, d\mu \int f \, d\mu \int g \, d\mu.$$

■

Finally, we shall need a two-parameter version of the so-called van der Corput trick, which is often helpful in dealing with multiple recurrence. (See, for example, [Be1] and [BeM2], Lemma A6.)

Proposition 5.7 *Assume that* $(x_{n,m})_{n,m\in\mathbb{N}}$ *is a double bounded sequence of vectors in a Hilbert space. If*

$$\lim_{H\to\infty} \frac{1}{H^2} \sum_{h_1,h_2=1}^{H} \left| \lim_{N-M\to\infty} \frac{1}{(N-M)^2} \sum_{n,m=M}^{N-1} \langle x_{n,m}, x_{n+h_1,m+h_2} \rangle \right| = 0,$$

then

$$\lim_{N-M\to\infty} \left\| \frac{1}{(N-M)^2} \sum_{n,m=M}^{N-1} x_{n,m} \right\| = 0.$$

We are now ready to start the proof of Theorem 5.3. Let $\mathcal{H} = L^2(X, \mathcal{B}, \mu)$. We are going to utilize the well-known decomposition $\mathcal{H} = \mathcal{H}_{\mathrm{d}} \oplus \mathcal{H}_{\mathrm{wm}}$, where the orthogonal subspaces $\mathcal{H}_{\mathrm{d}}$ and $\mathcal{H}_{\mathrm{wm}}$ correspond to the *discrete spectrum* and *weak mixing* of the unitary operator induced by T (i.e. $(Tf)(x) := f(Tx)$), and are defined by

$$\mathcal{H}_{\mathrm{d}} = \overline{\mathrm{Span}\{f : \exists \lambda : Tf = \lambda f\}},$$

$$\mathcal{H}_{\mathrm{wm}} = \mathcal{H}_{\mathrm{d}}^{\perp} = \Big\{ f \in \mathcal{H} : \forall g \ \lim_{N\to\infty} \frac{1}{N} \sum_{n=1}^{N} |\langle T^n f, g \rangle| = 0 \Big\}.$$

Remark 5.8 We shall actually need the following equivalent definition of $\mathcal{H}_{\mathrm{wm}}$:

$$\mathcal{H}_{\mathrm{wm}} = \Big\{ f \in \mathcal{H} : \forall g \ \lim_{H\to\infty} \frac{1}{H^2} \sum_{h_1,h_2=1}^{H} |\langle T^{h_1+h_2} f, g \rangle| = 0 \Big\}.$$

Let $f_i = \phi_i + \psi_i$, $\phi_i \in \mathcal{H}_d$, $\psi_i \in \mathcal{H}_d^\perp$, $i = 1, 2, 3$, be the corresponding decompositions of f_i. Substituting into the expression

$$\frac{1}{(N-M)^2} \sum_{n,m=M}^{N-1} f_1(T^n x) f_2(T^m x) f_3(T^{n+m} x), \qquad (*)$$

we shall get a representation of $(*)$ as a sum of eight expressions of the form

$$\frac{1}{(N-M)^2} \sum_{n,m=M}^{N-1} g_1(T^n x) g_2(T^m x) g_3(T^{n+m} x),$$

where each of the g_i's lies either in $\mathcal{H}_d$ or in $\mathcal{H}_d^\perp = \mathcal{H}_{wm}$. We shall show first that if at least one of the g_i's belongs to $\mathcal{H}_{wm}$, then

$$\lim_{N-M \to \infty} \left\| \frac{1}{(N-M)^2} \sum_{n,m=M}^{N-1} g_1(T^n x) g_2(T^m x) g_3(T^{n+m} x) \right\|_2 = 0.$$

Let, for instance, $g_3 \in \mathcal{H}_{wm}$ (the other six cases are verified in a similar fashion). Let $x_{n,m} = g_1(T^n x) g_2(T^m x) g_3(T^{n+m} x)$. We are going to apply Proposition 5.7. We have

$$\langle x_{n,m}, x_{n+h_1, m+h_2} \rangle = \int T^n g_1 T^m g_2 T^{n+m} g_3 T^{n+h_1} g_1 T^{m+h_2} g_2 T^{n+m+h_1+h_2} g_3 \, d\mu$$

$$= \int T^n (g_1 T^{h_1} g_1) T^m (g_2 T^{h_2} g_2) T^{n+m} (g_3 T^{h_1+h_2} g_3) \, d\mu.$$

By Proposition 5.6, this expression converges to

$$\int g_1 T^{h_1} g_1 \, d\mu \int g_2 T^{h_2} g_2 \, d\mu \int g_3 T^{h_1+h_2} g_3 \, d\mu.$$

Note that since $g_3 \in \mathcal{H}_{wm}$, one has, by Remark 5.8,

$$\lim_{H \to \infty} \frac{1}{H^2} \sum_{h_1, h_2 = 1}^{H} \left| \langle T^{h_1+h_2} g_3, g_3 \rangle \right| = 0.$$

Since $g_1, g_2 \in L^\infty$, this implies

$$\lim_{H \to \infty} \frac{1}{H^2} \sum_{h_1, h_2} \left| \int g_1 T^{h_1} g_1 \, d\mu \int g_2 T^{h_2} g_2 \, d\mu \int g_3 T^{h_1+h_2} g_3 \, d\mu \right| = 0,$$

and hence, in accordance with Proposition 5.7,

$$\lim_{N-M \to \infty} \left\| \frac{1}{(N-M)^2} \sum_{n,m=M}^{N-1} x_{n,m} \right\|_2$$

$$= \lim_{N-M \to \infty} \left\| \frac{1}{(N-M)^2} \sum_{n,m=M}^{N-1} g_1(T^n x) g_2(T^m x) g_3(T^{n+m} x) \right\|_2 = 0.$$

Now, to finish the proof of part (i) of Theorem 5.3, we have only to show that

$$\lim_{N-M\to\infty} \frac{1}{(N-M)^2} \sum_{n,m=M}^{N-1} f_1(T^n x) f_2(T^m x) f_3(T^{n+m} x)$$

exists whenever $f_1, f_2, f_3 \in \mathcal{H}_{\mathrm{d}}$. But this is almost obvious. Indeed, each $f \in \mathcal{H}_{\mathrm{d}}$ has a representation $f = \sum_\lambda a_\lambda f_\lambda$, where $f_\lambda(Tx) = \lambda f_\lambda(x)$, and it is enough to verify the convergence for finite approximations of the form $\tilde{f}_i = \sum_{k=1}^{L} a_{\lambda_k}^{(i)} f_{\lambda_k}$, for which the convergence statement is trivial.

We now turn our attention to part (ii) of Theorem 5.3. Let $A \in \mathcal{B}$ with $\mu(A) > 0$. Let $g = 1_A = f + h$, where $f \in \mathcal{H}_{\mathrm{d}}$ and $h \in \mathcal{H}_{\mathrm{d}}^\perp = \mathcal{H}_{\mathrm{wm}}$. Note that since g is bounded, f is bounded as well. In view of the proof of part (i) we have

$$\lim_{N-M\to\infty} \frac{1}{(N-M)^2} \sum_{n,m=M}^{N-1} g(T^n x) g(T^m x) g(T^{n+m} x)$$

$$= \lim_{N-M\to\infty} \frac{1}{(N-M)^2} \sum_{n,m=M}^{N-1} f(T^n x) f(T^m x) f(T^{n+m} x).$$

It follows that

$$\lim_{N-M\to\infty} \frac{1}{(N-M)^2} \sum_{n,m=M}^{N-1} \mu(A \cap T^n A \cap T^m A \cap T^{n+m} A)$$

$$= \lim_{N-M\to\infty} \frac{1}{(N-M)^2} \sum_{n,m=M}^{N-1} \int g(x) g(T^n x) g(T^m x) g(T^{n+m} x)\, d\mu$$

$$= \int g(x) \lim_{N-M\to\infty} \frac{1}{(N-M)^2} \sum_{n,m=M}^{N-1} g(T^n x) g(T^m x) g(T^{n+m} x)\, d\mu$$

$$= \int (f(x) + h(x)) \lim_{N-M\to\infty} \frac{1}{(N-M)^2} \sum_{n,m=M}^{N-1} f(T^n x) f(T^m x) f(T^{n+m} x)\, d\mu$$

$$= \lim_{N-M\to\infty} \frac{1}{(N-M)^2} \sum_{n,m=M}^{N-1} \int f(x) f(T^n x) f(T^m x) f(T^{n+m} x)\, d\mu.$$

(We used the fact that f is bounded and that the product of bounded functions from $\mathcal{H}_{\mathrm{d}}$ belongs to $\mathcal{H}_{\mathrm{d}}$.) Taking into account that the constant functions belong to $\mathcal{H}_{\mathrm{d}}$ and that $1_A = g = f + h$ with $f \in \mathcal{H}_{\mathrm{d}}$, $h \in \mathcal{H}_{\mathrm{d}}^\perp$ implies $\int f\, d\mu = \int g\, d\mu = \mu(A)$, we see that in order to prove (ii) it is enough to establish the following.

Proposition 5.9 *Let $(X, \mathcal{B}, \mu, T)$ be an ergodic measure-preserving system. If $f \in \mathcal{H}_d$, where f is bounded and non-negative, then*

$$\lim_{N-M\to\infty} \frac{1}{(N-M)^2} \sum_{n,m=M}^{N-1} \int fT^n fT^m fT^{n+m} f \, d\mu \geq \left(\int f \, d\mu \right)^4.$$

Proof. We shall need some basic facts about the eigenvalues and eigenfunctions of the unitary operators induced by measure-preserving transformations. (For details see [Ha] and [CFS].) Here is a summary of what we are going to use. (Warning: some of the facts below are true for ergodic transformations only.)

(a) The set $\Gamma = \{\lambda \in \mathbb{C} : \exists f \in L^2(X, \mathcal{B}, \mu) : Tf = \lambda f\}$ is a subgroup of the unit circle. Since we are dealing with Lebesgue spaces, this group is countable. Any eigenvalue $\lambda \in \Gamma$ has multiplicity one.

(b) The eigenfunctions corresponding to distinct eigenvalues are orthogonal. They have constant modulus and will be assumed to be normalized so that each of them will satisfy the condition $|f(x)| = 1$ a.e. We shall also assume that if $\lambda_1, \lambda_2 \in \Gamma$ and $f_{\lambda_1}, f_{\lambda_2}$ are the corresponding eigenfunctions, then $f_{\lambda_1} f_{\lambda_2} = f_{\lambda_1 \lambda_2}$.

We return to the proof of Proposition 5.9. Fix a set $\{f_\lambda\}_{\lambda \in \Gamma}$ of eigenfunctions so that the conditions described in item (b) above are satisfied. Let $f = \sum_{\lambda \in \Gamma} a_\lambda f_\lambda$ be the expansion of our function f in this basis. Note that $f_1 = 1$ a.e., $a_1 = \int f \, d\mu$, and also, for any $\lambda \neq 1$, $\int f_\lambda \, d\mu = 0$. Substituting $f = \sum a_\lambda f_\lambda$ into the integral

$$\int fT^n fT^m fT^{m+n} f \, d\mu$$

and changing the order of integration and summation, we shall arrive at the sum of terms of the form (where the sum will be taken over ρ, λ, τ and ν)

$$K_{\rho\lambda\tau\nu} = \int a_\rho f_\rho T^n (a_\lambda f_\lambda) T^m (a_\tau f_\tau) T^{n+m} (a_\nu f_\nu) \, d\mu$$

$$= \lambda^n \tau^m \nu^{n+m} \int a_\rho a_\lambda a_\tau a_\nu f_\rho f_\lambda f_\tau f_\nu \, d\mu$$

$$= (\lambda\nu)^n (\tau\nu)^m a_\rho a_\lambda a_\tau a_\nu \int f_{\rho\lambda\tau\nu} \, d\mu.$$

The contribution of such a term to the double Cesaro limit

$$\lim_{N-M\to\infty} \frac{1}{(N-M)^2} \sum_{n,m=M}^{N-1}$$

will be non-zero only if $\lambda\nu = 1$, $\tau\nu = 1$, and $f_{\rho\lambda\tau\nu} = f_1 = 1$ (which implies $\rho\lambda\tau\nu = 1$). The three conditions on ρ,λ,τ,ν imply $\lambda = \tau$, $\rho = \nu$, and hence

$a_\rho a_\lambda a_\tau a_\nu = a_\lambda^2 a_\nu^2$. Now, since $f \geq 0$, f is equal to its complex conjugate: $\bar{f} = f$, which gives

$$\sum \bar{a}_\lambda \bar{f}_\lambda = \sum a_\lambda f_\lambda$$

or

$$\sum \bar{a}_\lambda f_{\bar{\lambda}} = \sum a_{\bar{\lambda}} f_{\bar{\lambda}}.$$

From the uniqueness of the expansion we get $a_{\bar{\lambda}} = \bar{a}_\lambda$. Since $\lambda = \bar{\nu}$, we have

$$a_\rho a_\lambda a_\tau a_\nu = a_\lambda^2 a_\nu^2 = a_\rho^2 a_\nu^2 = \bar{a}_\nu^2 a_\nu^2 = |a_\nu|^4 \geq 0.$$

We showed that each time a term of the form $K_{\rho\lambda\tau\nu}$ gives a non-zero contribution to our double Cesaro limit, this contribution is non-negative. Also, at least one $K_{\rho\lambda\tau\nu}$ – namely, the one corresponding to $\rho = \lambda = \tau = \nu = 1$ – gives the contribution

$$\int a_1^4 f_1 \, d\mu = \left(\int f \, d\mu \right)^4 \int f_1 \, d\mu = \left(\int f \, d\mu \right)^4.$$

We are done. ∎

Theorem 5.3 can be easily derived from the following more general result which may be proved by a similar argument.

Theorem 5.10 *Let G be a countable abelian group and $(X, \mathcal{B}, \mu, \{T_g\}_{g\in G})$ a probability measure-preserving system. Let $\{F_n\}_{n=1}^\infty$ be a Følner sequence in $G \times G$. Then:*

(i) For any $f_1, f_2, f_3 \in L^\infty(X, \mathcal{B}, \mu)$

$$\lim_{n\to\infty} \frac{1}{|F_n|} \sum_{g,h\in F_n} f_1(T_g x) f_2(T_h x) f_3(T_{g+h} x)$$

exists in L^2.

(ii) For any $A \in \mathcal{B}$ with $\mu(A) > 0$

$$\lim_{n\to\infty} \frac{1}{|F_n|} \sum_{g,h\in F_n} \mu(A \cap T_g A \cap T_h A \cap T_{g+h} A) \geq \mu(A)^4.$$

Corollary 5.11 *For any countable abelian group G, any measure-preserving system $(X, \mathcal{B}, \mu, \{T_g\}_{g\in G})$, any $0 < \lambda < 1$, and any $A \in \mathcal{B}$ with $\mu(A) > 0$ the set*

$$\left\{ (g,h) \in G \times G : \mu(A \cap T_g A \cap T_h A \cap T_{g+h} A) > \lambda\mu(A)^4 \right\}$$

is syndetic.

It would be interesting to see whether Theorem 5.10 and Corollary 5.11 generalize further to higher order iterations and noncommutative group actions.

References

[B] J. Barrow-Green, *Poincaré and the Three Body Problem*, Amer. Math. Soc., Providence; London Math. Soc., London, 1997.

[Be1] V. Bergelson, A density statement generalizing Schur's theorem, *J. Combinatorial Theory (A)* **43** (1986), 338-343.

[Be2] V. Bergelson, Weakly mixing PET, *Ergodic Theory and Dynamical Systems*, **7** (1987), 337-349.

[Be3] V. Bergelson, Ergodic Ramsey Theory—an Update, Ergodic Theory of $\mathbb{Z}^d$-actions, M. Pollicott and K. Schmidt, editors, *London Math. Soc. Lecture Notes Series* **228**, Cambridge University Press (1996), 1-61.

[BeM1] V. Bergelson and R. McCutcheon, Recurrence for semigroup actions and a non-commutative Schur theorem, *Topological Dynamics and Applications* (Minneapolis, MN, 1995), *Contemporary Math.* **215**, Amer. Math. Soc., Providence, 1998, 205-222.

[BeM2] V. Bergelson and R. McCutcheon, An ergodic IP polynomial Szemerédi theorem, *Memoirs of AMS* (to appear).

[BeS] V. Bergelson and D. Shapiro, Multiplicative subgroups of finite index in a ring, *Proc. Amer. Math. Soc.* **119** (1993), 1127-1134.

[Bo1] L. Boltzmann, Entgegnung auf die wärmetheoretischen Betrachtungen des Hrn. E. Zermelo, *Annalen der Physik* **57** (1896), 773-784.

[Bo2] L. Boltzmann, Zu Hrn. Zermelo's Abhandlung "Über die mechanische Erklärung irreversibler Vorgänge", *Annalen der Physik* **59** (1896), 392-398.

[Bo3] L. Boltzmann, Über einen mechanischen Satz von Poincaré, *Wien Ber.* **106** (1897), p. 12.

[Br] S.G. Brush, *Kinetic Theory. Vol. 2. Irreversible Processes*, Pergamon Press, 1966.

[C] C. Carathéodory, Über den Wiederkehrsatz von Poincaré, *S. B. Preuss. Akad. Wiss.*, (1919), 580-584.

[CN] W. Comfort and S. Negrepontis, *The Theory of Ultrafilters*, Springer-Verlag, Berlin and New York, 1974.

[CFN] I. Cornfeld, S. Fomin, and Y. Sinai, *Ergodic Theory*, Springer-Verlag, 1982.

[D] L. Dickson, Lower limit for the number of sets of solutions of $x^l + y^l + z^l \equiv 0 \pmod{p}$, *J. Reine Angew. Math.* **135** (1908), 181-188.

[EE] P. Ehrenfest and T. Ehrenfest, *The Conceptual Foundations of the Statistical Approach in Mechanics* (translation of *Begriffliche Grundlagen der statistischen Auffassung in der Mechanik*), Dover, 1990.

[El] R. Ellis, Distal transformation groups, *Pac. J. Math.* **8** (1958), 401-405.

[F1] H. Furstenberg, Ergodic behavior of diagonal measures and a theorem of Szemerédi on arithmetic progressions, *J. d'Analyse Math.* **31** (1977), 204-256.

[F2] H. Furstenberg, *Recurrence in Ergodic Theory and Combinatorial Number Theory*, Princeton University Press, 1981.

[FK] H. Furstenberg and Y. Katznelson, An ergodic Szemerédi theorem for IP-systems and combinatorial theory, *J. d'Analyse Math.* **45** (1985), 117-168.

[FW] H. Furstenberg and B. Weiss, Topological dynamics and combinatorial number theory, *J. d'Analyse Math.* **34** (1978), 61-85.

[GRS] R. Graham, B. Rothschild, and J. Spencer, *Ramsey Theory*, Wiley, New York, 1980.

[Gi] J. Gillis, Note on a property of measurable sets, *J. London Math. Soc.* **11** (1936), 139-141.

[Gor] D. Goroff, Introduction to [P2].

[Gow1] W.T. Gowers, A new proof of Szemerédi's theorem for arithmetic progressions of length four, *Geom. Funct. Anal.* **8** (1998), 529-55.

[Gow2] W.T. Gowers, A new proof of Szemerédi's theorem, preprint.

[Ha] P. Halmos, *Lectures on Ergodic Theory*, Chelsea Publishing Co., New York, 1956.

[Hi1] D. Hilbert, Über die Irreducibilität ganzer rationaler Functionen mit ganzzahligen Coefficienten, *J. Math.* **110** (1892), 104-129.

[Hi2] H. Hilmy, Sur les théoremes de récurrence dans la dynamique générale, *Amer. J. Math.* **61** (1939), 149-160.

[Hi3] N. Hindman, Finite sums from sequences within cells of a partition of **N**, *J. Combinatorial Theory* (Series A) **17** (1974), 1-11.

[HiS] N. Hindman and D. Strauss, Algebra in the Stone-Čech compactification. Theory and applications, *de Gruyter Expositions in Mathematics* **27**, Walter de Gruyter, 1998.

[Ho] E. Hopf, *Ergodentheorie*, Chelsea, New York, 1948.

[Kac] M. Kac, *Probability and Related Topics in the Physical Sciences*, Interscience Publishers, 1959.

[KaM] M. Karpovsky and V. Milman, On subspaces contained in subsets of finite homogeneous spaces, *Discrete Mathematics* **22**, (1978), 273-280.

[Kh] A. Y. Khintchine, Eine Verschärfung des Poincaréschen "Wiederkehr-satzes", *Comp. Math.* **1** (1934), 177-179.

[O] J. Oxtoby, *Measure and Category*, Second edition, Springer-Verlag, 1980.

[Pa] W. Parry, *Topics in Ergodic Theory*, Cambridge Univ. Press, Cambridge, 1981.

[Pe] K. Petersen, *Ergodic Theory*, Cambridge Univ. Press, 1983.

[P1] H. Poincaré, Sur le problème des trois corps et les équations de la Dynamique, *Acta Mathematica* **13** (1890), 1-270.

[P2] H. Poincaré, *New Methods of Celestial Mechanics* (translation of *Les méthodes nouvelles de la mécanique céleste* I (1892), II (1893), and III (1899)), D. Goroff, editor, Amer. Inst. of Physics, New York, 1993.

[P3] H. Poincaré, *Calcul des probabilités*, *Gauthier-Villars*, 1912.

[R] H. Royden, *Real Analysis*, Third edition, Prentice Hall, 1988.

[Sch] I. Schur, Über die Kongruenz $x^m + y^m \equiv z^m \pmod p$, *Jahresbericht der Deutschen Math.-Ver.* **25** (1916),114-117.

[St] H. Steinhaus, Sur les distances des ensembles de mésure positive, *Fund. Math.* 1 (1920), 93-104.

[Sz] E. Szemerédi, On sets of integers containing no k elements in arithmetic progression, *Acta Arith.* **27** (1975), 199-245.

[Z1] E. Zermelo, Über einen Satze der Dynamik und die mechanischen Wärmetheorie, *Annalen der Physik* **57** (1896), 485-494.

[Z2] E. Zermelo, Über die mechanische Erklärung irreversibler Vorgänge, *Annalen der Physik* **59** (1896), 793-801.

Groups of Automorphisms of a Measure Space and Weak Equivalence of Cocycles

Sergey Bezuglyi
The Institute for Low Temperature Physics and Engineering
47, Lenin Ave
310164 Kharkov, Ukraine

This survey is devoted to a brief exposition of results proved mainly in papers [BG4, BG5, GS3, GS4]. We study ergodic countable approximately finite groups Γ of non-singular automorphisms of a measure space (X, μ) and cocycles $\alpha : X \times \Gamma \to G$ taking values in a l.c.s.c group G. The concept of weak equivalence of pairs (Γ, α) (which can be treated as a generalization of orbit equivalence of countable automorphism groups) was introduced and studied in these articles. All pairs (Γ, α) and $(\Gamma, \alpha \times \rho)$ can be classified by associated Mackey actions of G and $G \times \mathbb{R}$ where ρ is the Radon-Nikodym cocycle of Γ. The structure of cocycles up to weak equivalence is described. It is shown that the proved results can be applied to the solution of the outer conjugacy problem. Other applications of weak equivalence of cocycles are also considered.

Acknowledgment. The author was supported in part by INTAS-97 grant.

0   Introduction

Cocycles have been the subject of extensive investigations in the ergodic theory during the last thirty years. They appear naturally under solution of many problems because a cocycle over a single transformation of a measure space is represented by a measurable function taking values in a group G. An important role of cocycles is provided by the possibility to construct the new group actions which reflect the basic properties of the initial dynamical system. The constructions of a skew product action and the Mackey action associated to a cocycle have been known and explored for a long time [A, Ma]. In such a way, cocycles turn to be linked with actions of groups. It is well known that these constructions for two cohomologous cocycles over an automorphism group lead to isomorphic corresponding the skew products and Mackey actions but the converse statement is not true in general.

We will mainly study approximately finite (a.f.) groups of automorphisms Γ of a measure space. They can be defined as those that are orbit equivalent to a single automorphism. We call two cocycles over such a dynamical system weakly equivalent if they are cohomologous up to an automorphism from the normalizer of this dynamical system (see Definition 1.6). This equivalence relation is the subject of our investigations. One can treat this notion as either an equivalence relation on the set of all cocycles α over an a.f. group Γ or a generalization of the orbit equivalence relation if we consider cocycles over different automorphism groups. The latter means that we introduce the weak equivalence relation on the set of pairs (Γ, α). When we study non-singular groups Γ, then we need to take into account the Radon-Nikodym cocycle ρ and, therefore, consider the pairs $(\Gamma, \alpha \times \rho)$ where $\alpha \times \rho$ takes values into $G \times \mathbb{R}$.

All countable ergodic approximately finite groups were classified up to orbit equivalence due to H.A.Dye and W.Krieger [D1, D2, K1, K2]. It was proved that any two finite (infinite) measure preserving automorphisms are orbit equivalent and the class of orbit equivalence of a non-singular automorphism is defined by the associative flow. It is natural to find out whether there exists a complete systems of invariants for weak equivalence relation. The main result proved in reviewing papers can be formulated as follows: two cocycles (or two pairs) are weakly equivalent if and only if their Mackey actions are isomorphic. It is worth to note that the notion of weak equivalence of cocycles appeared for the first time in [GS1] (see also [GS3]) for cocycles with dense range in a group. The general case was studied in papers [BG4, BG5, GS4] for different classes of groups and independently in [F] where another technique was applied. In contrast to [F], our approach is based on the study of cocycle structure. Remark that the paper [AaHSch] is also devoted to the discussed problems.

The article is organized as follows. In Section 1, we first remind the main notions of orbit equivalence theory such as full groups of automorphisms, normalizers, cocycles, and approximative finiteness. We do this, first of all, for convenience of the reader who is not familiar with orbit equivalence. Then we discuss the definition of weak equivalence of pairs (Γ, α) and some constructions related to this definition where Γ is an ergodic countable group of automorphisms and α is a cocycle. In our exposition, we use the classification of cocycles offered by K. Schmidt [Sch1] that is mainly based on the notion of ratio set. This notion is well known for the Radon-Nikodym derivative (cocycle) and was used by W. Krieger for solution of orbit equivalence problem for nonsingular automorphisms [K1, K2]. According to [Sch1], one can distinguish regular, recurrent, transient cocycles. The very important class of cocycles consists of those with dense range in a group. It turns out that any two cocycles with dense range over an a.f. automorphism group Γ are cohomologous up to an element from the normalizer Γ (i.e. weakly equivalent). Section 2 contains the results related to the solution of weak equivalence problem for pairs (Γ, α) and $(\Gamma, \alpha \times \rho)$. We consider transient and regular cocycles separately because in these cases we do not need to assume that Γ is approximately finite. In general case, it is proved that any ergodic non-singular action of G is isomorphic to the Mackey action associated to a pair (Γ, α) (existence theorem). Then we describe the structure of cocycles over Γ and based on such a description, it is shown that the Mackey action is a complete invariant of weak equivalence (uniqueness theorem). To illustrate some applications of these results (see Section 3), we prove that the outer conjugacy problem for actions of countable amenable groups (solved earlier in [BG1, BG2, BG3, CK]) is a consequence of the uniqueness theorem. The other application is concerned actions of countable amenable groups induced from a subgroup in the sense of R.J. Zimmer. We find a sufficient condition under which an action is isomorphic to an induced action. It should be mentioned that there are other applications of the results mentioned above. It turns out the uniqueness theorem follows also from a generalization of the Connes-Krieger theorem to actions of discrete non-principal amenable groupoids (see [BG5]). Apart from, the uniqueness and existence theorems are used when an ergodic action of a normal subgroup $H \subset G$ is extended to an action of G [B].

1   Groups of automorphisms and cocycles

In this section, we present some basic facts on ergodic theory of countable automorphism groups of a measure space and cocycles over such groups. More detailed results may be found, for instance, in [HO, Sch1, Sch2, Ro, Z1].

1.1 Automorphism groups of a measure space. The term *measure space* will always stand for a triple $(X, \mathcal{B}, \mu)$ where $(X, \mathcal{B})$ is a standard Borel space and μ is a σ-finite nonatomic measure on $(X, \mathcal{B})$.

Let T be a Borel automorphism of $(X, \mathcal{B})$ and let μ be a measure on $(X, \mathcal{B})$; then T is called a *nonsingular* automorphism of $(X, \mathcal{B}, \mu)$ when $\mu(TA) = 0$ if and only if $\mu(A) = 0$, $A \in \mathcal{B}$, i.e. $T^{-1} \cdot \mu \sim \mu$.

We say that two objects defined on a measure space $(X, \mathcal{B}, \mu)$ (for example, automorphisms, sets, partitions, functions, etc.) are identical *mod* 0 if they agree on a conull subset, i.e. almost everywhere (a.e.). The corresponding terminology will be omitted in those cases where this does not lead to a misunderstanding.

Denote by $\mathrm{Aut}(X, \mathcal{B}, \mu)$ the set of identical *mod* 0 nonsingular automorphisms of $(X, \mathcal{B}, \mu)$. According to the above convention, we will not distinguish individual automorphisms and elements of $\mathrm{Aut}(X, \mathcal{B}, \mu)$ which also will be called the nonsingular automorphisms (see [Ro, Z1]). The set $\mathrm{Aut}(X, \mathcal{B}, \mu)$ is a group with the composition of automorphisms as a group operation. The identical automorphism of X is denoted by 1 or 1_X.

Let G be a locally compact second countable (l.c.s.c.) group with the identity e (this notation is fixed from now on). Let $a : G \times X \to X : (g, x) \mapsto a(g, x)$ be a Borel map such that (i) $a(gh, x) = a(g, a(h, x))$; (ii) $a(e, x) = x$; (iii) $x \mapsto a(g, x)$ is a one-to-one nonsingular map for each $g \in G$. Then a is called an *action* of G on $(X, \mathcal{B}, \mu)$. In this way, a homomorphism $\tau : g \mapsto (x \mapsto a(g, x)) : G \to \mathrm{Aut}(X, \mathcal{B}, \mu)$ is determined.

A countable subgroup $\Gamma \subset \mathrm{Aut}(X, \mathcal{B}, \mu)$ is called an *automorphism group*. The set $\Gamma x = \{\gamma x \; : \; \gamma \in \Gamma\}$ is the *orbit* of x with respect to Γ. A subset $A \in \mathcal{B}$ is called Γ-*invariant* if $\gamma A = A$ for all $\gamma \in \Gamma$. Γ is called *ergodic* if for any Γ-invariant Borel set $A \subset X$ either $\mu(A) = 0$ or $\mu(X - A) = 0$. We say that a Γ is *free* if the map $\gamma \mapsto \gamma x : \Gamma \to X$ is injective for a.e. $x \in X$.

In this paper, we will consider, as a rule, ergodic freely acting countable groups of automorphisms.

The following notions, introduced by H.A.Dye in [D1,D2], are of a crucial importance in the orbit theory of dynamical systems.

Definition 1.1. Let Γ be a countable automorphism group of $(X, \mathcal{B}, \mu)$. The set $[\Gamma] = \{R \in \mathrm{Aut}(X, \mathcal{B}, \mu) : Rx \in \Gamma x \text{ for a.e. } x \in X\}$ is called the *full group* generated by Γ. The set $N[\Gamma] = \{R \in \mathrm{Aut}(X, \mathcal{B}, \mu) : R[\Gamma]R^{-1} = [\Gamma]\}$ is called the *normalizer* of $[\Gamma]$. The full group generated by a single automorphism T is denoted by $[T]$.

Definition 1.2. A countable group $\Gamma \subset \mathrm{Aut}(X, \mathcal{B}, \mu)$ is said to be of *type I* if there exists a positive measure subset $A \in \mathcal{B}$ such that $\gamma A \cap A = \emptyset$, $\gamma \in \Gamma$, $\gamma \neq 1$, and $\cup_\gamma \gamma A = X$. A free countable group $\Gamma \subset \mathrm{Aut}(X, \mathcal{B}, \mu)$ is called *conservative* if for any positive measure subset $A \in \mathcal{B}$

$$\mu(\bigcup_{\gamma \in \Gamma}(A \cap \gamma^{-1}A)) > 0.$$

An ergodic group of automorphisms Γ is said to be of *type II* if there is a measure ν equivalent to μ such that $\gamma \cdot \nu = \nu$, $\gamma \in \Gamma$. If a Γ-invariant measure ν is finite (σ-finite), then Γ is said to be of *type II$_1$* (*II$_\infty$*). Γ is said to be of *type III* if there is no Γ-invariant measure equivalent to μ.

In [D1], the orbit equivalence relation on the set of countable automorphism groups was introduced.

Definition 1.3. Let Γ_i be a countable group of nonsingular automorphisms of $(X_i, \mathcal{B}_i, \mu_i)$, $i = 1, 2$. The groups Γ_1 and Γ_2 are said to be *orbit equivalent* if there exists a Borel isomorphism $\varphi : (X_1, \mathcal{B}_1, \mu_1) \to (X_2, \mathcal{B}_2, \mu_2)$ such that $\varphi\Gamma_1 x = \Gamma_2 \varphi x$ for μ_1-a.e. $x \in X_1$. Equivalently,

$$\varphi[\Gamma_1]\varphi^{-1} = [\Gamma_2], \quad \varphi \cdot \mu_1 \sim \mu_2.$$

It is easily seen that, for given countable group $\Gamma \subset \mathrm{Aut}(X, \mathcal{B}, \mu)$, the following properties are invariant with respect to orbit equivalence:
(i) Γ is ergodic;
(ii) Γ is conservative;
(iii) Γ is of type I;
(iv) Γ is of type II$_1$, II$_\infty$, or III (if Γ is ergodic).

In this context, H.A. Dye introduced and studied a very important class of automorphism groups (we give not original but an equivalent definition).

Definition 1.4. Let $\Gamma \subset \mathrm{Aut}(X, \mathcal{B}, \mu)$ be a countable ergodic automorphism group. Then Γ is called *approximately finite* (a.f.) (or *hyperfinite*) if Γ is orbit equivalent to the group $\{T^n : n \in \mathbb{Z}\}$ where $T \in \mathrm{Aut}(X, \mathcal{B}, \mu)$.

It turned out that, due to articles [D1, D2, K1, K2], all a.f. groups can be classified with respect to orbit equivalence.

Clearly, any action of $\mathbb{Z}$ is a.f. It is natural to ask for what other kinds of groups one can conclude that any their action is automatically a.f. The answer was given in [CFW]: *for any amenable group G, any action $\tau(G)$ on a measure space is a.f.*

1.2 Cocycles. The other central notion in the orbit theory is that of cocycles of dynamical systems. Since 70's, they have been studied in many papers from various points of view (see, for example, [AaHSch, BG1, BG2, FHM, FM, GS1, Ma, Sch1, Sch2, Z1]).

Let $(X, \mathcal{B}, \mu)$ and Γ be as above, and let G be a Borel second countable group.

Definition 1.5. A Borel map $\alpha : X \times \Gamma \to G$ is called a *cocycle* ("1-cocycle") over Γ if for any γ_1, γ_2 from Γ and for a.e. $x \in X$

$$\alpha(x, \gamma_2\gamma_1) = \alpha(\gamma_1 x, \gamma_2)\alpha(x, \gamma_1), \qquad (1.1)$$

$$\mu(\bigcup_{\gamma \in \Gamma}(\{x \in X : \gamma x = x\} \cap \{x \in X : \alpha(x, \gamma) \neq e\})) = 0$$

with e being the identity in G. The set of all cocycles is denoted by $Z^1(X \times \Gamma, G)$. Two cocycles $\alpha, \beta : X \times \Gamma \to G$ are called *cohomologous* (or Γ-cohomologous) (in symbols, $\alpha \sim \beta$) if there is a measurable map ("transfer function") $\varphi : X \to G$ such that for each $\gamma \in \Gamma$

$$\beta(x, \gamma) = \varphi(\gamma x)\alpha(x, \gamma)\varphi(x)^{-1}$$

for a.e. x. If $\alpha \sim 1$, i.e. $\alpha(x, \gamma) = \varphi(\gamma x)\varphi(x)^{-1}$ for some measurable $\varphi : X \to G$, then α is called a *coboundary*. The set of all coboundaries is denoted by $B^1(X \times \Gamma, G)$.

It follows from (1.2) that, for a.e. $x \in X$, the equality $\gamma x = \gamma_1 x$ implies $\alpha(x, \gamma) = \alpha(x, \gamma_1)$. This remark allows us to think of a cocycle $\alpha \in Z^1(X \times \Gamma, G)$ to be determined over the full group $[\Gamma]$ or, in other words, over the equivalence relation $\mathcal{R}_\Gamma$ on X generated by the partition into Γ-orbits.

If G is an abelian group, then $Z^1(X \times \Gamma, G)$ and $B^1(\Gamma \times X, G)$ are groups, and one can introduce the first cohomology group $H^1(X \times \Gamma, G) = Z^1(X \times \Gamma, G)/B^1(X \times \Gamma, G)$ [Sch2].

Example. 1) Let $\Gamma \in \mathrm{Aut}(X, \mathcal{B}, \mu)$ be a nonsingular transformation group and

$$\rho_\mu(x, \gamma) = \log \frac{d(\gamma^{-1} \cdot \mu)}{d\mu}(x).$$

Then ρ_μ is called the Radon-Nikodym cocycle. Clearly, $\rho_\mu \sim \rho_\nu$ if and only if ν is equivalent to μ.

2) All cocycles of an aperiodic $\mathbb{Z}$-action can be easily described. Specifically, for an action τ of $\mathbb{Z}$ on $(X, \mathcal{B}, \mu)$ and $\alpha \in Z^1(X \times \tau(\mathbb{Z}), G)$, set up $f(x) = \alpha(x, \tau(1))$. Then a consecutive application of the cocycle identity (1.1) allows one to express $\alpha(x, \tau(n))$ in terms of f for any n. Conversely, any measurable function $f : X \to G$ determines in this way a $\mathbb{Z}$-cocycle.

1.3 Weak equivalence. Let $\Gamma_i \subset \mathrm{Aut}(X_i, \mathcal{B}_i, \mu_i)$, $i = 1, 2$, be an ergodic countable group of automorphisms and G a l.c.s.c. group. Suppose that Γ_1 and Γ_2 are orbit equivalent. This means that there exists a one-to-one measurable map $\varphi : X_1 \to X_2$ such that $\varphi[\Gamma_1]\varphi^{-1} = [\Gamma_2]$ and $\varphi \cdot \mu_1 \sim \mu_2$. Let $\beta \in Z^1(X_2 \times \Gamma_2, G)$, then β can be "transferred" by means of φ onto Γ_1 as follows:

$$\varphi^{-1} \cdot \beta(x_1, \gamma_1) = \beta(\varphi x_1, \varphi \gamma_1 \varphi^{-1}), \quad (x_1, \gamma_1) \in X_1 \times \Gamma_1.$$

This relation defines a one-to-one correspondence between $Z^1(X_1 \times \Gamma_1, G)$ and $Z^1(X_2 \times \Gamma_2, G)$. It is clear that, in this case, Γ_1-cohomologous cocycles correspond to Γ_2-cohomologous ones and conversely.

Let us consider the set $\mathcal{A}$ of all pairs (Γ, α) where $\Gamma \subset \mathrm{Aut}(X, \mathcal{B}, \mu)$ is a countable ergodic transformation group and $\alpha \in Z^1(X \times \Gamma, G)$. Now we introduce the definition of an equivalence relation on the set $\mathcal{A}$. We will study this equivalence relation for the class of a.f. groups.

Definition 1.6. We call two pairs (Γ_1, α_1) and (Γ_2, α_2) from $\mathcal{A}$ to be *weakly equivalent* if there exists a measurable one-to-one map φ that provides an orbit equivalence of Γ_1 and Γ_2 and such that $\varphi^{-1} \cdot \alpha_2$ is Γ_1-cohomologous to α_1. If (Γ, α_1) and (Γ, α_2) are weakly equivalent, then we also call α_1 and α_2 weakly equivalent.

If we take $\alpha_i = \rho_i$, $i = 1, 2$, where ρ_i is the Radon-Nikodym cocycle of Γ_i, then Definition 1.6 is actually nothing more than the definition of orbit equivalence for automorphism groups. Therefore from this point of view, one can treat the notion of weak equivalence for pairs (Γ, α) as a generalization of orbit equivalence for transformation groups.

Let $\chi_{\mathbb{Z}}$ be the Haar measure on $\mathbb{Z}$ (the counting measure), and let τ denote the shift on $\mathbb{Z} : \tau(i) = i + 1$, $i \in \mathbb{Z}$. Let also $\Gamma \subset \mathrm{Aut}(X, \mathcal{B}, \mu)$ be a countable ergodic automorphism group. Consider the direct product group $\tilde{\Gamma} \subset \mathrm{Aut}(X \times \mathbb{Z}, \mu \times \chi_{\mathbb{Z}})$, $\tilde{\Gamma} = \Gamma \times \{\tau^n : n \in \mathbb{Z}\}$. If $\alpha \in Z^1(X \times \Gamma, G)$, then define $\tilde{\alpha}(x, i, \gamma, \tau^n) = \alpha(x, \gamma)$ which is a cocycle from $Z^1(X \times \mathbb{Z} \times \tilde{\Gamma}, G)$. The pair $(\tilde{\Gamma}, \tilde{\alpha})$ is called a *countable expansion* of (Γ, α).

Definition 1.7. The pairs (Γ_1, α_1) and (Γ_2, α_2) are called *stably weakly equivalent* if their countable expansions $(\tilde{\Gamma}_1, \tilde{\alpha}_1)$ and $(\tilde{\Gamma}_2, \tilde{\alpha}_2)$ are weakly equivalent.

One can easily show that if Γ_i, $i = 1, 2$, is either of type II_∞ or type III transformation group and $\alpha_i \in Z^1(X_i \times \Gamma_i, G)$, then (Γ_1, α_1) and (Γ_2, α_2) are weakly equivalent if and only if they are stably weakly equivalent.

For $A \in \mathcal{B}$ with $\mu(A) > 0$ we denote $[\Gamma]_A = \{R \in [\Gamma] : Rx = x$ for a.e. $x \in X - A\}$ where Γ is an ergodic countable automorphism group of $(X, \mathcal{B}, \mu)$. The subgroup $[\Gamma]_A$ is called *induced* on the set A. One can prove that there is a countable group $\Gamma_A \subset [\Gamma]$ of automorphisms of A such that $[\Gamma_A] = [\Gamma]_A$ [HO].

Let $B \subset X$, $\mu(B) > 0$, and $\Gamma \in \mathrm{Aut}(X, \mathcal{B}, \mu)$ be an ergodic countable group of automorphisms. For given $\alpha \in Z^1(X \times \Gamma, G)$, one can define a cocycle $\alpha_B \in Z^1(B \times \Gamma_B, G)$ as follows: $\alpha_B(x, \gamma_B) = \alpha(x, \gamma)$ where $\gamma \in \Gamma$ is found from the relation $\gamma_B x = \gamma x$.

If Γ is either of type III or type II_∞ and $\mu(B) = \infty$, $\Gamma \cdot \mu = \mu$, then (Γ, α) and (Γ_B, α_B) are weakly equivalent. If Γ is either of type II_1 or type II_∞ and $\mu(B) < \infty$, $\Gamma \cdot \mu = \mu$, then (Γ, α) and (Γ_B, α_B) are stably weakly equivalent. One can construct an example of pairs which are stably weakly equivalent but not weakly equivalent [F].

If a group $\Gamma \subset \mathrm{Aut}(X, \mathcal{B}, \mu)$ is of type III, that is Γ has a non-trivial Radon-Nikodym cocycle $\rho(\gamma, x)$, it is natural to consider instead of $\alpha \in Z^1(\Gamma \times X, G)$ the cocycle $\alpha_0 = (\alpha, \rho) \in Z^1(\Gamma \times X, G \times \mathbb{R})$. When we study the problem of weak equivalence, we will have to do this because to classify pairs (Γ, α) we need to take into account how the measure is transformed under the action of Γ.

1.4 Mackey actions. There is the well known construction of an action of the group G associated to a given pair (Γ, α) [Ma]. Namely, let $(X, \mathcal{B}, \mu)$ be a measure space, Γ an ergodic countable transformation group of $(X, \mathcal{B}, \mu)$, and $\alpha \in Z^1(X \times \Gamma, G)$, with G being a l.c.s.c. group. Define on $(X \times G, \mu \times \chi_G)$, where χ_G is the Haar measure on G, a group of automorphisms $\Gamma(\alpha)$, together with an action of G. Specifically, set up for $\gamma \in \Gamma$, $g \in G$ and a.e. $(x, h) \in X \times G$:

$$\gamma(\alpha)(x, h) = (\gamma x, \alpha(x, \gamma)h),$$
$$V(g)(x, h) = (x, hg^{-1}).$$

The automorphisms $\gamma(\alpha)$, $\gamma \in \Gamma$, constitute the group $\Gamma(\alpha)$, which is called the *skew product*. Evidently, $V(g)\gamma(\alpha) = \gamma(\alpha)V(g)$, $g \in G$, $\gamma \in \Gamma$. Therefore, if ξ is the measurable partition of $X \times G$ into ergodic components of $\Gamma(\alpha)$, then the G-action $V(G)$ generates an action $W(G)$ of G on the quotient space $((X \times G)/\xi, (\mu \times \chi_G)/\xi)$. Let $q : X \times G \to (X \times G)/\xi$ be the natural projection; then for $g \in G$

$$W(g)q(x, h) = q(V(g)(x, h)) = q(x, hg^{-1}).$$

The above action $W(G)$ will also be denoted by $W_{(\Gamma, \alpha)}(G)$ in order to indicate its connection with (Γ, α) which it is constructed from. We refer to $W_{(\Gamma, \alpha)}(G)$ as the *Mackey action* (or the action associated to the pair (Γ, α)). It is easy to see that the Mackey action $W_{(\Gamma, \alpha)}(G)$ is ergodic if and only if Γ is ergodic [Ma, Z1]. We also note that the Mackey action can be non-free even if Γ is free. Furthermore, the Mackey action can be either transitive or properly ergodic depending on properties of α.

The concept of the associated action generalizes that of the associated flow of automorphisms [K2, Ma, Ro]. They coincide when $G = \mathbb{R}$ and α is the Radon-Nikodym cocycle ρ.

If (Γ_1, α_1) and (Γ_2, α_2) are weakly equivalent, then the associated actions $W_{(\Gamma_1, \alpha_1)}(G)$ and $W_{(\Gamma_2, \alpha_2)}(G)$ are isomorphic. This fact is easily proved.

We will also use the notion of modulus of automorphisms from the normalizer $N[\Gamma]$. For $R \in N[\Gamma]$, define the skew product $R(\rho)$ acting on $X \times \mathbb{R}$:

$$R(\rho)(x, v) = (Rx, v + \rho(x, R)).$$

As above, we note that this transformation commutes with $V(u)$, $u \in \mathbb{R}$, and belongs to $N[\Gamma(\rho)]$. Therefore it generates an automorphism *mod* R that also commutes with the associated flow $W_{(\Gamma, \rho)}(\mathbb{R})$, i.e. *mod* R is in the centralizer $C(W_{(\Gamma, \rho)}(\mathbb{R}))$. In the case when Γ is of type II, *mod* R is just the translation on $\mathbb{R}$ by a logarithm of the constant by which R multiplies the Γ-invariant measure on X. Usually in such a situation exactly that constant is assigned to *mod* R and *mod* itself is considered as a map $N[\Gamma] \to \mathbb{R}_+^*$. It is easily seen that the map *mod* is a group homomorphism ([BG4, CK, H, K2]).

1.5 Ratio set. Under investigation of cocycles with values in a l.c.s.c. group G, it is important to know whether a cocycle can be reduced (by cohomologous transformation) to some cocycle with a countable set of values.

Theorem 1.8. [GS2] *Let H be a countable dense subgroup of a l.c.s.c. group G and let Γ be an ergodic a.f. group of automorphisms of* $\mathrm{Aut}(X, \mathcal{B}, \mu)$. *Then every cocycle $\alpha \in Z^1(X \times \Gamma, G)$ is cohomologous to some cocycle $\beta \in Z^1(X \times \Gamma, G)$ with values in H.*

We will remind briefly the classification of cocycles suggested by K. Schmidt in [Sch1].

As above, we assume that G is a l.c.s.c. group and $\Gamma \in \mathrm{Aut}(X, \mathcal{B}, \mu)$ an ergodic countable transformation group. Denote by $\overline{G} = G \cup \{\infty\}$ the one-point compactification of G (with $G = \overline{G}$ being implicit for G compact).

Definition 1.9. For Γ as above and $\alpha \in Z^1(X \times \Gamma, G)$, an element $f \in \overline{G}$ is called an *essential value* of α if for any neighborhood V_f of f in $\overline{G}$ and every $B \subset X$, $\mu(B) > 0$, one has

$$\mu(\bigcup_{\gamma \in \Gamma}(B \cap \gamma^{-1}B \cap \{x \in X : \alpha(x, \gamma) \in V_f\})) > 0.$$

The set of all essential values of α is denoted by $\bar{r}(\Gamma, \alpha)$; we also set up $r(\Gamma, \alpha) = \bar{r}(\Gamma, \alpha) \cap G$.

The proof of the following proposition is based on the definitions.

Proposition 1.10. *(i) If pairs (Γ_1, α_1) and (Γ_2, α_2) are (stably) weakly equivalent, then $\bar{r}(\Gamma_1, \alpha_1) = \bar{r}(\Gamma_2, \alpha_2)$. (ii) $r(\Gamma, \alpha)$ is a closed subgroup of G. (iii) If G is abelian, then α is a coboundary if and only if $\bar{r}(\Gamma, \alpha) = \{0\}$.*

In particular, we can take instead of α the Radon-Nikodym cocycle $\rho \in Z^1(X \times \Gamma, \mathbb{R})$ and define $\bar{r}(\Gamma, \rho)$. Since the list of all closed subgroups in $\mathbb{R}$ is well known, one can introduce the next definition.

Definition 1.11. Suppose that Γ is of type III. According to the cases $r(\Gamma, \rho) = 0$, $r(\Gamma, \rho) = \{n \log \lambda : n \in \mathbb{Z}\}$, $0 < \lambda < 1$, and $r(\Gamma, \rho) = \mathbb{R}$, the group Γ is said to be of type III_0, III_λ, $0 < \lambda < 1$, and III_1 respectively.

Now we distinguish a class of cocycles that will play an important role in the sequel.

Definition 1.12. A cocycle $\alpha \in Z^1(X \times \Gamma, G)$ is said to have a *dense range* in G if $r(\Gamma, \alpha) = G$.

Theorem 1.13. [GS1] *For an ergodic a.f. group Γ, a cocycle $\alpha \in Z^1(X \times \Gamma, G)$ has a dense range if and only if the group of automorphisms $\Gamma(\alpha)$ acts ergodically on the space $(X \times G, \mu \times \chi_G)$. In other words, the as-*

sociated action $W_{(\Gamma,\alpha)}(G)$ for a cocycle α with dense range is trivial.

In particular, it follows from Theorem 1.13 that for a.f. Γ of type III_1 the associated flow $W_{(\Gamma,\rho)}(\mathbb{R})$ is trivial.

The other interesting result follows from [GS1, Z1]: *Let Γ be an ergodic a.f. automorphism group and α a cocycle over Γ with dense range in a l.c.s.c. group G, then G is amenable.* As a consequence of this fact, we will have to restrict our investigation by the class of amenable groups (or amenable actions of groups [Z1]).

The next theorem gives a method of investigation of cocycles with dense range in G. Let $\ker(\alpha)$ denote the full subgroup $\{\gamma \in [\Gamma] : \alpha(x,\gamma) = e$ for a.e. $x \in X\}$ of $[\Gamma]$.

Theorem 1.14. [GS1] *Given a cocycle $\alpha \in Z^1(X \times \Gamma, G)$ with dense range over an ergodic a.f. countable group Γ, there exists a cocycle β cohomologous to α such that $\ker(\beta)$ is an ergodic full subgroup in $[\Gamma]$.*

If G is a countable group, then for any $\alpha \in Z^1(X \times \Gamma, G)$ with dense range in G, we get that $\ker(\alpha)$ is automatically ergodic. Moreover, for given $g \in G$, there exists $\gamma_g \in [\Gamma]$ such that $\alpha(x,\gamma_g) = g$ a.e. One can show that $\gamma_g \in N[\ker(\alpha)]$, $g \in G$, and $[\Gamma]$ is generated by $\ker(\alpha)$ and $\{\gamma_g \mid g \in G\}$ [BG2].

For a countable group G, it is natural to introduce the following definition. We say that an element $g_0 \in G$ belongs to the range $E(\Gamma, \alpha)$ of $\alpha \in Z^1(X \times \Gamma, G)$ if

$$\mu(\bigcup_{\gamma \in \Gamma}\{x \in X : \alpha(x,\gamma) = g_0\}) > 0.$$

Proposition 1.15. [BG5] *For a countable group G, the following statements are equivalent:*
(i) $E(\Gamma, \alpha) = r(\Gamma, \alpha)$;
(ii) the group of automorphisms $\ker(\alpha) = \{\gamma \in \Gamma : \alpha(x,\gamma) = e$ for a.e. $x \in X\}$ is ergodic.

Definition 1.16. Let Γ be a conservative transformation group of $(X, \mathcal{B}, \mu)$. A cocycle $\alpha \in Z^1(X \times \Gamma, G)$ is called *recurrent* if for every set $B \in \mathcal{B}$ with $\mu(B) > 0$ and every neighborhood V of the identity in G

$$\mu(\bigcup_{\gamma \in \Gamma, \gamma \neq e} (B \cap \gamma^{-1}B \cap \{x \in X : \alpha(x,\gamma) \in V\})) > 0.$$

Otherwise α is called *transient*.

Equivalently, a cocycle α is transient if and only if there exist a measurable set $B_0 \subset X$, $\mu(B_0) > 0$, and a neighborhood V_0 of the identity in G such that

$$\mu\left(\bigcup_{\gamma \in \Gamma, \gamma \neq e} (B_0 \cap \gamma^{-1}B_0 \cap \{x \in X : \alpha(x,\gamma) \in V_0\})\right) = 0.$$

This means that, for a transient cocycle α, the cocycle α_{B_0} over the group $\Gamma_{B_0} \subset [\Gamma]$ does not take values in V_0.

For a transient cocycle α, we have $\bar{r}(\Gamma, \alpha) = \{0, \infty\}$.

Proposition 1.17. [Sch1] *The skew product $\Gamma(\alpha)$ is conservative if and only if α is a recurrent cocycle. $\Gamma(\alpha)$ is of type I if and only if α is transient.*

For the rest of this section we assume that G is an abelian l.c.s.c. group and α is a cocycle from $Z^1(X \times \Gamma, G)$. Since $r(\Gamma, \alpha)$ is a closed subgroup in G, one can define a new cocycle $\hat{\alpha} \in Z^1(X \times \Gamma, \hat{G})$ with values in $\hat{G} = G/r(\Gamma, \alpha)$ as follows:

$$\hat{\alpha}(\gamma, x) = \alpha(\gamma, x) + r(\Gamma, \alpha).$$

Then $r(\Gamma, \hat{\alpha}) = \{\hat{0}\}$ with $\hat{0}$ being the identity in $\hat{G}$.

Definition 1.18. A cocycle $\alpha \in Z^1(X \times \Gamma, G)$ is called *regular* if $\bar{r}(\Gamma, \hat{\alpha}) = \{\hat{0}\}$ and *non-regular* if $\bar{r}(\Gamma, \hat{\alpha}) = \{\hat{0}, \infty\}$.

Proposition 1.19. [Sch1] *For given $\alpha \in Z^1(X \times \Gamma, G)$, the following statements are equivalent:*
(i) α is a regular cocycle;
(ii) α is Γ-cohomologous to a cocycle α_1 with values in $r(\Gamma, \alpha)$.

Definition 1.20. A cocycle $\alpha \in Z^1(X \times \Gamma, G)$ is called *lacunary* if there exists a neighborhood V_0 of the identity in G such that

$$\mu\left(\bigcup_{\gamma \in \Gamma}\{x \in X : \alpha(x,\gamma) \in V_0 - \{0\}\}\right) = 0.$$

Proposition 1.21 [Sch1]. *A cocycle $\alpha \in Z^1(X \times \Gamma, G)$ is cohomologous to a lacunary cocycle if and only if there exists a neighborhood V_1 of the identity in G such that $r(\Gamma, \alpha) \cap V_1 = \{0\}$.*

2   Results on weak equivalence and structure of cocycles

In this section, we discuss the main results concerning the weak equivalence problem of cocycles. Note that we are unable to mention all results proved in articles [BG4, BG5, GS3, GS4]. It seems to be reasonable to show the main ideas used there. We start with transient and regular cocycles.

2.1 Transient cocycles. We assume that Γ is a freely acting ergodic countable group of non-singular automorphisms of $(X, \mathcal{B}, \mu)$, and G is a l.c.s.c. group with the Haar measure χ_G. Let $\alpha \in Z^1(X \times \Gamma, G)$ be a transient cocycle. It follows from Proposition 1.17 that $\Gamma(\alpha)$ is of type I automorphism group of $(X \times G, \mu \times \chi_G)$. Therefore, the partition ξ of $(X \times G, \mu \times \chi_G)$ into $\Gamma(\alpha)$-orbits is measurable. Then, the quotient space $\Omega = (X \times G)/\xi$ can be regarded as a measurable subset of positive measure in $X \times G$ which intersects every $\Gamma(\alpha)$-orbit exactly at one point. Hence $X \times G = \cup_{\gamma \in \Gamma} \gamma(\alpha)\Omega$, and the Mackey action $W(G) = W_{(\Gamma, \alpha)}(G)$ can be written as follows:

$$W_{(\Gamma, \alpha)}(g)(x, h) = \gamma(\alpha)(x, hg^{-1}), \quad g \in G, \quad (x, h) \in \Omega,$$

where $\gamma(\alpha) \in \Gamma(\alpha)$ is found by the relation $(x, hg^{-1}) \in \gamma(\alpha)^{-1}\Omega$. The Mackey action is free and ergodic in this case.

It follows from the Definition 1.16 that there exists $B \subset X$, $\mu(B) > 0$, such that $\alpha(x, \gamma_B) \notin V_0$ where $(x, \gamma_B) \in B \times [\Gamma]_B$, $\gamma_B \neq 1$, and V_0 is a neighborhood of the identity in G. Similarly, one can construct the Mackey action $W_B(G) = W_{(\Gamma_B, \alpha_B)}(G)$ associated to (Γ_B, α_B) where $[\Gamma_B] = [\Gamma]_B$.

Recall the definition of the return cocycle. Let U be a free ergodic Borel action of a l.c.s.c. group H on $(X, \mathcal{B}, \mu)$. A Borel subset $E \subset X$ is called a complete lacunary section for $U(H)$ if $\mu(X - U(H)E) = 0$ and there is a neighborhood V of the identity in H such that $U(V)x \cap E = \{x\}$ for all $x \in E$ [FHM]. There arises a countable measurable equivalence relation on E, i.e. a countable automorphism group Γ_E acting on E [FM]. Every Γ_E-orbit is a part of an $U(H)$-orbit. The return cocycle u_E over Γ_E is defined as follows: $u_E(x, \gamma_E) = h$ if $\gamma x = U(h)x$.

Proposition 2.1. [BG5] *Let $\alpha \in Z^1(X \times \Gamma, G)$ be a transient cocycle over a countable freely acting ergodic group $\Gamma \in \text{Aut}(X, \mathcal{B}, \mu)$, and let $W_B(G)$ be the Mackey action corresponding to the pair (Γ_B, α_B) defined as above. Then (Γ, α) is weakly equivalent to the pair (Γ_E, u_E) where E is a complete lacunary section of $W_B(G)$ and u_E is the return cocycle.*

Based on this proposition, we can prove the following criterion of weak equivalence for transient cocycles. Note that in the next statement we do not assume approximative finiteness of automorphism groups.

Theorem 2.2. [BG5, GS2] *Let (Γ_1, α_1) and (Γ_2, α_2) be two pairs with the transient cocycles α_1 and α_2 over ergodic freely acting automorphism groups Γ_1 and Γ_2 respectively. Then their Mackey actions $W_{(\Gamma_1, \alpha_1)}(G)$ and*

$W_{(\Gamma_1,\alpha_1)}(G)$ *are isomorphic if and only if the pairs* (Γ_1,α_1) *and* (Γ_2,α_2) *are stably weakly equivalent.*

We observe that the cocycle $\alpha_0 = \alpha \times \rho \in Z^1(X \times \Gamma, G \times \mathbb{R})$ is transient if and only if $\alpha \in Z^1(X \times \Gamma, G)$ is transient [BG4]. It follows from this remark that Theorem 2.2 is true if we replace G by $G_0 = G \times \mathbb{R}$ and α by $\alpha_0 = \alpha \times \rho$.

2.2 Regular cocycles. Let Γ and G be as above. In this subsection, we study the case of transitive associated actions. We show (Theorem 2.4) that such actions are generated by regular cocycles only. All statements are proved in [BG5].

Theorem 2.3. *Assume that* $\alpha \in Z^1(X \times \Gamma, G)$ *and the associated action* $W_{(\Gamma,\alpha)}(G)$ *is isomorphic to the transitive action of* G *on the quotient space* G/H *where* H *is a proper closed subgroup in* G. *Then* α *is cohomologous to a cocycle* β *taking values in* H. *If* G *is either abelian or countable, then* $r(\Gamma,\alpha) = r(\Gamma,\beta) = H$.

From this theorem and Proposition 1.19, we deduce the following statement.

Theorem 2.4. *Let* $\Gamma \in \mathrm{Aut}(X,\mathcal{B},\mu)$ *be a countable ergodic group and* $\alpha \in Z^1(X \times \Gamma, G)$ *where* G *is an abelian l.c.s.c. group. Then the Mackey action* $W_{(\Gamma,\alpha)}(G)$ *is transitive if and only if* α *is regular.*

The property of regularity of cocycles is invariant with respect to weak equivalence. Note also that the regularity of $\alpha_0 = \alpha \times \rho$ implies the regularity of α and ρ. The converse statement is false.

Theorem 2.5. *Let* (Γ_1,α_0^1) *and* (Γ_2,α_0^2) *be two pairs with regular cocycles* $\alpha_0^i \in Z^1(X_i \times \Gamma_i, G \times \mathbb{R})$, $i = 1,2$. *Then these pairs are weakly equivalent if and only if* $r(\Gamma_1,\alpha_0^1) = r(\Gamma_2,\alpha_0^2)$ *(or, equivalently, their Mackey actions* $W_{(\Gamma_1,\alpha_0^1)}(G \times \mathbb{R})$ *and* $W_{(\Gamma_2,\alpha_0^2)}(G \times \mathbb{R})$ *are isomorphic).*

2.3 Cocycles with dense range. A few statements on such a class of cocycles was formulated in Section 1. Now we consider more general situation when a cocycle has a dense range in a closed subgroup $H \subset G \times \mathbb{R}$ or even in a measurable field of such subgroups. We first consider cocycles with values in an abelian l.c.s.c. group G.

Proposition 2.6. [BG4] *Let* H_0 *be a closed subgroup of* $G_0 = G \times \mathbb{R}$ *with* G *being an abelian group. Then there exists a pair* (Γ,α_0) *such that*

$r(\Gamma, \alpha_0) = H_0$ *where* $\alpha \in Z^1(X \times \Gamma, G)$ *and* Γ *is an a.f. ergodic subgroup in* $\mathrm{Aut}(X, \mathcal{B}, \mu)$.

We give the sketch of the proof. Let $\{h_0(n)\}_{n=1}^{\infty}$ be a dense sequence of elements from H_0 such that every $h_0(n)$ occurs in it infinitely often. Write down $h_0(n) = (h_G(n), h_{\mathbb{R}}(n))$, $n \in \mathbb{N}$, where $h_G(n) \in H_G$ and $h_{\mathbb{R}}(n) \in H_{\mathbb{R}}$ are the projections of $h_0(n)$. Set $Y = \{0,1\}^{\mathbb{N}}$ and denote $A_n(i) = \{y \in Y \mid y_n = i\}$, $i = 0, 1$, $n \in \mathbb{N}$. Define the probability product-measure $\nu = \otimes_n \nu_n$ on Y such that

$$\nu_n(A_n(1)) = e^{h_{\mathbb{R}}(n)} \nu_n(A_n(0)), \quad n \in \mathbb{N}.$$

Take the automorphism $\delta_n \in \mathrm{Aut}(Y, \nu)$, $n \in \mathbb{N}$, such that $\delta_n(\{y_k\}) = \{y_k'\}$ where $y_k' = y_k$ if $k \neq n$ and $y_n' = y_n + 1 \pmod 2$. One has $\rho(y, \delta_n) = h_{\mathbb{R}}(n)$, $y \in A_n(0)$, $n \in \mathbb{N}$. Denote by Γ the group of automorphisms of (Y, ν) generated by δ_n, $n \in \mathbb{N}$. It is well known that Γ is ergodic and a.f. [HO]. Define a cocycle $\alpha \in Z^1(Y \times \Gamma, H_G)$ by its values on the generators of Γ:

$$\alpha(y, \delta_n) = \begin{cases} h_G(n) & \text{if } y \in A_n(0) \\ -h_G(n) & \text{if } y \in A_n(1) \end{cases}$$

Then $\alpha_0(y, \delta_n) = h_0(n)$, $y \in A_n(0)$, $n \in \mathbb{N}$. One can show that $\ker(\alpha_0)$ is an ergodic subgroup. It follows from this fact that $r(\Gamma, \alpha_0) = H_0$.

The following proposition is extremely important under investigation of cocycles with dense range taking values in a l.c.s.c. group. Based on this fact, we can prove one of the main results of weak equivalence theory, Theorem 2.8 (see below).

Proposition 2.7. *Let* $\Gamma \in \mathrm{Aut}(X, \mathcal{B}, \mu)$ *be an ergodic a.f. group and let* $\alpha_0 \in Z^1(X \times \Gamma, G \times \mathbb{R})$ *be a cocycle with values in a closed subgroup* $H \subset G \times \mathbb{R}$ *and with dense range in* H. *Take* $(g, r) \in H$ *and let* A *and* B *be the subsets in* X *of positive measure such that* $\mu(B) = e^r \mu(A)$. *Then for any neighborhood* V *of* (g, r) *there exists an automorphism* $\gamma \in [\Gamma]$ *such that* $\gamma(A) = B$ *mod 0, and* $\alpha_0(x, \gamma) \in V$ *for a.e.* $x \in A$.

Theorem 2.8 [GS1] (uniqueness theorem for cocycles with dense range). *Let the cocycles* α *and* β *belong to* $Z^1(X \times \Gamma, G)$ *and* α_0 *and* β_0 *take values and have dense ranges in the same closed subgroup* $H \subset G \times \mathbb{R}$ *where* Γ *is an ergodic a.f. automorphism group. Then there exist an automorphism* $\theta \in N[\Gamma]$, *mod* $\theta = \mathrm{id}$, *and a Borel map* $f : X \to G \times \mathbb{R}$ *such that* $f(\gamma x) \alpha_0(x, \gamma) f(x)^{-1} = \beta_0(\theta x, \theta \gamma \theta^{-1})$ *for all* $\gamma \in [\Gamma]$ *and a.e.* $x \in X$, *i.e. the pairs* (Γ, α_0) *and* (Γ, β_0) *are weakly equivalent.*

Consider an important particular case when G is countable. For simplicity, assume that Γ is an ergodic a.f. measure preserving group. Then the conclusion of the above theorem is as follows: there exists $\theta \in N[\Gamma]$ such that $\alpha(x, \gamma) = \beta(\theta x, \theta \gamma \theta^{-1})$ a.e. [BG2].

Let a Borel non-free action U of a l.c.s.c. group G on $(X, \mathcal{B}, \mu)$ be given. Then the Borel field of stabilizers $x \mapsto G_x = \{g \in G \mid U(g)x = x\}$ arises naturally. Under the study of non-free Mackey actions, we will have to construct a field of cocycles with dense range in a given field of closed subgroups. The method of this problem solution is similar to Proposition 2.6. Moreover for such Borel fields of cocycles, one can prove the uniqueness theorem analogous to Theorem 2.8 (see [BG5, GS1]).

2.4 Structure of cocycles and weak equivalence. We start with a few results concerning the structure of cocycles over an ergodic a.f. group. To clarify this structure, we will consider only cocycles with values in abelian l.c.s.c. and countable amenable groups G. Note that sometimes it is more convenient to use the language of measurable groupoids (see [FHM] for definitions)

Let Γ be an ergodic a.f. countable group of measure preserving automorphisms of $(X, \mathcal{B}, \mu)$, and let $\alpha \in Z^1(X \times \Gamma, G)$ be a recurrent cocycle with values in an abelian group G. Suppose $r(\Gamma, \alpha) = H$ is a proper (may be trivial) subgroup of G. As in Section 1, consider the cocycle $\hat{\alpha}(x, \gamma) = \alpha(x, \gamma) + H$ with values in $\hat{G} = G/H$. It is proved in [BG4] that, up to weak equivalence, the cocycle $\hat{\alpha}$ is lacunary.

Theorem 2.9. *Take a pair (Γ, α) as above. Then $(X, \mathcal{B}, \mu)$ is isomorphic to $(X_0 \times Y, \mu_0 \times \nu)$ and $[\Gamma]$ is generated by two automorphisms Q_0 and S_0 acting on $(X_0 \times Y, \mu_0 \times \nu)$ as follows: $Q_0(x_0, y) = (Qx_0, U(x_0)y)$, $S_0(x_0, y) = (x_0, Sy)$ where Q and S are ergodic, $Q_0 \in N[S_0]$. The cocycle $\hat{\alpha}$ has the properties: $\hat{\alpha}(x_0, y, Q_0) = \varphi(x)$, $\hat{\alpha}(x_0, y, S_0) = \hat{0}$ where $\varphi(x_0) \in G - V_0$ (V_0 is a neighborhood of $\hat{0} \in \hat{G}$).*

In other words, this theorem shows that if $H = \{0\}$, then any α can be represented (up to weak equivalence) as a cocycle constructed by a transient cocycle and a cocycle with dense range. The other corollary is that the Mackey action $W_{(\Gamma, \alpha)}(G)$ is isomorphic to $W_{(Q, \alpha(\varphi))}(G)$ where $\alpha(\varphi)$ is a cocycle defined by Q and the function $\varphi(x_0)$. This also means that H is the stabilizer of $W_{(\Gamma, \alpha)}(G)$ for any point. Remark that a statement analogous to Theorem 2.9 is also true when we take a non-singular automorphism group Γ and α is replaced by $\alpha \times \rho$ (see [BG4]).

Assume now that G is a countable amenable group and $(\Gamma, \alpha \times \rho)$ is defined as above. The automorphism group $\Gamma(\alpha \times \rho)$ together with the action V of

$G \times \mathbb{R}$ (see Section 1) generate a measurable groupoid $\mathcal{G}$.

Consider on $\mathbb{R}$ the measure $d\mu_1(u) = \exp(-u)d\chi_{\mathbb{R}}(u)$, where $\chi_{\mathbb{R}}$ is the Lebesgue measure. Then, $\mu \times \chi_G \times \mu_1$ is the $\Gamma(\alpha_0)$-invariant measure. Apart from, for all $(h, v) \in G \times \mathbb{R}$, one has $\rho(x, g, u, V(h, v)) = v$. Define a homomorphism (cocycle) $\beta_0 = \beta \times \rho$ from $\mathcal{G}$ into $G \times \mathbb{R}$:

$$\beta_0(x, g, u, \gamma(\alpha \times \rho)) = (e, 0), \quad \gamma(\alpha \times \rho) \in \Gamma(\alpha \times \rho),$$

$$\beta_0(x, g, u, V(h, v)) = (h, v), (h, v) \in G \times \mathbb{R}.$$

One can show that the action of $G \times \mathbb{R}$ associated to $(\mathcal{G}, \beta_0)$ is isomorphic to $W_{(\Gamma, \alpha \times \rho)}(G \times \mathbb{R})$. Therefore, we can freely replace $(\Gamma, \alpha \times \rho)$ by $(\mathcal{G}, \beta_0)$ when it is convenient.

Suppose the associated action $W = W_{(\Gamma, \alpha \times \rho)}(G \times \mathbb{R})$ is defined on (Ω, m) and $\omega \mapsto H_\omega = \{(h, u) \in G \times \mathbb{R} \mid W(h, u)(\omega) = \omega\}$ is the Borel field of stabilizers. We must distinguish two possible cases: (I) H_ω is discrete in the product topology of $G \times \mathbb{R}$ a.e.; (II) H_ω has the form $(H_\omega^G, \mathbb{R})$ a.e.

In [BG5], the structure of the groupoid $\mathcal{G}$ and the cocycle $\beta \times \rho = \beta_0$ was studied and the following theorem was proved.

Theorem 2.10. *The pair $(\mathcal{G}, \beta_0)$ (or otherwise $(\Gamma, \alpha \times \rho)$) is weakly equivalent to (Γ', β_0') such that the latter has the following structure: 1) the ergodic group of automorphisms $\Gamma' \subset \mathrm{Aut}(X_0 \times Y_0, \mu_0 \times \nu_0)$ is generated by Q_0 and S_0 acting by the formulae: $Q_0(x_0, y_0) = (Qx_0, U(x_0)y_0), S_0(x_0, y_0) = (x_0, Sy_0), Q_0 \in N[S_0]$; 2) the cocycle $\beta_0' = (x_0 \mapsto \beta_{0x_0}')$ is transient on Q_0 and has a dense range in the field of groups $x_0 \mapsto H_{0x_0}$, where H_{0x_0} is the stabilizer of the associated action $W(G_0)$, $x_0 \in X_0$; 3) Q and S are ergodic automorphisms, the type of S depends on the structure of the field of groups $x_0 \mapsto H_{0x_0}$ and can be any, except III_0.*

The next theorem will be referred to as the existence theorem later on. Roughly speaking, this theorem says that any action of l.c.s.c. amenable group G is, in fact, the Mackey action associated to a pair (Γ, α).

Theorem 2.11 (existence theorem). *Let a (non-free) amenable ergodic non-singular action W of the group $G \times \mathbb{R}$ be given on a measure space $(X, \mathcal{B}, \mu)$. Then there exists a pair $(\Gamma, \alpha \times \rho)$ for which the associated action $W_{(\Gamma, \alpha \times \rho)}(G \times \mathbb{R})$ is isomorphic to $W(\Gamma \times \mathbb{R})$ where Γ is a countable ergodic a.f. group of automorphisms and α is a cocycle over Γ with values in G.*

Remark that such a pair $(\Gamma, \alpha \times \rho)$ is not unique. Moreover, the group Γ can be chosen, for instance, measure preserving.

The next theorem is a general statement that has been referred to as the uniqueness theorem. Note that, in fact, this name is assigned to a number

of statements that deal with a kind of cocycles and groups (some of them
are mentioned above). In particular, the case of abelian groups was studied
in [BG4]. The main difficulty which differs non-abelian Mackey actions from
the abelian case is that for such non-free Mackey actions, one has to study
Borel fields of groups and cocycles.

Theorem 2.12 [GS4] (uniqueness theorem) *Suppose that the Mackey ac-
tions* $W_{(\Gamma,\alpha\times\rho)}(G \times \mathbb{R})$ *and* $W_{(\Gamma,\beta\times\rho)}(G \times \mathbb{R})$ *are isomorphic where* α, β *are
recurrent cocycles over* Γ *with values in a l.c.s.c. amenable group* G. *Then
there exist an automorphism* $\theta \in N[\Gamma]$ *and a Borel map* $f : X \to G$ *such that*
$\beta(\theta x, \theta\gamma\theta^{-1} = f(\gamma x)\alpha(x,\gamma)f(x)^{-1}$ *for all* $\gamma \in \Gamma$ *and a.e.* $x \in X$.

3 Applications of weak equivalence

3.1 Outer conjugacy. In this subsection, we consider another technique of
solution of the outer conjugacy problem for actions of countable amenable
groups, which differs from that known earlier [BG1, BG2, BG3, CK]. This
new method is based on the uniqueness theorem for pairs (Γ, α_0). We also
apply this technique to the outer conjugacy problem for actions of fields of
countable amenable groups.

Let T be an ergodic non-singular automorphism of the measure space
$(X_0, \mathcal{B}_0, \mu_0)$, and suppose that G is a countable amenable group. Consider an
action $\tau : G \to \mathrm{Aut}(X_0, \mathcal{B}_0, \mu_0)$ of G on $(X_0, \mathcal{B}_0, \mu_0)$. Suppose $\tau(G) \subset N[T]$
and $\tau(g) \notin [T]$ for all $g \neq e$ where e is the identity in G. The latter is not
an essential restriction. We assume it solely for convenience of our exposition
(see Remark 3.4 below). We will consider here the most complicated case
when T is of type III_0; all other types of automorphisms can be studied similarly.

Definition 3.1. The actions τ_1 and τ_2 of G on $(X_0, \mathcal{B}_0, \mu_0)$ with $\tau_i(G) \subset
N[T]$, $i = 1, 2$, are called outer conjugate if there exists an automorphism
$R \in N[T]$ such that

$$\tau_1(g) = R^{-1}\tau_2(g)Rt, \quad g \in G,$$

where $t = t(g) \in [T]$.

It follows from [K2] that T is orbit equivalent to an a.f. ergodic group
of type III_0 automorphisms $\mathcal{G}(Q, \phi) \subset \mathrm{Aut}(X \times Y, \mu \times \nu)$ which is generated
by $Q_0(x,y) = (Qx, U(x)y)$, $S_0(x,y) = (x, Sy)$ where $x \mapsto U(x) \in N[S]$ is
a measurable field of automorphisms, Q and S are ergodic automorphisms,
$S \circ \nu = \nu$, and $\nu(Y) = \infty$. Besides, $\rho(x, y, Q_0) = \phi(x) > 0$.

It was proved in [BG1] that the automorphisms $\tau(g)$, $g \in G$, can be brought, by multiplying by elements from $[\mathcal{G}]$, to the form

$$\tau(g)(x,y) = (a(g)x, V_x(g)y), \quad g \in G. \tag{3.1}$$

After this, τ is not an action of G but the automorphisms $\tau(g)$, $g \in G$, form an "action" of G module $[\mathcal{G}]$. Apart from $a(g) \in N[Q]$, $V_x(g) \in N[S]$. Set

$$H = \{g \in G : a(g) = \mathbf{1}\}. \tag{3.2}$$

Then H is a normal subgroup of G (maybe trivial).

Denote by Γ the group of automorphisms of $(X \times Y, \mu \times \nu)$ generated by $\tau(G)$ and $\mathcal{G}(Q, \phi)$. It follows from [BG3, CFW] that Γ is approximately finite. Define a cocycle α on Γ by its values on the generators of Γ as follows:

$$\alpha(x, y, \tau(g)t) = g, \quad t \in [\mathcal{G}], \quad g \in G. \tag{3.3}$$

Then $\alpha_0 = \alpha \times \rho \in Z^1(X \times Y \times \Gamma, G \times \mathbb{R})$, and let $W(G_0) = W_{(\Gamma, \alpha_0)}(G_0)$ denote the Mackey action of $G_0 = C \times \mathbb{R}$ associated to (Γ, α_0).

One can easily find the action $W(G_0)$. For this, construct the associated flow $W_{\mathcal{G}}(\mathbb{R})$ by $(\mathcal{G}, \rho)$. It is known that $W_{\mathcal{G}}(\mathbb{R})$ is isomorphic to a special flow built by the basis automorphism Q and the ceiling function $\phi(x)$. $W_{\mathcal{G}}(\mathbb{R})$ acts on $X(\phi) = \{(x, u) \in X \times \mathbb{R} : 0 \leq u < \phi(x), x \in X\}$ [K2]. Remind that we defined in Section 1 the map *mod* from the normalizer of an automorphism group onto the centralizer of associated flow.

Lemma 3.2. *The action $W(G_0)$ associated to (Γ, α_0) is isomorphic to the action of $G_0 = G \times \mathbb{R}$ on the set $X(\phi)$ generated by $W_{\mathcal{G}}(\mathbb{R})$ and the group $\{\mathrm{mod}\ \tau(g) : g \in G\} \subset C\{W_{\mathcal{G}}(\cdot)\}$.*

Proof. Take $g_0 = (g, v) \in G \times \mathbb{R}$; we will show that

$$W(g_0) = \mathrm{mod}\ \tau(g)W_{\mathcal{G}}(v).$$

The generators of $\Gamma(\alpha_0)$ act on $(X \times Y \times G \times \mathbb{R}, \mu \times \nu \times \chi_{\times}\chi_{\mathbb{R}})$ as follows:

$$\tau(g)(\alpha_0)(x, y, h, u) = (a(g)x, V_x y, gh, u + \rho(x, y, \tau(g))),$$

$$Q_0(\alpha_0)(x, y, h, u) = (Qx, U(x)y, h, u + \phi(x)),$$
$$S_0(\alpha_0)(x, y, h, u) = (x, Sy, h, u).$$

It can be directly checked that $X(\phi) \times \{e\} \times \{0\}$ intersects every $\Gamma(\alpha_0)$-orbit exactly at one point (e is the identity in G). Therefore, $X(\phi)$ can be considered as the set, on which $W(G_0)$ acts. It is evident that $W(e, v) = W_{\mathcal{G}}(v)$, $v \in \mathbb{R}$. To find $W(g, 0)$, $g \in G$, we determine which point from $X(\phi)$ belongs to the $\Gamma(\alpha_0)$-orbit running through the point (x, g^{-1}, u), $(x, u) \in X(\phi)$ (recall that $W(g, 0)$ is the projection of $V(g)$: $(x, y, h, u) \mapsto (x, y, hg^{-1}, u)$ along the $\Gamma(\alpha_0)$-orbits). Thus, $\tau(g^{-1})(\alpha_0)(x', e, u') = (x, g^{-1}, u)$, where $(x', u') = (\text{mod } \tau(g))(x, u)$, and hence, $W(g, u) = \text{mod } \tau(g) W_{\mathcal{G}}(v)$.

Lemma 3.3. *The group $H_0 = \{(h, -p) \in G \times \mathbb{R} \mid h \in H$, mod $\tau(h) = W_{\mathcal{G}}(p)\}$ is the stabilizer of the associated action $W(G_0)$ where H is defined in* (3.2).

Remark 3.4. (1) In Lemma 3.2, we have found the form of the (Γ, α_0)-associated action of G_0 where Γ is constructed by $\tau(G)$ and $\mathcal{G}$, and α is defined in (3.3). We now formulate a statement on the structure of the associated action of G_0 for (Γ, α_0), where Γ is built by $\tau(G)$ and $[T]$ when T is not of type III$_0$. If T is of type II$_1$, then $W_{(\Gamma, \alpha_0)}(G_0)$ is isomorphic to the transitive action of $\mathbb{R}$ on itself. If T is of type II$_\infty$, then $W_{(\Gamma, \alpha_0)}(G_0)$ is isomorphic to the action of $G \times \mathbb{R}$ on $(\mathbb{R}, \chi_{\mathbb{R}})$, where $\{e\} \times \mathbb{R}$ acts by shifts on itself and $G \times \{0\}$ acts by shifts on mod $\tau(g)$, $g \in G$. If T is of type III$_1$, then the associated action $W_{(\Gamma, \alpha_0)}(G_0)$ is trivial. If T is of type III$_\lambda$ ($0 < \lambda < 1$), then $W_{(\Gamma, \alpha_0)}(G_0)$ is isomorphic to the action of $G \times \mathbb{R}$ on $[0, -\log \lambda)$ such that $\{e\} \times \mathbb{R}$ has the transitive periodic action with the period $-\log \lambda$, and $G \times \{0\}$ acts by shifts on mod $\tau(g)$, $g \in G$, where mod $\tau(g) \in \mathbb{R}/\mathbb{Z} \log \lambda$.

(2) If we do not assume that $\tau(G) \cap [T] = \{1\}$, then the cocycle α from (3.3) should be replaced by a cocycle with values in $\hat{G} = G/G'$, where $G' = \{g \in G : \tau(g) \in [T]\}$. Then it is sufficient to solve the outer conjugacy problem for actions of $\hat{G}$ (see [BG2]).

Theorem 3.5. [BG3, BG5] *Let $\mathcal{G}(Q, \phi) = \mathcal{G}$ be an ergodic a.f. group of the type III$_0$ automorphisms of $(X_0, \mu_0) = (X \times Y, \mu \times \nu)$, and let τ_1, τ_2 be actions of a countable amenable group G such that $\tau_i(G) \subset N[\mathcal{G}]$ and $\tau_i(G) \cap [\mathcal{G}] = \{1\}$, $i = 1, 2$. The actions τ_1 and τ_2 are outer conjugate if and only if there exists an automorphism $\gamma \in C\{W_{\mathcal{G}}(\cdot)\}$ such that*

$$\gamma(\text{mod } \tau_1(g))\gamma^{-1} = \text{mod } \tau_2(g), \quad g \in G.$$

Proof. Clearly, we need to prove only the sufficiency of the theorem condition. Without loss of generality (see e.g. [BG3]) we may suppose that mod $\tau_1(g) = \text{mod } \tau_2(g)$, $g \in G$.

Let (Γ_i, α_0^i), $i = 1, 2$, be the pair defined by $\tau_i(G)$ and $\mathcal{G}$ as above. From Lemma 3.2 it follows that the associated actions $W_1(G_0)$ and $W_2(G_0)$ are, in fact, the same action; let $W(G_0)$ denote this action. Hence, by Theorem 2.12 (see also [BG5, Theorem 6.3], (Γ_1, α_0^1) and (Γ_2, α_0^2) are weakly equivalent. This means that there exists a one-to-one measurable map $\theta \in \text{Aut}(X_0, \mu_0)$ such that $\theta[\Gamma_1]\theta^{-1} = [\Gamma_2]$ and

$$\theta^{-1} \cdot \alpha_0^2(x_0, \gamma_1) = f(\gamma_1 x_0)\alpha_0^1(x_0, \gamma_1)f(x_0)^{-1}, \quad \gamma_1 \in [\Gamma_1], \qquad (3.4)$$

where $f(x_0) = (\xi(x_0), \eta(x_0))$ is a measurable function from X_0 into G_0. It follows from (3.4) that

$$\theta^{-1} \cdot \alpha^2(x_0, \gamma_1) = \xi(\gamma_1 x_0) \alpha^1(x_0, \gamma_1) \xi(x_0)^{-1}. \tag{3.5}$$

Let β be a cocycle on $X_0 \times \{\theta^n : n \in \mathbb{Z}\}$ which is specified by the equality

$$\beta(x_0, \theta) = f(x_0), \quad x_0 \in X_0,$$

where f satisfies (3.4). The automorphism $\theta(\beta) \in \mathrm{Aut}(X_0 \times G_0, \mu_0 \times \chi_{G_0})$ transforms $\Gamma_1(\alpha_0^1)$-orbits onto $\Gamma_2(\alpha_0^2)$-orbits, as it is seen from (3.4). Therefore, on the quotient space $X(\phi)$ it induces an automorphism which we denote by $\overline{\mathrm{mod}}\ \theta$ and which belongs evidently to $C\{W\}$. In particular, $\overline{\mathrm{mod}}\ \theta \in C\{W_{\mathcal{G}}(\cdot)\}$. By using the methods of [GDB], we obtain that in $N[\Gamma_1]$ there is an automorphism σ such that $\sigma \cdot \alpha_0^1$ is cohomologous to α_0^1 and $\mathrm{mod}\ \sigma = (\overline{\mathrm{mod}}\ \theta)^{-1}$. Set $\theta_1 = \theta\sigma$; then

$$\theta_1[\Gamma_1]\theta_1^{-1} = [\Gamma_2], \quad \overline{\mathrm{mod}}\ \theta_1 = 1. \tag{3.6}$$

Besides, equalities (3.4) and (3.5) remain valid if θ is replaced by θ_1. In this case, the functions f and ξ will also be replaced by certain functions f_1 and ξ_1 which are evidently defined by f, ξ and the value of α_0^1 on σ. Hence,

$$\theta_1^{-1} \cdot \alpha^2(x_0, \gamma_1) = \xi_1(\gamma_1 x_0) \alpha^1(x_0, \gamma_1) \xi_1(x_0)^{-1}, \quad \gamma_1 \in [\Gamma_1]. \tag{3.7}$$

Since $\theta_1 \in \mathrm{Aut}(X \times Y, \mu \times \nu)$ and $\overline{\mathrm{mod}}\ \theta_1 = 1$, then θ_1 can be brought, by multiplying by an automorphism from $[\mathcal{G}]$, to the form $(x, y) \mapsto (x, R_x y)$ [BG1]. We denote this automorphism by θ_1, as earlier. It follows from (3.6) that: (1) the automorphism R_x $(x \in X)$ preserves the measure ν on Y; (2) the function $\xi_1(x_0)$ takes values in $\tilde{H}_0 = \{h_0 \in H_0 : h_0 = (h, 0)\}$, where $H_0 \subset G \times \mathbb{R}$ is the stabilizer of W. Indeed, it follows from (3.7) that the automorphism $(x_0, g, u) \mapsto (\theta_1 x_0, \xi_1(x_0)g, u)$ maps $\Gamma_1(\alpha_0^1)$-orbits onto $\Gamma_2(\alpha_0^2)$-orbits, so that the identical map is induced on the quotient space $X(\phi)$. Therefore, by Lemma 3.2, $\xi_1(x_0) \in H_0$, and since θ_1 is measure preserving, we obtain that $\xi_1(x_0) \in \tilde{H}_0$.

Take an automorphism $\tilde{S}_1 \in \mathrm{Aut}(X \times Y, \mu \times \nu)$ whose full group $[\tilde{S}_1]$ is generated by S_0 and $\tau_1(\tilde{H}_0)$. In view of ergodicity of S on (Y, ν), the cocycle α^1 considered on $[\tilde{S}_1]$ has a dense range in $\tilde{H}_0$. Therefore, there exists an automorphism $t_0 = (x \to t(x)) \in [\tilde{S}_1]$ such that $\alpha^1(x, y, t(x)) = \xi_1(x, y)$ (remind that $\xi_1(x, y)$ is piecewise constant). Thus, the right-hand side of (3.7) equals to $t_0^{-1} \cdot \alpha^1(x_0, \gamma_1)$, i.e.

$$t_0\theta_1^{-1} \cdot \alpha^2(x_0, \gamma_1) = \alpha^1(x_0, \gamma_1), \quad \gamma_1 \in [\Gamma_1].$$

Denote $\tilde{\theta}_1 = \theta_1 t_0^{-1}$; then,

$$\alpha^2(\tilde{\theta}_1 x_0, \tilde{\theta}_1 \gamma_1 \tilde{\theta}_1^{-1}) = \alpha^1(x_0, \gamma_1), \quad \gamma_1 \in [\Gamma_1]. \tag{3.8}$$

For a.e. $x_0 \in E(g) = \{x_0 \in X_0 \mid \alpha^1(x_0, \gamma_1) = g\}$, $g \in G$, we get from (3.3) and (3.8) that

$$\gamma_1 x_0 = \tau_1(g) t_1 x_0, \quad t_1 \in [\mathcal{G}],$$

$$\tilde{\theta}_1 \gamma_1 \tilde{\theta}_1^{-1}(\tilde{\theta}_1 x_0) = \tau_2(g) t_2 \tilde{\theta}_1 x_0, \quad t_2 \in [\mathcal{G}].$$

Therefore, for a.e. $x \in \tilde{\theta}_1 E(g)$, $g \in G$,

$$\tilde{\theta}_1 \tau_1(g) \tilde{\theta}_1^{-1} x_0 = \tau_2(g) t x_0, \quad t \in [\mathcal{G}]. \tag{3.9}$$

Since $\mathcal{G}$ is ergodic, equality (3.9) can be extended to the whole measure space (X_0, μ_0).

The above method can be applied for the solution of outer conjugacy problem for actions of fields of countable amenable groups. Let $\mathcal{G} = \mathcal{G}(Q, \phi)$ be an ergodic a.f. type III_0 group of automorphisms of a measure space (Y, ν). Let also (X, μ) be a measure space, and suppose that a Borel field $x \mapsto G_x$ of countable amenable groups is defined on X (see all definitions in [K2, S]).

Definition 3.6. Let τ_1 and τ_2 be two actions of a field of groups $x \mapsto G_x$, $x \in X$, such that $\tau_i(G_x) \subset N[\mathcal{G}]$ and $\tau_i(G_x) \cap [\mathcal{G}] = 1$, $x \in X$, $i = 1, 2$. Then τ_1 and τ_2 are called outer conjugate, if there exists a measurable field of automorphisms $x \mapsto R(x) \in N[\mathcal{G}]$ such that $\tau_1(g) = R(x)^{-1} \tau_2(g) R(x) t$, where $g \in G_x$, $t = t(g, x)$, $x \in X$.

For a.e. $x \in X$, denote by Γ_x the countable ergodic a.f. group of automorphisms of (Y, ν) generated by $\mathcal{G}$ and $\tau(G_x)$. Define a cocycle $\alpha(x) : Y \times \Gamma_x \to G_x$:

$$\alpha(x)(y, \tau(g)t) = g, \quad g \in G_x, \quad t \in [T].$$

In such a way, we get the measurable field of pairs $x \mapsto (\Gamma_x, \alpha_0(x))$, $x \in X$. Denote by $W_x = W_{(\Gamma_x, \alpha_0(x))}(G_x \times \mathbb{R})$ the action of $G_x \times \mathbb{R}$ associated to $(\Gamma_x, \alpha_0(x))$. The measurable field $x \mapsto W_x$, $x \in X$, may be regarded as ergodic components of a global action W of the field of groups $x \mapsto G_x \times \mathbb{R}$ associated to $x \mapsto (\Gamma_x, \alpha_0(x))$. It follows from the uniqueness theorem for such fields of pairs that if the measurable fields $x \mapsto W_x^1$ and $x \mapsto W_x^2$ of associated actions of $x \mapsto G_x \times \mathbb{R}$ are isomorphic, then $x \mapsto (\Gamma_x^1, \alpha_0^1(x))$ and $x \mapsto (\Gamma_x^2, \alpha_0^2(x))$ are weakly equivalent.

The structure of $x \mapsto W_x$, $x \in X$, can be found as in Lemma 3.2. One can generalize the proof of Theorem 3.5 to actions of measurable fields of groups.

Theorem 3.7. [BG5] *Let τ_1 and τ_2 be measurable actions of a Borel field of groups $x \mapsto G_x$, $x \in X$, such that $\tau_i(G_x) \subset N[\mathcal{G}]$, $\tau_i(G_x) \cap [\mathcal{G}] = 1$, $i = 1, 2$, where $\mathcal{G} = \mathcal{G}(Q, \phi)$ is an a.f. ergodic type III_0 group of automorphisms. The actions τ_1 and τ_2 are outer conjugate if and only if there exists a measurable field $x \mapsto \theta(x)$ of automorphisms such that $\theta(x) \in C\{W_{\mathcal{G}}(\cdot)\}$ and*

$$\mod \tau_1(g_x) = \theta(x)^{-1}(\mod \tau_2(g_x))\theta(x), \quad g_x \in G_x, \quad x \in X.$$

3.2 Induced actions. To illustrate another application of the mentioned results, we will find a condition under which a non-free action of a countable amenable group is induced from a subgroup in the sense of R.J.Zimmer [Z2].

Let G be an amenable countable group and let K be a subgroup of G. Consider an ergodic non-singular action T of K on a measure space (Ω, m). Define a cocycle $\alpha : \Omega \times K \to G$, setting $\alpha(\omega, T(k)) = k$, $k \in K$. Let $W_{(T(K),\alpha)}(G)$ be the action of G associated to $(T(K), \alpha)$.

Definition 3.8. [Z2] The action $W_{(T(K),\alpha)}(G)$ is called *induced* from an action T of a subgroup K.

If $T(K)$ is ergodic, then $W_{(T(K),\alpha)}(G)$ is also ergodic, and if $T(K)$ is a non-free action, then $W_{(T(K),\alpha)}(G)$ is also non-free.

Consider now another approach to the notion of induced actions. For $K \subset G$, denote by $[g]_K = Kg$, $g \in G$. Choose a section $\theta : G/K \to G$ such that $\theta([e]_K) = e$, where e is the identity in G. Define for $p \in G$,

$$\gamma([g]_K, p) = \theta([g]_K p^{-1})p\theta([g]_K)^{-1}. \tag{3.10}$$

Then $\gamma : G/K \times G \to K$ is a cocycle over the natural action of G on G/K. Define the action λ of G on $\Omega \times G/K$ as follows:

$$\lambda(p)(\omega, [g]_K) = (T(\gamma([g]_K, p))\omega, [g]_K p^{-1}), \quad p \in G, \tag{3.11}$$

where $T : K \to \text{Aut}(\Omega, m)$ is as above. It is easy to check that $\lambda(G)$ is ergodic on $\Omega \times G/K$. It was proved in [Z2] that $\lambda(G)$ is isomorphic to the action $W_{(T(K),\alpha)}(G)$ induced from the action T of the subgroup $K \subset G$.

Define now a class of non-free actions of countable groups. Let $T(K) \subset \text{Aut}(\Omega, m)$ be a non-free ergodic action of K, and let H be a normal subgroup of K.

Definition 3.9. An action T of K is called *H-simple* if the stabilizer $H_\omega = \{k \in K : T(k)\omega = \omega\}$ coincides with H a.e. on Ω. The action $T(K)$ is called *simple*, if it is H-simple for a normal subgroup $H \subset K$.

Proposition 3.10. [BG5] *Let T be an H-simple action of K on (Ω, m) and let $W_{(T(K),\alpha)}(G) \subset \text{Aut}(X, \mu)$ be the action induced from $T(K)$. Then,* *the stabilizer* $G_x = \{g \in G : W_{(T(K),\alpha)}(g)x = x\}$ *is conjugated to H in G for a.e. $x \in X$.*

The proof of this proposition is straightforward. In fact, one can show that $G_{\pi(\omega,g)} = g^{-1}Hg$ where $\pi : \Omega \times G \to (\Omega \times G)/\xi$ is the projection from $\Omega \times G$ onto the quotient space with respect to the partition ξ into $T(K)(\alpha)$-orbits. It turns out that the converse statement is also true.

Theorem 3.11. [BG5] *Let W be a non-free non-singular ergodic action of a countable amenable group G on a measure space (X_0, μ_0) and let $x_0 \to H_{x_0}$ be the Borel field of stabilizers of $W(G)$. Assume that $H_{x_0} \neq \{e\}$ and all the stabilizers are pairwise conjugated in G a.e. on X_0. Then, there exist a subgroup $K \subset G$ and a normal subgroup H in K, such that W is isomorphic to the action $W_{(T(K),\alpha)}(G)$ induced from an H-simple action T of K.*

Proof. According to Theorem 2.11 (the existence theorem) and Theorem 6.4 of [BG5], the action W of G can be regarded as that associated to a pair (Γ, β), where Γ is a group of measure preserving automorphisms of $(X \times Y, \mu \times \nu)$ generated by the automorphisms $Q_0(x,y) = (Qx, U(x)y)$, $S_0(x,y) = (x, S_x y)$ with $Q_0 \in N[S_0]$, and a cocycle β on Q_0 and S_0 is defined on the generators: $\beta(x, y, Q_0) = \phi(x) \notin H_x$, $\beta(x, y, S_x)$ has a dense range in H_x, $x \in X$.

We have that for any two stabilizers H_{x_0} and $H_{x'_0}$, there exists element $g = g(x_0, x'_0) \in G$ such that $H_{x'_0} = gH_{x_0}g^{-1}$ a.e. $(x_0, x'_0) \in X_0 \times X_0$. From this, we obtain that the field of stabilizers $x_0 \to H_{x_0}$ $(x_0 \in X_0)$ is piecewise constant, because G is countable. Hence, there exists a subgroup $H \subset G$ such that the set $B = \{x \in X : H_x = H\}$ has a positive measure. Therefore, we may pass from (Γ, β) to a stably weakly equivalent pair $(\Gamma_B, \beta_B) = (\Gamma', \beta')$ such that a cocycle β' on S'_0 has a dense range in H and β' on Q'_0 takes the value $\phi'(x) \notin H$. The function $\phi'(x)$ is also piecewise constant, and denote by $\{\phi'\}$ the set of the essential values of ϕ'. Let K be a subgroup of G which is generated by elements of H and the set $\{\phi'\}$. Note that K is a proper subgroup of G, because we again may pass to a stably weakly equivalent pair, using piecewise constancy of ϕ'. It is convenient to use the former notations, $(\Gamma, \beta), \phi, Q_0$ and S_0, assuming validity of the above properties.

It is easily seen that H is a normal subgroup of K.

In Section 2, we have described the structure of elements from $W_{(\Gamma, \beta)}(G)$. They act on the measure space which is obtained as a quotient space of $X \times G/H$ by the measurable partition into orbits of the automorphism $Q([\phi]_H) :$ $(x, [g]_H) \to (Qx, [\phi(x)g]_H)$, $[g]_H = Hg \in G/H$. Thus, if π is the natural projection on the quotient space, then $X_0 = \pi(X \times G/H)$.

Because β takes its values in $K \subset G$, the associated action $W_1(K) =$

$W^1_{(\Gamma,\beta)}(K)$ of K can be assigned to (Γ,β). Evidently, $X \times K/H \subset X \times G/H$ and $X \times K/H$ is a $Q([\phi]_H)$-invariant set of positive measure. Therefore,

$$W_1(k) = W_{(\Gamma,\beta)}(k)|_{X \times K/H}, \quad k \in K.$$

Besides, one has that $W_{(\Gamma,\beta)}(p)\pi(x, [g]_H) = \pi(x, [g]_H p^{-1})$, $p \in G$.

Construct now the action of G induced from the action W_1 of K. Note that since the stabilizer of any point for $W_1(K)$ is the normal subgroup H, then $W_1(K)$ is an H-simple action.

Let $\theta : G/K \to G$ be a section of G over G/K and let $\gamma : G/K \to K$ be the cocycle defined in (3.10). Set up, as in (3.11), for $p \in G$

$$\lambda(p)(\pi(x, [k]_H), \cdot [g]_K) = (W_1(\gamma([g]_K, p))\pi(x, [k]_H), [g]_K p^{-1})$$

$$= (\pi(x, [k]_H \gamma([g]_K, p)^{-1}), [g]_K p^{-1}). \tag{3.12}$$

Since λ is isomorphic to the action of G induced from $W_1(K)$, we will consider λ below.

To prove isomorphism of $\lambda(G)$ and $W_{(\Gamma,\beta)}(G)$, we define a map $\Phi : G/H \to K/H \times G/H$ as follows. For $g \in G$, we have $g = k\theta([g]_K)$; then set $\Phi([g]_H) = ([k]_H, [g]_K)$. Clearly, Φ is one-to-one. It follows from (3.12) that for $p \in G$,

$$\lambda(p)(\pi(x, [k]_H), [g]_K) = (\pi(x, [k]_H \theta([g]_K) p^{-1} \theta([g]_K p^{-1})^{-1})), [g]_K p^{-1}), \tag{3.13}$$

$$Hgp^{-1} = Hk\theta([g]_K) p^{-1} \theta([g]_K p^{-1})^{-1} \theta([g]_K p^{-1}), \tag{3.14}$$

where $g = k\theta([g]_K)$. Since $\theta([g]_K) p^{-1} \theta([g]_K p^{-1})^{-1} \in K$, then we get from (3.14), that

$$\Phi([gp^{-1}]_H) = ([k\gamma([g]_K, p)^{-1}]_H, [gp^{-1}]_K), \ p \in G. \tag{3.15}$$

Hence, (3.13), (3.14), and (3.15) give us the following equality:

$$\Phi^* W_{(\Gamma,\beta)}(p)\pi(x, [g]_H) = \lambda(p)\Phi^*\pi(x, [g]_H),$$

where $\Phi^*\pi(x, [g]_H) = \pi(x, \Phi([g]_H))$.

We remark that Theorem 3.11 remains valid, if it is only assumed that the action $W(G)$ is amenable.

Corollary 3.12. *If G has a countable number of subgroups, then any its non-free amenable action is induced. For example, this fact is true for actions of a nilpotent group with a finite number of generators.*

References

[AaHSch] Aaronson J, Hamachi T., and Schmidt K., Associated actions and uniqueness of cocycles, *"Algorithms, Fractals, and Dynamics"*, 1995, 1 - 25.

[A] Anzai H., Ergodic skew product transformations on the torus, *"Osaka Math. J."*, **3** (1951), 83 -97.

[B] Bezuglyi S.I., *H*-cocycles and ergodic actions of group extensions, *"Dokl. NAN Ukraine"* (to appear).

[BG1] Bezuglyi S.I. and Golodets V.Ya., Groups of measure space transformations and invariants of outer conjugation for automorphisms from normalizers of type III full groups, *"J. Funct. Anal."*, **60** (1985), 341 - 369.

[BG2] Bezuglyi S.I. and Golodets V.Ya., Outer conjugacy for actions of countable amenable groups on a measure space, *"Izv. Acad. Sci. USSR, Math. Ser."*, **50** (1986), 643 - 660 (in Russian).

[BG3] Bezuglyi S.I. and Golodets V.Ya., Type III_0 transformations of a measure space and outer conjugacy of countable amenable groups of automorphisms, *"J. Operator Theory"*, **21** (1989), 3 - 40.

[BG4] Bezuglyi S.I. and Golodets V.Ya., Weak equivalence and the structure of cocycles of an ergodic automorphism, *"Publ. RIMS Kyoto Univ."*, **27** (1991), 577 - 625.

[BG5] Bezuglyi S.I. and Golodets V.Ya., Weak equivalence of cocycles of an ergodic automorphism and the Connes-Krieger theorem for actions of discrete non-principal amenable groupoids, Preprint, Heidelberg University, 1991.

[CFW] Connes A., Feldman J., and Weiss B., An amenable equivalence relation is generated by a single transformation, *"Ergod. Theory and Dyn. Syst."*, **1** (1981), 431 - 450.

[CK] Connes A. and Krieger W., Measure space automorphisms, the normalizers of their full groups, and approximate finiteness, *"J. Funct. Anal."*, **24** (1977), 336 - 352.

[D1] Dye H.A., On groups of measure preserving transformations, I, *"Amer. J. Math."*, **81** (1959), 119 - 159.

[D2] Dye H.A., On groups of measure preserving transformations, II, *"Amer. J. Math."*, **85** (1963), 551 - 576.

[F] Fedorov A.L., The Krieger theorem for cocycles (in Russian), Preprint, 1985.

[FHM] Feldman J., Hahn P., and Moore C.C., Orbit structure and countable sections for actions of continuous groups, *"Adv. Math."*, **28** (1978), 186 - 230.

[FM] Feldman J. and Moore C.C., Ergodic equivalence relations, cohomology, and von Neumann algebras.I, *"Trans. Amer. Math. Soc."*, **234** (1977), 289 - 324.

[GDB] Golodets V.Ya., Danilenko A.I. and Bezuglyi S.I. Cocycles of an ergodic approximable dynamical system and automorphisms compatible with cocycles, *"Advances in Soviet Math."*, **19** (1994), 73 - 96.

[GS1] Golodets V.Ya. and Sinelshchikov S.D., Existence and uniqueness of cocycles of an ergodic automorphism with dense ranges in amenable groups, Preprint, Inst. Low Temp. Phys. and Engin., 1983.

[GS2] Golodets V.Ya. and Sinelshchikov S.D., Structure of automorphisms of measurable groupoids and comparison of transient cocycles (Russian), *"Dokl. AN USSR"*, ser.A., No.5 (1987), 3 - 5.

[GS3] Golodets V.Ya. and Sinelshchikov S.D., Amenable ergodic actions of groups and images of cocycles, *"Soviet Math. Dokl."*, **41** (1990), 523 - 526.

[GS4] Golodets V.Ya. and Sinelshchikov S.D., Classification and structure cocycles of amenable ergodic equivalence relations, *"J. Funct. Anal."*, **121** (1994), 455 - 485.

[H] Hamachi T., The normalizer group of an ergodic automorphism of type III and the commutant of an ergodic flow, *"J. Funct. Anal."*, **40** (1981), 387 - 450.

[HO] Hamachi T. and Osikawa M., Ergodic groups of automorphisms and Krieger's theorems, *"Sem. Math. Sci. Keio Univ."*, **3** (1981), 1 - 113.

[K1] Krieger W., On non-singular transformations of a measure space, I, II, *"Z. Wahrscheinlichkeitstheorie verw. Geb."*, **11** (1969), 83 - 97, 98 - 119.

[K2] Krieger W., On ergodic flows and isomorphism of factors, *"Math. Ann."*, **223** (1976), 19 - 70.

[M] Mackey G.W., Virtual groups and group actions, *"Math. Ann."*, **166** (1966), 187 - 207.

[Ro] Rohlin V.A., Selected problems of the metrical theory of dynamical systems (Russian), *"Uspekhi Math. Nauk"*, **4** (1949), No.2, 57 - 128.

[S] Sutherland C.E., A Borel parametrization of Polish groups, *"Publ. RIMS"*, Kyoto Univ., **21** (1985), 1067 - 1086.

[Sch1] Schmidt K., Lecture on Cocycles of Ergodic Transformation Groups, Univ. of Warwick, 1976, 232 p.

[Sch2] Schmidt K., Algebraic Ideas in Ergodic Theory, *"Regional conference series in mathematics"*, v.76, 1990.

[Z1] Zimmer R.J., Ergodic Theory and Semisimple Groups, Birkhäuser, Boston, 1984.

[Z2] Zimmer R.J., Induced and amenable ergodic actions of Lie groups, *"Ann. Sci., Ecole Norm. Sup."*, **11** (1978), 407 - 428

A Descriptive View of Ergodic Theory

Matthew Foreman
Department of Mathematics
University of California, Irvine
Irvine, CA 92697

1   Introduction

This article is intended for two purposes: to survey a small portion of Ergodic Theory for set theorists and to illustrate with some examples the potential relevance of Descriptive Set Theory to ergodic theorists.

As such there is a serious risk that both audiences will find the portions they know to be trivial and the unfamiliar parts incomprehensible (or worse). The author begs the indulgence and patience of all readers on this point. The attempt is to write a manuscript of use to the union of the two audiences, rather than the intersection; the latter a set of woefully small measure.

The painful choices for such a project are clear from the very beginning; no more so than in choosing which notation to use for the non-negative integers. After some reflection the decision was made to use "ω" rather than (say) "N", in part out of laziness. It is the author's sincere hope that this simple matter will not present an obstacle to any reader.

The main point of this paper is that logical language and techniques add a new invariant for distinguishing between classes of (say) measure preserving transformations, and provide a conceptual framework for negative classification results.

This paper focuses primarily on the distinction between Borel and non-Borel sets. The essence of this distinction being that Borel sets are those sets closely tied to countable amounts of "information" and thus showing that a class is non-Borel shows that it is inherently abstract. (The distal transformations are an example of this.) Logical tools discern much finer distinctions than this, for example the level of the Borel hierarchy, or for countable collections of objects the level in the arithmetic hierarchy. However for the purposes of this paper, the Borel/non-Borel distinction suffices.

To give another type of example: consider a class $\mathcal{C}$ of objects that are a subset of a Polish space X. Then the isomorphism relation on $\mathcal{C}$ is a subset of $X \times X$ and hence it makes sense to ask if it is Borel.

If $\mathcal{C}$ admits complete invariants that are calculable by a Borel function, then clearly the answer is "yes". (The class of Bernoulli shifts give an example of this.) However, the converse is false. (In the current context, the ergodic transformations with discrete spectrum are an example where the isomorphism relation is Borel, but there is no complete set of invariants consisting of members of a Polish space.)

The statement that the equivalence relation of isomorphism is not Borel is a strong anti-classification statement. To anthropomorphize a bit: the existence of a complete Borel invariant means that there is a countable list of "questions" that one can ask the individual elements of $\mathcal{C}$ and two elements of $\mathcal{C}$ are isomorphic iff they give the same answers to this list of questions.

To say that the equivalence relation is Borel is to say that there is a countable protocol such that given a pair of elements x, y of C, one can ask each one a countable list of questions, compare answers and see if they are isomorphic. In this case, however, the protocol uses x's answers to determine the questions asked of y and vice versa.

Saying that the equivalence relation is not Borel then precludes even this kind of invariant. At the time of this writing it is open whether the isomorphism relation on the ergodic measure preserving transformations is Borel.

The sections of this paper are organized as follows:

In **Section 2** "A Logical View of the Complexity of Sets", we introduce the Baire and Cantor spaces and review some of their universal properties. After reviewing the ordinals, we define the Borel hierarchy and define the analytic and co-analytic sets. We prove some basic closure properties of these pointclasses. The final topic is logical notation (e.g., Π^1_1, Σ^0_α etc.) and methods of calculating complexity.

In **Section 3**, we begin by giving the basic elements for analyzing the structure of measure preserving transformations, the ideas of isomorphism and factor. We draw out the connection between unitary operators and measure preserving transformations and give a criterion for when a unitary operator is induced by a measure preserving transformation.

This criterion shows that for a fixed measure space the group of measure preserving transformations is a Polish group. A concrete complete metric is given for this group.

A basic requirement for taking the set theoretic point of view is to find the appropriate setting for studying the complexity of the class of objects at hand. The rest of this section is devoted to carrying this out, for classes of homeomorphisms, and for measure preserving transformations. This is done in such a way that the complexity is invariant under the various choices made. Part of this "coding" requires a discussion of partitions, and generators.

In **Section 4**, some examples of calculations of complexity are given, in the form of a survey of a portion of ergodic theory, noting how "effective" various results are. This begins with ergodic transformations, ergodic decompositions and Bernoulli shifts. A classical application of descriptive set theory is noted in the theorems of Halmos and Rohlin distinguishing between the weakly and strongly mixing transformations. Entropy is introduced and briefly discussed.

Krieger's theorem relating entropy to finite generators is asserted to be Borel, and Ornstein's theorem that entropy is a complete invariant of Bernoulli shifts is stated. In the other direction, a result of Ornstein and Shields is presented to the effect that the K-automorphisms are quite complicated.

Homeomorphisms of compact spaces are considered, and it is asserted that the collection of minimal transformations is a simple Borel set, but that distal transformations do not form a Borel set at all.

In preparation for the Halmos-von Neumann theorem (and to illustrate some points about the effectivity of the spectral theorem), it is shown that it is possible to calculate the eigenvalues of linear operators in a Borel way.

Section 5 is an introduction to equivalence relations, under the ordering of reducibility. This subject has been highly developed in Descriptive Set Theory, and is only touched on here, in as much as it immediately relates to ergodic theory.

The partial ordering of Borel reducibility is introduced, along with various Borel equivalence relations, such as E_0. The dichotomy theorem between *smooth* and E_0 is stated, as well as the interpretation of the equivalence relation of isomorphism on Bernoulli shifts as smooth and the Ornstein-Shields result as a reduction of E_0 to isomorphism on the K-automorphisms.

The "+" operation on equivalence relations is introduced, and it is shown that every Borel equivalence relation with countable classes is reducible to Eq^+, the + operation applied to the equivalence relation of equality. The spectral theorem for normal operators is interpreted in this language.

After this the Halmos-von Neumann theory is reviewed for the benefit of the set theorists, and it is shown that isomorphism relation on the discrete spectrum transformations has the same complexity as "Eq^+".

In **Section 6**, the structure of Π_1^1 (co-analytic) sets is explained. This is intimately related to the notion of reducibility given in the previous section. The notion of "complete Π_1^1" is given and canonical examples of complete Π_1^1 sets are described. The definition of a Π_1^1-norm is given along with the crucial Boundedness Theorem.

In **Section 7**, a particular equivalence relation is considered, the conjugacy action of a Polish group on itself.

The section begins by giving a canonical example of a non-Borel equivalence relation. This relation is then shown to be reducible to the conjugacy relation on the automorphism group of the rational numbers. As a corollary, it follows that there is a Polish group where the conjugacy relation is not Borel. On the other hand, it is explained why the conjugacy relation on the unitary operators is Borel. An example is given that shows that for measure preserving transformations, unitary equivalence does not imply isomorphism.

Section 8 is devoted to a brief exegesis of the Furstenberg structure theorem. Some basic ergodic theorems are stated, and the notions of weakly mixing and compact transformations are given. Weak mixing is characterized in terms of eigenvalues and it is shown that a transformation is compact iff it has discrete spectrum. The Hilbert-Schmidt operators are introduced and it is shown that the compact functions are spanned by Hilbert-Schmidt operators with invariant kernels.

The relativized theory is introduced and the notions of compact and weak mixing are generalized. The Furstenberg structure theorem is then outlined and stated. The definition of measure distal and generalized discrete spectrum are given.

Skew products are introduced and the notion of cocycle is given. A characterization of compact extensions is given in terms of cocycles into homogeneous spaces, as well as a criterion for isomorphism for *normal* compact extensions.

Section 9 is a discussion of a theorem of Ferenc Beleznay and the author showing that the collection of measure distal transformations is not a Borel set. Moreover, for all countable ordinals α there is a measure distal transformation whose approximating tower takes at least α steps. It is remarked that there are sets of integers such that the associated measure preserving transformation is isomorphic to any given distal transformation.

It is shown that the function that assigns to a measure distal transformation the least length of an approximating tower induces a $\coprod_1^1$-norm. Using this, it is shown that the function that assigns to every ergodic transformation the maximal distal factor less than or equal a given ordinal α, is a Borel function.

The proof of this theorem is outlined. Some of the methods for concretely calculating the distal extensions are given. A characterization of measure distal in terms of the ordertypes of maximal approximating towers is proved. This is essential for showing that the Furstenberg norm is a $\coprod_1^1$-norm.

The reduction of WO to the measure distal flows is described. Crucial to this is the ability to precisely calculate the maximal compact extension. The examples constructed of transfinite norm are given as limits of transformations of finite norm. In order for the limiting process to converge properly, skew products of finite dimensional tori are considered with fractional coefficients. The apparently novel results that these are ergodic and the compact extensions behave properly are described.

In **Section 10**, the situation for topologically distal transformations is described. The theory for distal transformations is analogous (and prior) to the theory for measure distal transformations and the analogous theorems go through. However, instead of controlling compact extensions via cocycles, iterated "generic" skew products are used.

Section 11 asks some questions.

Some elementary mathematical prerequisites are presumed of the reader, and some unfamiliar notations may be used. Some that come to mind are, in no particular order:

- If α is an irrational multiple of $2\pi i$, then $\{e^{n\alpha} : n \in \mathbb{Z}\}$ is dense in the unit circle $\mathbb{T}$.

- Basic notions of probability theory such as Bayes' Law. One idea used frequently in the paper is that of *independence*. If $(X, \mathcal{B}, \mu)$ is a measure space and $\mathcal{P}, \mathcal{Q}$ are two measurable partitions of X, then $\mathcal{P}$ and $\mathcal{Q}$ are *independent* iff for every $A \in \mathcal{P}$ and $B \in \mathcal{Q}, \mu(A \cap B) = \mu(A)\mu(B)$. Two sets A and B are said to be independent iff the partitions $\mathcal{P} = \{A, X \backslash A\}$ and $\mathcal{Q} = \{B, X \backslash B\}$ are independent.

- The word *essential* will be used to mean "except for a set of measure zero".

- Familiarity with product spaces. We will frequently be considering a Polish space X (i.e., a topological space homeomorphic to a complete separable metric space) and considering its countable product. We will denote this countable product X^ω or $X^\mathbb{Z}$, among other ways. Basic open sets for the product topology are given by cylinders determined by finite functions into a basis for X. Specifically, if we are viewing this countable product as $X^\mathbb{Z}$ and $\mathcal{U}$ is a basis for the topology for X, then basic open sets are determined by functions $s : S \to \mathcal{U}$ (where $S \subset \mathbb{Z}$ is finite) by taking the cylinder set $[s] = \{f \in X^\mathbb{Z} : \forall n \in S, f(n) \in s(n)\}$.

- The measure theoretic analogues of the previous statements: If $(X, \mathcal{B}, \mu)$ is a measure space with measure algebra $\mathcal{B}$, then the product space $X^\mathbb{Z}$ has measure algebra generated by cylinder sets determined by a function $s : S \to \mathcal{B}$ for a finite $S \subset \mathbb{Z}$. The measure of the cylinder set is $\prod_{n \in S} \mu(s(n))$. Clearly, if $s : S \to \mathcal{B}$ and $t : T \to \mathcal{B}$ are such that $S \cap T = \emptyset$, then the cylinder sets determined by s and t are independent in the sense of measure.

- Maharam's Theorem, which has as a corollary that any two non-atomic, separable measure spaces are essentially isomorphic.

- Some familiarity with (left) Haar measure, the unique left translation invariant probability measure on a compact group.

- We will assume that for every measure space $(X, \mathcal{B}, \mu)$ we have a fixed canonical basis $\{e_i : i \in \omega\}$ for $L^2(X)$. This is not a limiting assumption in light of Maharam's theorem.

- Since the unit circle with the operation of multiplication is canonically isomorphic to the unit interval with 0 and 1 identified and the operation of addition, there are two possibilities for notation. We will use each according to what is convenient in context. For addition on the unit interval we will use the notation $\oplus$ when we are trying to be careful.

- The fact that every compact metric space is homeomorphic to a closed subset of the Hilbert Cube.

- We will use "OR" for the class of ordinals.

- $\mathcal{G}_\delta$ subsets of Polish spaces are Polish spaces with the induced topology.

- For sets $B \subset X \times Y$ and $x \in X$, we will use B_x for the set $\{y \in Y : (x, y) \in B\}$.

- The paper will adopt the logician's convention that the natural number N is the set of smaller natural numbers, i.e., $N = \{0, 1, 2, \ldots, N - 1\}$.

- The paper will not attempt to give real proofs of any statement. Instead there will be sketches that are hopefully complete enough to convey some basic ideas.

The author would like to thank the many people who read the paper and helped prevent him from embarrassing himself. These include V. Bergelson, A. Kechris, and A. Louveau. Thanks also go to many people who helped him mathematically, especially those in Jerusalem. Particular debt is owed to Benjamin Weiss for his patient midwifery over the years.

The author apologizes in advance for brazenly cribbing not only the ideas, but the notation and organization of various other expositions. In particular much of this paper was taken directly from Walters' book, *An Introduction to Ergodic Theory* [Wal82] and Furstenberg's book *Ergodic Theory and Combinatorial Number Theory* [Fur81]. The rationalization for this is the author's feeling that it is important to collect all of this information in one place; the author hopes that the victims will realize that this larceny reflects sincere admiration.

Acknowledgment. Research and preparation of this paper were partially supported by NSF Grant DMS 98-03126.

2 A Logical View of the Complexity of Sets

In this section we discuss "logical notation" for the Borel sets and the projective hierarchy, and give some examples and explanation showing its utility.

As a general heuristic, quantification over countable sets yields Borel sets and quantification over Polish spaces yields projective sets, but constructions that require quantification over arbitrary well orderings (or use of uncountable axiom of choice) take one out of the realm of definable sets and into the realm of non-measurable sets.

The general context for this will be a Polish space, i.e., a complete separable metric space. In the examples, the Polish space will usually be the space of measure preserving transformations with the Halmos metric, or the space of unitary operators with the weak operator topology.

For the rest of this paper we will assume that we are working in a Polish space. We will frequently tacitly assume as well that our space has no isolated points.

2.1   The Baire Space and the Cantor Space

Two Polish spaces play a special role in descriptive set theory, the Baire space and the Cantor space.

Definition 1 *Let ω have the discrete topology. The* Baire Space *is the product of countably many copies of ω with the product topology. Similarly, if we view the number* **2** *as the set* $\{0, 1\}$ *endowed with the discrete topology, the* Cantor Space *is the product of countably many copies of* **2** *with the product topology.*

Since the product of countably many copies of ω can be viewed as the collection of functions from ω to ω it is often denoted ${}^{\omega}\omega$ (or less frequently ω^{ω}). The Cantor set is denoted either ${}^{\omega}2$ or 2^{ω}.

These spaces are familiar in very conventional contexts. The Cantor set defined in this manner is homeomorphic to the usual "middle thirds" Cantor set. (A 0 or a 1 indicates whether you are going right or left in the middle thirds Cantor set.) The continued fractions expansion of irrational numbers gives a homeomorphism between ${}^{\omega}\omega$ and the irrational real numbers. Occasionally we will want our space to reflect more structure on ω, such as a group structure. In this case we will write, e.g., $2^{\mathbb{Z}}$ or $2^{\mathbb{Q}}$.

Part of the reason for the significance of the Baire and Cantor sets is their universal character:

Theorem 2 *Let X be a Polish Space.*

1. *There is a continuous surjection of ${}^{\omega}\omega$ onto X.*

2. *If X is uncountable, there is a continuous injection of ${}^{\omega}2$ into X.*

The Baire and Cantor spaces are also convenient for other reasons: If X is either the Baire space or the Cantor space, then X is homeomorphic to either a finite or countable product of X with itself. If X is the Baire space and ω is again given the discrete topology, then $\omega \times X \approx X$. These facts allow easy "coding" into elements of the Baire space and the applicability of recursion theoretic technique.

2.2   Ordinals

Basic to understanding the logical conception of the Borel sets is an understanding of the countable ordinals. The "official" definition of an ordinal is a transitive set, all of whose elements are transitive. (A set x is *transitive* iff for all $y \in x$ and all $z \in y$, we have $z \in x$).

However this definition somewhat obscures the main points: Every ordinal is well-ordered and every well-ordering is isomorphic to a unique ordinal. The set of countable ordinals is itself well-ordered, and in fact forms the first uncountable ordinal, ω_1. This allows transfinite induction on the countable ordinals. The ordinals come in two "flavors", limit and successor ordinals, typically leading to two cases in an induction.

The natural numbers form an initial segment of the ordinals, and the first infinite ordinal ω is the set of natural numbers. (We will use ω and $\mathbb{N}$ interchangeably for the set of natural numbers which, for us, includes the number 0.) The next ordinal is $\omega + 1$ which is the well ordering obtained by putting one point after the order type of ω. The second limit ordinal is $\omega + \omega$, which is the well ordering obtained by putting a "copy" of the natural numbers after the natural numbers.

There are (non-commutative) operations of addition, multiplication and exponentiation on the ordinals. The first few ordinals look like:

$$0, 1, 2, 3, 4, \ldots, \omega, \omega+1, \omega+2, \ldots, \omega+\omega, \omega+\omega+1, \ldots, \omega\times 3, \ldots, \omega\times 4, \ldots \omega\times\omega,$$
$$\ldots, \omega^3, \ldots, \omega^4, \ldots, \omega^\omega, \ldots, \ldots$$

Any elementary set theory book such as [Lév79], [JW96] or [End77] will contain information on the ordinals.

We will use the notation "OR" for the class of ordinals.

2.3   Borel Sets

We define the hierarchy of Borel sets by induction on the countable ordinals. We let:

$$\Sigma_1^0 = \{\text{Open subsets of X}\}$$

and

$$\Pi_1^0 = \{X \backslash U : U \in \Sigma_1^0\} = \{\text{Closed subsets of X}\}.$$

More generally, for a countable ordinal $\alpha > 0$:

$$\Sigma_\alpha^0 = \{\bigcup_n A_n : \text{for each } n \text{ there is a } \xi < \alpha, A_n \in \Pi_\xi^0\}$$

and

$$\Pi_\alpha^0 = \{X \backslash A : A \in \Sigma_\alpha^0\}$$

Then the Borel sets are the union over the countable ordinals α of the Σ^0_α, which is the same as the union of the Π^0_α:

$$\{\text{Borel sets}\} = \bigcup_{\alpha < \omega_1} \Sigma^0_\alpha = \bigcup_{\alpha < \omega_1} \Pi^0_\alpha.$$

The intersection of a "sigma" class and the corresponding "pi" class is called the "delta" class:

$$\Delta^0_\alpha = \Sigma^0_\alpha \cap \Pi^0_\alpha.$$

This notation is a systematic way of codifying the usual Borel hierarchy defined in any first year analysis course. For example, a moments' thought will show that the $\mathcal{F}_\sigma$ sets are the Σ^0_2 sets, the $\mathcal{G}_\delta$ sets are the Π^0_2 sets, etc.

The *Borel Rank* of a Borel set B is the least ordinal α such that B lies in $\Sigma^0_\alpha \cup \Pi^0_\alpha$.

A classical fact is:

Theorem 3 *In $^\omega 2$ (and hence in any uncountable Polish space) all of the following inclusions are proper:*

$$
\begin{array}{ccc}
 & \Sigma^0_\alpha & \\
 \subset & & \subset \\
\Delta^0_\alpha & & \Delta^0_{\alpha+1} \\
 \subset & & \subset \\
 & \Pi^0_\alpha &
\end{array}
$$

2.4   Quantifiers

If X is a Polish space and ω is given the discrete topology, then the product space $X \times \omega$ is still a Polish space. Moreover, if $\langle A_n : n \in \omega \rangle$ is a collection of Borel subsets of X, then so is

$$A = \{(x, n) : x \in A_n\} \subset X \times \omega$$

If each $A_n \in \Pi^0_\alpha$, then $A \in \Pi^0_\alpha$. Thus a typical element of $\Sigma^0_{\alpha+1}$ is:

$$\{x : \exists n (x, n) \in A\} = \bigcup_n A_n.$$

Similarly, $\Pi^0_{\alpha+1}$ sets B can be constructed by taking a Σ^0_α set $A \subset X \times \omega$ and letting

$$B = \{x : \forall n (x, n) \in A\} \in \Pi^0_{\alpha+1}$$

The moral of these trivial remarks is that "number quantification" (i.e., quantifying over the natural numbers) corresponds to countable unions (in the case of existential quantification) and intersections (in the case of universal quantification) and that this is *a very useful tool for calculating Borel rank.*

Sometimes quantification that appears to be over uncountable sets can be reduced to quantification over countable sets. A common example of this are statements "$\forall \epsilon > 0$" or "$\exists \delta > 0$" which can frequently be replaced by the equivalent statement "for all rational $\epsilon > 0$" or "there is a rational $\delta > 0$".

Quantifying over uncountable sets gives a very different result as we will see in the next section.

2.5 Analytic and Co-analytic Sets

In this section we give the definition of the analytic and co-analytic sets. The collection of analytic and co-analytic sets properly extends the collection of Borel sets. Analytic and co-analytic sets retain many of the nice properties of Borel sets such as having the property of Baire and being universally measurable sets.

Definition 4 *We define a set $A \subset X$ to be* analytic *iff it is the image of a Borel subset B of a Polish space Y under a continuous map $f : Y \to X$. A set B is* co-analytic *iff $X \backslash B$ is analytic.*

Many natural sets (such as the stopping times of a stochastic process) arise as analytic sets. In 1917, Suslin gave an example of an analytic set that was not Borel, thus rectifying a mistake of Lebesgue ([Sus17]).

Fix such a set $B \subset Y$ and a continuous map $f : Y \to X$. Define $C = \{(x, y) : x = f(y) \text{ and } y \in B\}$. Since the graph of f is easily seen to be a closed subset of $Y \times X$, the set C is a Borel subset of the Polish space $X \times Y$ and the image of B under the map f is the projection of C onto the X-axis. Thus we get the following characterization of the analytic sets:

Proposition 5 *The analytic subsets of a set X are precisely the collection of projections of Borel subsets of $X \times Y$, where Y is a Polish space.*

This leads to the following remark: $A \subset X$ is analytic iff there is a Polish space Y and a Borel subset $B \subset X \times Y$ such that:

$$A = \{x : \exists y \in Y (x, y) \in B\}.$$

By Theorem 2 one can assert even more: $A \subset X$ is analytic iff there is a Borel set $B \subset X \times {}^\omega\omega$ such that:

$$A = \{x : \exists y \in {}^\omega\omega \ (x, y) \in B\}.$$

In other words, *the analytic sets are exactly the projections of Borel sets in the product of X with the Baire space.*

By taking complements we see that the co-analytic sets C can be defined by taking a Borel set $B \subset X \times {}^\omega\omega$ and letting C be the collection of $x \in X$ such that the "line" above x lies inside B. In other words, the co-analytic sets are those built from Borel subsets of a Polish spaces by universal quantification over a Polish space Y; i.e., $C = \{x : (\forall y \in Y)(x, y) \in B\}$.

This method of constructing analytic and co-analytic sets makes the following theorem easy to prove:

Theorem 6 *The collection of analytic sets is closed under countable unions and intersections. The same assertion is true for the co-analytic sets.*

⊢   We give the proof here to illustrate some of the techniques. Note that the assertion for co-analytic sets follows from the assertion for analytic sets by taking complements.

Suppose that B is a countable union of analytic sets. There are Borel sets $A_n \subset X \times {}^\omega\omega$ such that $x \in B$ iff $\forall n \exists f (x, f) \in A_n$. Define a set $C \subset X \times (\omega \times {}^\omega\omega)$ by setting $(x, n, f) \in C$ iff $(x, f) \in A_n$. Then C is clearly Borel and $x \in B$ iff $\exists (n, f)(x, (n, f)) \in C$.

The second assertion is somewhat more surprising: one would expect the intersection to introduce a universal quantifier that obstructs analyticity. Let Y be the product of countably many copies of ${}^\omega\omega$. (So Y is homeomorphic to ${}^\omega\omega$.) Thus elements of Y can be viewed as countable sequences $\langle g_n \rangle$ where each $g_n \in {}^\omega\omega$.

If B is a countable intersection of analytic sets then there are Borel sets $A_n \subset X \times {}^\omega\omega$ such that $x \in B$ iff $\forall n \exists g_n (x, g_n) \in A_n$. Define a Borel set $C \subset Y$ by setting $g \in C$ iff for all n, $g_n \in A_n$. Then C is Borel and $x \in B$ iff $\exists g \in Y \ (x, g) \in C$.   ⊣

The following theorem of Suslin gives the exact relation between Borel sets and analytic sets:

Theorem 7 *A set $B \subset X$ is Borel iff it is both analytic and co-analytic.*

This theorem gives a "quick and dirty" method of showing that a set is Borel.

2.6   Notation and Random Observations

The collection of analytic sets is denoted Σ_1^1 and the collection of co-analytic sets is denoted Π_1^1.

The co-analytic sets are not closed under projections; there are co-analytic sets $C \subset {}^\omega\omega \times {}^\omega\omega$ such that the projection of C to the first coordinate is not a co-analytic set. In fact the operation of taking complements and projections gives a hierarchy of collections of subsets of X that does not terminate at a finite stage. These are written in logical notation as follows:

$$A \in \Sigma^1_{n+1}$$

iff there is a Π^1_n set $B \subset (X \times {}^\omega\omega)$ such that for all $x \in X$

$$x \in A \Leftrightarrow \exists f (x, f) \in B.$$

Moreover, a set $B \in \Pi^1_n$ iff there is a Σ^1_n set $A \subset X$ such that $B = X \backslash A$. Again we let $\Delta^1_n = \Sigma^1_n \cap \Pi^1_n$.

The following theorem is in analogy with the Borel sets.

Theorem 8 *In ${}^\omega 2$ (and hence in any uncountable Polish space) all of the following inclusions are proper:*

$$
\begin{array}{ccc}
 & \Sigma^1_n & \\
\subset & & \subset \\
\Delta^1_n & & \Delta^1_{n+1} \\
\subset & & \subset \\
 & \Pi^1_n &
\end{array}
$$

This hierarchy is called the *projective hierarchy*. Showing that the sets in the projective hierarchy are nicely behaved (i.e., have the property of Baire, are universally measurable, etc.) requires Large Cardinal assumptions.

For the next definition we will use the jargon of "point classes" ([Mos80]), i.e., collections of subsets of some Polish spaces. Typical examples are the collection of Borel sets of a given rank, and the collection of analytic sets.

Definition 9 *Let Γ be a pointclass. A set $U \subset Y \times X$ is* universal *for Γ on X iff $U \in \Gamma$ and $\Gamma \cap P(X) = \{U_y : y \in Y\}$.*

The basic ideas in theorems 3 and 8 are the same, one builds a universal set for a given level of the hierarchy and then diagonalizes against that set. The following proposition is used for both theorems.

Proposition 10 *Let Γ be a pointclass which has a universal set $U \subset X \times X$ on X such that Γ is closed under preimages of continuous functions. Then there is a set $D \subset X$ such that $D \in \Gamma$ but $X \backslash D \notin \Gamma$.*

⊢ To see this proposition, we note that since Γ is closed under continuous preimages, $D = \{x \in X : (x, x) \in U\}$ is in Γ. Suppose that $C = X \backslash D \in \Gamma$. Since U is universal, there is some $z \in X, C = U_z$. But then $z \in C$ iff

$(z, z) \in U$ iff $z \in D$, which contradicts the fact that C is the complement of D.   ⊣

Note that each level of the Borel and Projective hierarchies is closed under continuous preimages. Hence the following theorem, which is most easily proved using logical methods, suffices to show Theorems 3 and 8.

Theorem 11 *Let* Γ *be either the collection of* Σ^0_α *sets for some countable ordinal* α *or the collection of* Σ^1_n *sets for some* $n \in \omega$ $(n > 0)$. *Let* X *be a perfect Polish space. Then there is a set* $U \subset X \times X$ *that is universal for* Γ *on* X.

The reader is referred to either [Mos80] or [Kec95] for a proof of Theorem 11 as well as detailed information on the Borel, Projective and other hierarchies of sets.

2.7 Some Pseudo-historical Remarks

A non-logician might be puzzled by the notation here. It was first proposed formally by Addison [Add59]. As an aid to memory let me make the following (somewhat apocryphal) remarks.

The use of Σ for unions is natural as unions are sums of a sort and similarly for Π for intersections. The "tilde" under the symbol indicates that there are no restrictions on what unions are taken: in different contexts the unions and intersections can be restricted to be effective or computable in various senses.

The superscript 0 or 1 stems from a tradition of "type theory" that dates at least back to logicians such as Russell and Whitehead. In an effort to circumvent Russell's Paradox, they classified sets according how many iterations of the powerset operation one needed to find them.

The natural numbers were quite safe, so they were considered type 0; functions from the natural numbers to the natural numbers (or equivalently subsets of the natural numbers) were type 1. Functions from type 1 objects to type 1 objects were type 2 etc.

The superscript i in Σ^i_j refers to the type being quantified over. Since Borel sets are built quantifying over the natural numbers the superscript is 0. The projective sets are built by quantifying over $^\omega\omega$, i.e., type 1 objects. Hence they are denoted Σ^1_n, and Π^1_n.

3   The Polish Spaces

The examples we will consider will come either from ergodic theory or from topological dynamics. In order to use the descriptive complexity as an invariant we must be able to view the objects of study as members of an appropriate Polish space. This will usually involve considering three basic spaces, the group of unitary operators, the group of measure preserving transformations and collections of homeomorphisms of a compact metric spaces. (The latter is not a group, as we will vary the spaces we consider.)

3.1   Some Structural Ideas for Measure Preserving Transformations

Let X be a set and B a σ-algebra of subsets of X and μ a probability measure defined on the sets in B. Then (X, B, μ) is a *standard measure space* (or *Lebesgue space*) iff X is measure isomorphic to the unit interval with Lebesgue measure defined on the Lebesgue measurable sets.

Let (X, B, μ) and (Y, C, ν) be two standard measure spaces. A function $T : X \to Y$ is a measure preserving transformation iff

- There is a set of μ measure one on which T is a bijection.

- T and T^{-1} are measurable.

- For all $A \in B$, $\nu(A) = \mu(T^{-1}(A))$.

We will typically identify two measure preserving transformations if they agree on a set of measure one.

Remark 12 *We will be assuming throughout this paper that the transformations we are considering are aperiodic. Since we will be concentrating (for the most part) on the ergodic transformations, this is not a serious limitation.*

A four tuple (X, B, μ, G) is a *measure preserving system* iff G is a group acting on X by measure preserving transformations. In an abuse of notation, if G is generated by a single element T, we will frequently write (X, B, μ, T), instead of $(X, B, \mu, \langle T \rangle)$.

If (X, B, μ, T) and (Y, C, ν, S) are measure preserving systems, then they are *isomorphic* iff there are measure one sets X', Y' and an invertible measure preserving $f : X' \to Y'$ such that the diagram:

$$
\begin{array}{ccc}
X' & \xrightarrow{T} & X' \\
\downarrow f & & \downarrow f \\
Y' & \xrightarrow{S} & Y'
\end{array}
$$

commutes.

An important tool for the study of measure preserving systems is the notion of a "factor".

Definition 13 (Y, C, ν, S) *is a factor of* (X, B, μ, T) *(and* (X, B, μ, T) *is an extension of* (Y, C, ν, S)*) iff there is an essential set* $X' \subset X$ *and a measurable map* $\pi : X' \to Y$ *such that for all* $A \in C$, $\mu(\pi^{-1}(A)) = \nu(A)$ *and the following diagram commutes:*

$$
\begin{array}{ccc}
X' & \xrightarrow{T} & X' \\
\downarrow \pi & & \downarrow \pi \\
Y & \xrightarrow{S} & Y
\end{array}
$$

Clearly there is a canonical $1-1$ correspondence between T-invariant sub-σ-algebras of B and factors of (X, B, μ, T).

Thus, one way of viewing factors is that there is a measure preserving system (X, B, μ, T) generating a phenomenon that is only observable in the invariant σ-algebra C. The resulting observable dynamical system is the factor (Y, C, ν, S).

3.2 Unitary Operators and Measure Preserving Transformations

Example 14 *Let* $\mathbb{H}$ *be a countably infinite dimensional Hilbert space. Let* $\mathcal{U}(\mathbb{H})$ *be the group of unitary operators on* $\mathbb{H}$. *Then the weak and the strong operator topologies coincide on* $\mathcal{U}(\mathbb{H})$ *and make* $\mathcal{U}(\mathbb{H})$ *into a Polish space.*

A measure preserving transformation $T : X \to Y$ naturally induces a unitary transformation U_T from $L^2(Y)$ to $L^2(X)$, where U_T is defined by $U_T(f) = f \circ T$. (We will sometimes abuse notation by writing $T(f)$ instead of $U_T(f)$.) The next theorem (see e.g., [Wal82]) gives a criterion for when a unitary operator is induced by a measure preserving transformation:

Theorem 15 *Let* (X, B, μ) *and* (Y, C, ν) *be Lebesgue spaces. Let* U *be a unitary operator from* $L^2(Y)$ *onto* $L^2(X)$. *Then there is a measure preserving transformation* $T : X \to Y$ *such that* $U = U_T$ *iff*

- $U : L^\infty(Y) \to L^\infty(X)$

- *For* $f, g \in L^\infty(Y)$, $U(fg) = U(f)U(g)$

⊢   Sketch: χ is the characteristic function of a set $A \in C$ iff $\chi^2 = \chi$. Hence U takes characteristic functions to characteristic functions. Define $\phi(A) = A'$ iff $U(\chi_A) = \chi_{A'}$. The linearity of U shows that ϕ is a Boolean algebra isomorphism and the continuity shows that it is σ-complete. Hence $\phi : C \to B$ is a complete isomorphism that preserves measure, and thus is induced by a measure preserving transformation T.   ⊣

Since it suffices to verify the hypothesis of Theorem 15 on the characteristic functions of a dense subset of B, one can check that the collection of unitary operators that satisfy Theorem 15 form a $\underline{\Pi}_2^0$ subset of $\mathcal{U}(L^2(X))$. Thus we see:

Example 16 *Let $\mathcal{H}$ be the group of all measure preserving transformations of (X, B, μ). Then $\mathcal{H}$ is isomorphic to a $\underline{\Pi}_2^0$ subset of $\mathcal{U}(L^2(X))$. In particular, $\mathcal{H}$ is a Polish group when given the induced topology, and $\mathcal{H}$ is therefore isomorphic to a closed subgroup of $\mathcal{U}(L^2(X))$.*

Halmos defined a concrete complete metric on $\mathcal{H}$: Let $\{E_n : n \in \omega\}$ be a collection of measurable subsets of X that generate B as a σ-algebra. (One can think of these as the rational intervals in $X = [0, 1]$.)

Definition 17 *The* Halmos Metric *on the group of measure preserving transformations of X is defined as follows: If S and T are measure preserving transformations from X to X, then:*

$$d(S, T) = \sum_{n \in \omega} 2^{-n}[\mu(T(E_n) \Delta S(E_n)) + \mu(T^{-1}(E_n) \Delta S^{-1}(E_n))].$$

This defines a complete metric on the group of measure preserving transformations.

Note that two measure preserving transformations on X are isomorphic iff they are conjugate in the group of measure preserving transformations. For this reason we will use the words "conjugate" and "isomorphic" as synonyms.

3.3 Homeomorphisms

For considering topological transformations, we use the fact that every compact metric space is homeomorphic to a compact subset of the Hilbert Cube. Moreover, this homeomorphism can be chosen canonically from a basis for the topology for X.

Example 18 *Let $\mathbb{I}$ be the Hilbert cube, and $\mathcal{K}_0$ be the space of compact subsets of $\mathbb{I}^2$ with the Hausdorff metric. If X is a compact subset of $\mathbb{I}$ and $T : X \to X$ is a homeomorphism, then T can be viewed as an element of $\mathcal{K}_0$, by identifying it with its graph.*

3.4 Coding Issues

As in the previous example, there may be more than one choice for viewing the objects of study as members of Polish spaces. To wit: we could consider the infinite product of the Hilbert Cube with itself, $\mathbb{I}^{\mathbb{Z}}$, and view a homeomorphism T of a compact metric space X as an element of the space $\mathcal{K}_\omega$ of compact subsets of $\mathbb{I}^{\mathbb{Z}}$, by identifying T with the graph of its iterates as a subset of $X^{\mathbb{Z}}$.

The following proposition from [BF95] shows that this change is irrelevant:

Proposition 19 *There are Borel functions* $\pi : \mathcal{K}_\omega \to \mathcal{K}_0$ *and* $e : \mathcal{K}_0 \to \mathcal{K}_\omega$ *such that for all* $x \in \mathcal{K}_\omega, x$ *codes a homeomorphism of a compact metric space* X *iff* $\pi(x)$ *codes a homeomorphism of* X *and for all* $y \in \mathcal{K}_0, y$ *codes a homeomorphism of a compact space* X *iff* $e(y)$ *codes a homeomorphism of* X. *Moreover,* x *and* $\pi(x)$ *code the same homeomorphism. Similarly,* y *and* $e(y)$ *code the same homeomorphism.*

We can also code measure preserving transformations canonically as measures on the Baire Space. For this we need the notion of a *generator*.

Let $\mathcal{P}$ be a measurable partition of the space X, and let T be a measure preserving transformation. Let $T^n \mathcal{P}$ be the partition of X that is the image of $\mathcal{P}$ under T^n. Let $\bigvee_k^l T^n \mathcal{P}$ be the partition generated by $\bigcup_{n=k}^l T^n \mathcal{P}$, i.e., the smallest partition refining each $T^n \mathcal{P}$ for n between k and l.

Definition 20 *A partition* $\mathcal{P}$ *is called a* generator *iff* $\bigcup_{n=-\infty}^\infty T^n \mathcal{P}$ *generates* B *as a* σ-algebra.

The collection of measurable partitions itself forms a Polish space; there are many equivalent ways of doing this. One is to view a partition as sequence $\langle f_n \rangle$ of elements of $L^2(X)$ such that $f_n^2 = f_n$, $f_n \perp f_m$ for $n \neq m$ and $\sum \| f_n \|_2^2 = 1$. This is a $\underline{\Pi}_2^0$ condition on the space of sequences of elements of $L^2(X)$ and hence gives a Polish structure on the collection of partitions. Since T acts by measure preserving transformations, T is a homeomorphism of the space of partitions. Moreover the map $(T, \mathcal{P}) \mapsto T\mathcal{P}$ is a jointly continuous function of T and $\mathcal{P}$.

Measure preserving transformations always have a countable generator:

Theorem 21 *[DGS76] Every measure preserving transformation has a countable generator.*

We note that this theorem is true in considerable generality: Weiss showed that if $T : X \to X$ is a Borel map (where X is a Polish space), then there is a countable generator for T. Moreover, the generator is computed effectively. In particular, in the context of Theorem 21 the generator can be computed in a continuous way from T.

If $\mathcal{P} = \{A_n : n \in \omega\}$ is a generator for the measure preserving transformation T, then almost every point x can be uniquely identified by its "history", i.e., by the function $f_x(i) = j$ iff $T^i(x) \in A_j$ ([DGS76]). This gives a map $\phi : X \to {}^Z\omega$ that is defined almost everywhere.

We can then define a measure on ${}^Z\omega$ by setting $\nu(C) = \mu(\phi^{-1}(C))$. If $s : [-N, N] \to \omega$ and O is the basic open set $\{f : f \restriction [-N, N] = s\}$, then O is measurable and $\nu(S) = \mu(\bigcap_{n=-N}^N A_{s(n)})$, hence ν defines a measure on the Borel sets of ${}^Z\omega$ and the Borel sets generate the ν-measurable sets.

Thus if $\sigma : {}^{\mathbb{Z}}\omega \to {}^{\mathbb{Z}}\omega$ is the "shift" map $(\sigma(f)(n) = f(n+1))$, and C is (the completion of) the collection of Borel sets in ${}^{\mathbb{Z}}\omega$, then the system $({}^{\mathbb{Z}}\omega, C, \nu, \sigma)$ is isomorphic to the original system (X, B, μ, T).

Note that in the triple $({}^{\mathbb{Z}}\omega, C, \nu, \sigma)$, ν is the only element that varies with the transformation T. We have just outlined a proof of the following result:

Proposition 22 *Every measure preserving transformation is isomorphic to the shift map on ${}^{\mathbb{Z}}\omega$, with an invariant measure.*

The effective version of this theorem shows us that the coding leaves the complexity invariant. (We state the Borel versions of the coding results. More care yields finer results, such as continuity.) Let $\mathbb{M}$ be the collection of shift invariant measures on ${}^{\mathbb{Z}}\omega$. Recall that $\mathcal{H}$ is the group of measure preserving transformations.

Theorem 23 *There are Borel functions $\gamma : \mathcal{H} \to \mathbb{M}$ and $\tau : \mathbb{M} \to \mathcal{H}$ such that for all $T \in \mathcal{H}, \gamma(T)$ is isomorphic to T, and for all $\mu \in \mathbb{M}$ the system $({}^{\mathbb{Z}}\omega, C, \mu, \sigma)$ is isomorphic to $\tau(\mu)$.*

We finish by remarking that many of the constructions of ergodic theory are effective. For example there are effective versions of Rohlin's theorem.

4   Some Examples of the Technique

Let T be a measure preserving transformation on (X, B, μ). Then T is *ergodic* iff the only T-invariant subsets of X have measure zero or one. This is easily seen to be equivalent to the statement that every measurable function f such that $f = f \circ T$ almost everywhere, is constant almost everywhere. The following criterion for ergodicity can be verified using the weak ergodic theorem (see section 8 for a statement of this theorem):

Proposition 24 *Let C be a generating set for the σ-algebra B. Then the following are equivalent:*

1. *T is ergodic.*

2. *For some $\epsilon > 0$, for all $U, V \in C$, there is a k, $\mu(T^k(U) \cap V) > \epsilon\mu(U \cap V)$.*

3. *For all $\epsilon \in (0, 1)$ and all $U, V \in C$, there is a k such that $\mu(T^k(U) \cap V) > \epsilon\mu(U \cap V)$.*

If C is taken to be countable, then this condition is easily seen to be a $\underline{\Pi}_2^0$ condition. Moreover, a Rohlin tower argument shows that the orbit of every measure preserving transformation is dense ([Hal60]). Hence:

Corollary 25 *The collection of ergodic transformations T form a dense $\mathcal{G}_\delta$ subset of the group of measure preserving transformations.*

4.1   Ergodic Decompositions

For background, we include the ergodic decomposition theorem which shows that every measure preserving transformation is an "integral" of ergodic measure preserving transformations.

Let X be a Polish space and (X, B, μ, T) a measure preserving system, where B is the completion of the Borel sets by μ. Let $\mathcal{M}(X)$ be the collection of Borel measures on X (with the weak topology).

Theorem 26 *Let T be a measure preserving transformation on the Lebesgue space (X, B, μ, T). Then there is a Lebesgue space (Ω, C, ν) and a C-measurable function $\mu : \Omega \to \mathcal{M}(X)$ such that:*

- *for all $w \neq w' \in \Omega$, $\mu(w)$ and $\mu(w')$ are mutually singular and T invariant.*

- *For all $Y \in B, \mu(Y) = \int \mu(w)(Y) d\nu(w)$.*

- *For almost all $w \in \Omega, (X, B, \mu(w), T)$ is ergodic.*

Using this theorem as a rationale, the focus on studying the structural properties of measure preserving systems has been on ergodic transformations. (However, as remarked in the comments after the section on the Halmos-von Neumann theorem, intractably complicated transformations can be built from very simple ergodic measure preserving systems.)

4.2   Bernoulli Shifts

Let I be a finite or countable set with the discrete topology, and for each $i \in I$ assume we have a positive p_i so that $\sum_i p_i = 1$. (In other words, I is an atomic measure space for a probability measure.)

Let $I^{\mathbb{Z}}$ be the collection of functions from $\mathbb{Z}$ to I with the product topology, and the product probability measure μ. Let σ be the shift map.

Then $(I^{\mathbb{Z}}, B, \mu, \sigma)$ is a measure preserving system, called a *Bernoulli Shift*. (In actual usage, anything isomorphic to a Bernoulli shift if called a Bernoulli shift.) This example is also called an *iid* for *independent identically distributed.*

If I has n elements then this gives a good model for repeatedly flipping an n-faced coin ($n \leq \omega$) where the probability of getting face i is given by p_i.

The next theorem gives a nice example of the use of the Suslin theorem to show that a collection is Borel:

Theorem 27 *(Feldman [Fel74]) The collection of transformations isomorphic to a Bernoulli Shift on a finite set I is a Borel set.*

⊢ Outline of proof: We show that the collection of Bernoulli Shifts is Σ_1^1 and Π_1^1.

We use Theorem 23 (the coding result), to see that it is Σ_1^1. To show that a transformation is Bernoulli, it suffices to find a finite generator $\mathcal{P}$ such that for all $n \in \mathbb{Z}$, $T^n \mathcal{P}$ is independent of $\bigvee_{k=0}^{n-1} T^k \mathcal{P}$. Thus T is isomorphic to a Bernoulli Shift iff there is a finite partition $\mathcal{P}$ that generates independently. This is clearly a Σ_1^1 statement.

The quantifier "there is a partition $\mathcal{P}$" is an existential quantifier over a Polish space, so it suffices to show that the collection of $\mathcal{P}$ that are independent generators for T form a Borel set.

Without loss of generality, our space $X = [0,1]$. So $\mathcal{P}$ generates iff for all $k \in \omega, \epsilon > 0$ there is an N such that every element of the partition $\{(0, \frac{1}{k+1}), (\frac{1}{k+1}, \frac{2}{k+1}), \ldots, (\frac{k}{k+1}, 1)\}$ can be approximated within ϵ by a union of elements of $\bigvee_{l=-N}^{N} T^l(\mathcal{P})$. This is a Π_2^0 condition.

The condition that $T^n \mathcal{P}$ is independent of $\bigvee_{k=0}^{n-1} T^k \mathcal{P}$ is a closed condition on the space of partitions.

To see that the collection of Bernoulli shifts is Π_1^1, we use ideas defined in section 4.4: the idea of entropy and the notion of being finitely determined. It is then a theorem [Orn74] that that a transformation T is isomorphic to a finite Bernoulli shift iff T has finite entropy and every finite generator is finitely determined.

We postpone the proof that this is a Π_1^1 condition to the section on entropy. We content ourselves for now to note that this statement is of the form T has finite entropy (which will be a Borel condition) conjoined with:

$$\forall \mathcal{P}(\mathcal{P} \text{ is a finite generator for } T \implies \mathcal{P} \text{ is finitely determined}).$$

Supposing (as we will show later) that the statement $\mathcal{P}$ is finitely determined is a Π_1^1 condition on $\mathcal{P}$, this statement is of the form:

$$(\forall x \in X)(\text{Borel condition 1} \implies (\forall y \in Y)(\text{Borel condition 2})),$$

where X and Y are Polish spaces and "Borel condition 1" and "Borel condition 2" are Borel subsets of the appropriate spaces. Applying elementary logic, this is equivalent to:

$$(\forall x \in X)(\forall y \in Y)(\text{Borel condition 1} \implies \text{Borel condition 2})$$

or:

$$(\forall (x,y) \in X \times Y)(\text{Borel condition 1} \implies \text{Borel condition 2})$$

Since the Borel sets are closed under Boolean combinations, this is a universal quantifier over a Polish space, followed by a Borel condition; thus a Π_1^1-set.⊣

4.3   The Theorems of Halmos and Rohlin

The first example of the use of descriptive set theory to distinguish between classes of transformations is due to Halmos and Rohlin [Hal60]:

Definition 28 *Let* (X, B, μ, T) *be a measure preserving system.*

- T *is strongly mixing iff for all* $U, V \in B$ *we have*

$$\lim_{n \to \infty} \mu(T^n(U) \cap V) = \mu(U)\mu(V).$$

- T *is weakly mixing iff for all* $U, V \in B$, *we have*

$$\lim_{D} \mu(T^n(U) \cap V) = \mu(U)\mu(V),$$

where $\lim_{D}$ *means limit with respect to the filter of sets of natural numbers of density one.*

Thus strong mixing says that the iterates of A become asymptotically independent of B and weak mixing says that this holds in density. The density limit is equivalent (morally and rigorously) to the limit in L^2.

Example 29 *Every Bernoulli shift is strongly mixing.*

⊢ Note that it suffices to check weak or strong mixing on a base for the measure space. A base for the measure space I^Z consists of cylinder sets of the form $\{f : f \restriction [-N, N] = s\} = [s]$, where s is a function from $[-N, N] \to I$. Given cylinder sets determined by $s : [-N, N] \to I$ and $t : [-M, M] \to I$, with $M \geq N$, we see that $T^M([s])$ is independent of $[t]$, and hence that every Bernoulli shift is strongly mixing.   ⊣

Halmos and Rohlin showed:

Theorem 30 *In the group of measure preserving transformations,*

- *(Rohlin) [Roh48] The class of strongly mixing measure preserving transformations is a meager* $\underline{\Pi}^0_3$ *set.*

- *(Halmos)[Hal60] The class of weakly mixing transformations is a dense* $\underline{\Pi}^0_2$ *set.*

The latter statement follows because (as we will see in later sections) T is weakly mixing iff the product transformation $T \times T : X \times X \to X \times X$ is ergodic. This is a $\underline{\Pi}^0_2$ condition and the collection of T satisfying it are dense.

Corollary 31 *There is a weakly mixing transformation that is not strongly mixing.*

4.4 Entropy

The entropy of a partition is intended to be the expected value of the amount of "information" obtained by knowing in which element of the partition a randomly chosen element of the space is.

The information function on the space should depend only on the measures of the partition. The information from landing in a smaller measure set should be more than the information from landing in a large measure set. Moreover, if $\mathcal{P}_0$ and $\mathcal{P}_1$ are independent partitions, the information from knowing what element of $\mathcal{P}_0$ a point x is in, should give no information about element of $\mathcal{P}_1$ x is in. Rephrasing this if Q is the coarsest partition of X refining $\mathcal{P}_0$ and $\mathcal{P}_1$, then the information function for Q should be the sum of the information functions for $\mathcal{P}_0$ and $\mathcal{P}_1$.

Since the elements of Q are sets of the form $A \cap B$ with $A \in \mathcal{P}_0$ and $B \in \mathcal{P}_1$, and the measure of $A \cap B$ is the product of the measures of A and B, we see that we need a function that changes products to sums.

The following proposition explains the use of logarithms in the definition of entropy.

Proposition 32 *([Par69], [Smo71]) Suppose that f is a function on $(0,1]$ such that:*

- $f(1) = 0$,

- *f is monotonically decreasing and continuous,*

- $f(ab) = f(a) + f(b)$.

Then for some base k, $f(x) = -\log_k(x)$, all $x \in (0,1]$.

Definition 33 *Let $\mathcal{P}$ be a countable measurable partition of X. The information function $I_P : X \to \mathbb{R}^+$ is given by*

$$I_P(x) = -\sum_{A \in \mathcal{P}} \log(\mu(A)) \chi_A(x),$$

where $\chi_A(x)$ is the characteristic function of A.

The entropy $H(\mathcal{P})$ of the partition $\mathcal{P}$ is the expected value of the partition function:

$$H(\mathcal{P}) = \sum_{A \in \mathcal{P}} \mu(A) \log(\mu(A))$$

where $0 \log 0$ is defined to be 0 for our purposes.

The entropy function is a surjective continuous function from the space
of partitions to the non-negative real numbers. It has many nice conceptual
properties: there is a notion of conditional information and entropy. Here the
idea is that one is given two partitions $\mathcal{P}$ and $\mathcal{Q}$ and computes the conditional
information function:

$$I(\mathcal{P}|\mathcal{Q}) = - \sum_{P \in \mathcal{P}, Q \in \mathcal{Q}} log(\frac{\mu(P \cap Q)}{\mu(Q)}) \chi_{P \cap Q}.$$

Note that at a point $x \in Q$, this formula gives the information of $x \in P$,
relative to the measure space (Q, ν_Q), where $\nu_Q(A) = \frac{\mu(A \cap Q)}{\mu(Q)}$. This is viewed
informally as the amount of information gained by knowing that $x \in P$ given
that we know that $x \in Q$. The conditional entropy function is then the
expected value of the conditional information, and is given by the formula:

$$H(\mathcal{P}|\mathcal{Q}) = \int I(\mathcal{P}, \mathcal{Q}).$$

An easy calculation shows that:

$$H(\mathcal{P}|\mathcal{Q}) = \sum_{Q \in \mathcal{Q}} \mu(Q) H(\mathcal{P}|Q),$$

where $H(\mathcal{P}|Q)$ is the entropy of $\mathcal{P}$ with respect to ν_Q.

Some computation then shows that the entropy of the partition $\mathcal{P} \bigvee \mathcal{Q}$ is
equal to the conditional entropy of $\mathcal{P} \bigvee \mathcal{Q}$ given $\mathcal{P}$ plus the entropy of $\mathcal{P}$. In
symbols:

$$H(\mathcal{P} \bigvee \mathcal{Q}) = H(\mathcal{P}) + H(\mathcal{P} \bigvee \mathcal{Q}|\mathcal{P})$$

Conceptually this is saying that the expected information of performing
the "experiment" $\mathcal{P} \bigvee \mathcal{Q}$ is the same as the expected information for performing the experiment $\mathcal{P}$ and, knowing the results of the first experiment,
performing the second experiment $\mathcal{Q}$. It is easy to check that if $\mathcal{P}$ and $\mathcal{Q}$ are
independent, then $H(\mathcal{P} \bigvee \mathcal{Q})$ is the sum of $H(\mathcal{P})$ and $H(\mathcal{Q})$.

Expanding inductively, we see that $H(\bigvee_{k=0}^{n}(T^k(\mathcal{P}))$ can be viewed as the
cumulative information gained by repeated performing the experiment $\mathcal{P}$ at
times $k = 0, 1, \ldots, n$. The average amount of information gained is then
$\frac{1}{n} H(\bigvee_{k=0}^{n}(T^k(\mathcal{P}))$.

Note that $H(\bigvee_{k=0}^{n} T^k(\mathcal{P})) = H(\bigvee_{k=1}^{n+1}(\mathcal{P}))$, since the distributions of the
partitions $\bigvee_{k=0}^{n} T^k(\mathcal{P})$ and $\bigvee_{k=1}^{n+1}(\mathcal{P})$ are the same. Thus:

$$H(\bigvee_{k=0}^{n+1}(\mathcal{P})) = H(\bigvee_{k=1}^{n+1}(\mathcal{P})) + H(\mathcal{P}|\bigvee_{k=1}^{n+1}(\mathcal{P}))$$

$$= H(\bigvee_{k=0}^{n}(\mathcal{P})) + H(\mathcal{P}|\bigvee_{k=1}^{n+1}(\mathcal{P}))$$

This in turn shows that $H(\bigvee_{k=0}^{n+1}(\mathcal{P})) - H(\bigvee_{k=0}^{n}(\mathcal{P}))$ is a decreasing sequence, and hence:

$$h(T,\mathcal{P}) = \lim_{n\to\infty} \frac{1}{n} H(\bigvee_{k=0}^{n}(T^k(\mathcal{P})))$$

converges monotonically and gives the *Entropy of T with respect to the partition $\mathcal{P}$*. Note that this is a Borel (Baire Class 1) function of T and $\mathcal{P}$.

Definition 34 *The* entropy *of a measure preserving transformation T is defined to be:*

$$h(T) = \sup\{h(T,\mathcal{P}) : \mathcal{P} \text{ is a finite partition}\}.$$

Note that infinite entropy is possible.

From the definitions we have checked:

Proposition 35 *The entropy function is a Borel function from the space of measure preserving transformations to the extended positive real numbers.*

The following facts can be found in any good book on entropy:

Theorem 36 *Let T be a measure preserving transformation. Then:*

1. *If $\mathcal{P}$ is a generator for T, then $h(T) = h(T,\mathcal{P})$. In particular, the entropy of T is bounded by $\log|\mathcal{P}|$. This bound is achieved if T is a Bernoulli shift in $|\mathcal{P}|$ elements.*

2. *(Krieger)[Kri70] If $h(T) < \log(k)$ and T is ergodic, then there is a generator for T with less than or equal to k elements.*

Following the proof in [WeiBA] one can check:

Proposition 37 *There is a Borel function $\boldsymbol{k}$ from $\{T : h(T) < \infty\}$ to the space of partitions of X such that from all T, $\boldsymbol{k}(T)$ is a generator for T.*

Note that by Theorem 23, we can view this as a Borel function from the collection of measures on $^{Z}\omega$ that give the shift map finite entropy to the space of partitions of $^{Z}\omega$, yielding generators.

Krieger's theorem can be strengthened: since the partition produced in Krieger's theorem is finite, it shows that the transformation T is isomorphic to a transformation S on k^Z with an invariant measure μ. The "Jewett-Krieger Theorem" says that there is a shift invariant closed set $K \subset k^Z$ on which there is a unique shift invariant ergodic measure μ, and T is isomorphic to shift on this ergodic measure. Moreover, as in Proposition 37, K can be computed effectively from T.

The most remarkable theorem involving entropy is the theorem of Ornstein:

Theorem 38 *Let* (X, B, μ, T) *and* (Y, C, ν, S) *be two Bernoulli shifts. Then* (X, B, μ, T) *is isomorphic to* (Y, C, ν, S) *iff* $h(T) = h(S)$.

Thus entropy is a complete invariant of the collection of measure preserving transformations isomorphic to a Bernoulli shift. This result extends to Bernoulli shifts on an infinite alphabet as well.

This theorem has some initially surprising consequences:

Example 39 *Let* (X, B, μ, T) *be the Bernoulli shift on a 4 sided coin with probabilities* $\{\frac{1}{4}, \frac{1}{4}, \frac{1}{4}, \frac{1}{4}\}$ *and* (Y, C, ν, S) *be the Bernoulli shift on a 5 sided coin with probabilities* $\{\frac{1}{2}, \frac{1}{8}, \frac{1}{8}, \frac{1}{8}, \frac{1}{8}\}$. *Then* T *and* S *have the same entropy and hence are isomorphic. (This result was first shown by Meshalkin, by giving an explicit isomorphism between the two shifts.)*

In fact Theorem 38 extended the theorem of Sinai that every measure preserving system (X, B, μ, T) has a Bernoulli factor (Y, C, ν, S) with the same entropy as T. The Sinai theorem had as a corollary that any two Bernoulli shifts of the same entropy are embeddable as factors of each other.

Following the proof of [Shi73] one can check:

Proposition 40 *There is a Borel function* S: $\mathcal{H} \to \mathcal{H}$ *such that for all measure preserving transformations* T, $S(T)$ *is a factor of* T *and has the same entropy.*

We now pay the debt incurred in the proof of Theorem 27. First, since the entropy function is Borel, the collection of transformations with finite entropy is a Borel set. What we must show is that the definition of *finitely determined* is $\amalg_1^1$. This uses the machinery developed by Ornstein in the proof of Theorem 38. We refer the reader to the survey by Ornstein and Weiss for more information on this topic [OW91]. The following definitions are taken from [Shi73].

If $\mathcal{P} = \langle P_1, \ldots, P_n \rangle$ is an ordered partition, we define the distribution of $\mathcal{P}$ to be the sequence $d(\mathcal{P}) = \langle \mu(P_1), \mu(P_2), \ldots, \mu(P_n) \rangle$. For each n, we define the semi-metric *distribution distance* on the space of ordered partitions with n elements by setting:

$$|d(\mathcal{P}) - d(\mathcal{Q})| = \sum_{i=1}^{n} |\mu(P_i) - \mu(Q_i)|.$$

The *partition distance* of $\mathcal{P}$ and $\mathcal{Q}$ is the metric on the space of ordered partitions with n elements defined by:

$$|\mathcal{P} - \mathcal{Q}| = \sum_{i=1}^{n} \mu(P_i \Delta Q_i).$$

Both the distribution distance and the partition distance are easily checked to be continuous functions from the space of pairs of partitions into the real numbers.

Let $T, \bar{T}$ be two measure preserving transformations on $X = [0,1]$ and $\mathcal{P}, \mathcal{Q}$ be two partitions. The *process distance* $\bar{d}$ between $(T, \mathcal{P})$ and $(\bar{T}, \mathcal{Q})$ is defined to be:

$$\bar{d}((T, \mathcal{P}), (\bar{T}, \mathcal{Q})) = \sup_n \inf_{S \in \mathcal{H}} \frac{1}{n} \sum_{k=0}^{n-1} |T^i(\mathcal{P}) - S\bar{T}^i(\mathcal{Q})|.$$

In this form this is not clearly Borel function. However, given an $S \in \mathcal{H}$, the number $\frac{1}{n} \sum_{k=0}^{n-1} |T^i(\mathcal{P}) - S\bar{T}^i(\mathcal{Q})|$ is determined by the distribution of $\bigvee_{i=0}^{n-1} S\bar{T}^i(\mathcal{Q})$ relative to $\bigvee_{i=0}^{n-1} T^i(\mathcal{P})$. The infimum can be achieved by an infimum over the possible distributions, and for this we can consider a canonical countable set. This shows that the function $\bar{d}$ is a Borel function.

Definition 41 *Let $\mathcal{P}$ be a k-set partition. Then $\mathcal{P}$ is* finitely determined *relative to an ergodic transformation T iff for every $\epsilon > 0$ there is an $n > 0$ and a $\delta > 0$ such that if $\bar{T}$ is any ergodic transformation with $h(\bar{T}) \geq h(T, \mathcal{P})$, and $\bar{P}$ is any k-set partition such that*

1. $|d(\bigvee_{i=0}^{n-1} \bar{T}^i \overline{\mathcal{P}}) - d(\bigvee_{i=0}^{n-1} T^i \mathcal{P})| \leq \delta$,

2. $0 \leq h(T, \mathcal{P}) - h(\bar{T}, \overline{\mathcal{P}}) \leq \delta$,

then $\bar{d}((T, \mathcal{P}), (\bar{T}, \overline{\mathcal{P}})) < \epsilon$.

While the significance of this definition is far from clear at first sight, for our purposes the important point is its complexity. We can check that this is a $\coprod_1^1$ condition on the pairs (T, P), as all of the various functions involved in the definitions are Borel (e.g., $h, \bar{d}$ etc.) The only non-Borel quantification is over all ergodic transformations $\bar{T}$.

4.5 K-Automorphisms

Property K (for Kolmogorov) transformations (or *K-automorphisms*) are those for which "knowledge of the past tends to give no information about the future." Clearly Bernoulli transformations are of this sort, but the class of K-automorphisms is much broader as will be seen in the section on equivalence relations.

We refer the reader to [OS73] for information about K-automorphisms. Much of this section is taken from that paper.

Definition 42 (X, B, μ, T) *is a K-automorphism iff there is a finite generator* $\mathcal{P}$ *such that every set in* $\bigcap_{n=0}^{\infty} B(\bigcup_{i=-\infty}^{-n} T^i(\mathcal{P}))$ *has either measure zero or measure one, where* $B(\bigcup_{i=-\infty}^{-n} T^i(\mathcal{P}))$ *is the σ-algebra generated modulo μ by* $\bigcup_{i=-\infty}^{-n} T^i(\mathcal{P})$.

Proposition 43 *The collection of K-automorphisms is a Borel subset of $\mathcal{H}$.*

It is a theorem of Rohlin and Sinai that T is a K-automorphism iff all non-trivial partitions $\mathcal{P}$ are not measurable with respect to $B(\bigcup_{-\infty}^{-1} T^i(\mathcal{P}))$. This is easily seen to be equivalent to the statement that for all non-trivial partitions $\mathcal{P}, h(T, \mathcal{P}) \neq 0$. This is clearly a $\amalg_1^1$ statement.

To show that the class of K-automorphisms is Σ_1^1, we will need the following notion of ϵ-independence: Given partitions $\mathcal{P}$ and $\mathcal{Q}$, we say that $\mathcal{P}$ is ϵ-independent of $\mathcal{Q}$ iff there is a collection $\mathcal{C}$ of sets in $\mathcal{Q}$ of total measure at least $1 - \epsilon$ such that for all $Q \in \mathcal{C}$:

$$\sum_{P \in \mathcal{P}} |\frac{\mu(P \cap Q)}{\mu(Q)} - \mu(P)| < \epsilon.$$

The property that every element of $\bigcap_{n=0}^{\infty} B(\bigcup_{i=-\infty}^{-n} T^i(\mathcal{P}))$ has either measure zero or measure one is equivalent to the statement that there is a generator $\mathcal{P}$ such that for all positive m and $\epsilon > 0$ there is an integer N such that $\bigvee_{i=-(n+1)}^{n+m} T^i \mathcal{P}$ is ϵ-independent of $\bigvee_{i=1}^{k} T^i \mathcal{P}$ for all $n \geq N$ and all $k > 0$.

(Intuitively, if we view our transformation as the shift transformation on $^Z\omega$, this is saying that every basic open interval becomes asymptotically independent of all intervals of length m in the distant past.)

Since the only non-Borel quantification in this condition is an existential quantifier over the space of partitions, the fact that the definition of K-automorphism can be written in Σ_1^1 form now follows easily.

4.6 Homeomorphisms of Compact Metric Spaces

These techniques can be used to classify homeomorphisms of compact metric spaces.

Recall that $\mathcal{K}_0$ be the space of compact subsets of $\mathbb{I}^2$ with the Hausdorff metric, and that we are viewing homeomorphisms of compact metric spaces as elements of $\mathcal{K}_0$.

Recall that a homeomorphism $T : X \to X$ is minimal iff for every $x \in X$, the orbit $\{T^n(x) : n \in \omega\}$ is dense.

Proposition 44 *[BF95] The following are Borel subsets of $\mathcal{K}_0$:*

1. *The collection of homeomorphisms of a compact metric space.*

2. *The collection of minimal homeomorphisms of a compact metric space.*

⊢   The former statement can be rewritten in terms of a basis for the metric space, and hence involves only quantification over countable sets. Since X is compact, it is easy to verify that the latter statement is equivalent to the result that for all basic open sets O, there is a $k \in \omega$ such that $\bigcup_{i=1}^{k} T^{-i}(O)$ is a cover of X. Again, using the compactness of X, one can show that this is a Borel condition.   ⊣

We recall a classical definition:

Definition 45 *Let T be a homeomorphism of a compact metric space (X, d). Points $x \neq y \in X$ are distal iff there is a $\delta > 0$ such that for all $n \in \mathbb{Z}$, $d(T^n x, T^n y) > \delta$. Otherwise x, y are proximal. T is called distal (or proximal) iff every pair of distinct points in X is distal (resp. proximal).*

The reader is referred to Glasner's article in this volume for a more complete explication of these ideas.

Clearly proximal and distal are antithetical ideas, and each class is given by a $\coprod_1^1$ definition. However they are quite different in complexity:

Theorem 46 *(Beleznay-Foreman [BF95]) In the space $\mathcal{K}_0$:*

1. *The collection of distal homeomorphisms is not Borel.*

2. *The collection of proximal transformations is Borel.*

⊢   The first statement can be strengthened to state that the collection of distal transformations is a maximally complicated $\coprod_1^1$-set, in the sense described in section 6. This result is outlined in section 9.

To see the second statement: The points x, y are proximal iff for the transformation $T^* = T \times T : X \times X \to X \times X$, the T^*-orbit of (x, y) gets arbitrarily close to the diagonal. Hence is suffices to show that for a closed subset K of a compact space X the collection of T such that every T-orbit gets arbitrarily close to K is a Borel set.

Note that if $d(T^n x, K) < \epsilon$, then there is an open neighborhood O of x such that for all $y \in O, d(T^n y, K) < \epsilon$. Thus the following are equivalent:

- For all $\epsilon > 0$ and $x \in X$, there is an n, so that $d(T^n x, K) < \epsilon$.

- For all $\epsilon > 0$ there is a finite open cover $\mathcal{U}$ of X such that for each $O \in \mathcal{U}$, there is an n, so that $T^n O$ is contained in a ball of radius ϵ around K.

An argument similar to Proposition 44 shows that the latter condition is Borel.   ⊣

4.7   Measure Distal Transformations

Parry, in [Par68] defined a class of transformations that have zero entropy, and showed that the topologically distal transformations fall into that class. Later, in analogy to the topologically distal homeomorphisms, Furstenberg described a collection of measure preserving transformations called the *measure distal* transformations. Zimmer [Zim76b], using the term *generalized discrete spectrum* to describe the class, and showed that it exactly coincided with Parry's original class.

The collection of measure distal transformations also turns out to be a complete Π_1^1-set, although for reasons somewhat different than in the topological case.

This example is described later in Section 9, "An Extended Example."

4.8   Eigenvalues of Linear Operators

Let $L : H \to H$ be an arbitrary linear operator, where H is a countable dimension Hilbert space. We show that the statement that λ is an eigenvalue of L is Borel. This is shown in somewhat more generality then we will need later, when we apply it only to the unitary operators.

One obstacle to stating the result is that $\mathcal{L}$, the space of linear operators on H with the weak operator topology, is not a Polish space. However, its Borel structure is a standard Borel structure, and we can use the techniques and language of descriptive set theory. In particular, it makes sense to speak of Borel sets as if they were Borel sets in a Polish space. (See [Kec95], page 80, for an explanation of this.)

Proposition 47 *(Foreman) Let $\mathcal{L}$ be the space of linear operators on H with the weak operator topology. Then $\{(L, \lambda) : \lambda$ is an eigenvalue of $L\} \subset \mathcal{L} \times \mathbb{C}$ is a Borel set.*

Since the map $(L, \lambda) \mapsto L - \lambda I$ is a Borel map from $\mathcal{L}$ to $\mathcal{L}$ it suffices to show that the collection of L that have non-trivial kernel is a Borel set.

Fix an orthogonal basis $\{e_i : i \in \omega\}$ for H. Then

$$ker(L) \neq (0)$$
$$\text{iff}$$
$$\exists h \in H (h \neq 0 \wedge L(h) = 0)$$
$$\text{iff}$$
$$(\forall P)(\text{ if } P \text{ is a projection to } ker(L), \text{ then } (\exists i)P(e_i) \neq 0)$$

The first statement is clearly Σ_1^1, so we must show that the second statement is Π_1^1.

Note that P is a projection to $ker(L)$ iff

1. for all $i, P^2(e_i) = e_i$, and P is linear on the rational combinations of $\{e_i\}$.

2. For all rational combinations h of $\{e_i\}, \| P(h) \| \leq \| h \|$.

3. $\sup\{\| P(h) \|: h$ is a rational combination of $\{e_i\}$ and $\| h \| \leq 1\} = 1$

   Note that these three clauses say that P is an idempotent linear operator of norm 1, i.e., a projection.

4. For all $i, L(P(e_i)) = 0$. (So P maps into the kernel of L.)

5. There is a rational number $\delta > 0$ such that for all rational combinations h of $\{e_i\}, \delta \| L(h) \| \geq \| h - P(h) \|$. (Using the inverse function theorem this is equivalent to P mapping onto the kernel of L.)

Since conditions 1-5 only involve quantification over a specified countable set (e.g., rational combinations of $\{e_i\}$), the condition that P is a projection to $ker(L)$ is Borel.

We note that it follows easily that $\{(L, \lambda, x) : L(x) = \lambda x\}$ is Borel.

Moreover, the Luzin-Novikov uniformization theorem yields the following consequence. Suppose that B is a Borel subset of $\mathcal{L}$ such that every $L \in B$ has a countable nonempty set of eigenvalues. Then there is a Borel map $E : \omega \times \mathcal{L} \to \mathbb{C}$ such that for all $L \in B, \langle E(n, L) : n \in \omega \rangle$ is an enumeration of the eigenvalues of L.

5 Equivalence Relations

In this section, we discuss examples of analytic equivalence relations and mention some of their theory. We will say that an equivalence relation E on a Polish space X is Borel (or analytic, co-analytic etc.) iff it is Borel (or analytic, co-analytic etc.) when viewed as $E \subset X \times X$.

Definition 48 *Let X and Y be standard Borel spaces and $E \subset X \times X, F \subset Y \times Y$ be equivalence relations. Let B be a Borel subset of X that is a union of E equivalence classes. Then $E \restriction B$ is Borel (resp. continuously) reducible to F iff there is a Borel (resp. continuous) function $f : X \to Y$ such that for all $x_1, x_2 \in B$ we have:*

$$x_1 E x_2 \text{ iff } f(x_1) F f(x_2).$$

If E is Borel reducible to F, we write $E \leq F$.

The reducing function f is supposed the "reduce" the question of E-equivalence of two elements of B to the question of F-equivalence. If the function f is sufficiently transparent (e.g., Borel or continuous, depending on the context), then this shows that the relation F contains all of the information and is at least as complicated as E is on the set B.

Many "classifications" are of this form:

Example 49 *(See Theorem 38) Let F be the equality relation on the non-negative real numbers. Let $\mathbb{B} \subset \mathcal{H}$ be the collection of Bernoulli shifts. Let E be the conjugacy (isomorphism) relation on $\mathcal{H}$. Let $h : \mathbb{B} \to \mathbb{R}$ be the entropy function. Then h is a reduction of $E \restriction B$ to F.*

Equivalence relations that are reducible to the identity relation on a Polish space are called *smooth* (see [HKL90]) Harrington, Kechris and Louveau showed that there is a Borel equivalence relation that is intimately tied with smooth equivalence relations:

Definition 50 *Let $^{\omega}2$ be the Cantor set. Define E_0 to be the equivalence relation on $^{\omega}2$ given by $f \equiv g$ iff there is an n such that for all $m > n, f(m) = g(m)$. The* Vitali *equivalence relation on the interval $[0, 1]$ is defined by setting $s \equiv t$ iff $s - t \in \mathbb{Q}$.*

The following can be found in [DJK94]:

Proposition 51 *There is a Borel isomorphism between $^{\omega}2$ and $[0, 1]$ that reduces E_0 to the Vitali equivalence relation.*

Part of the significance of E_0 is that standard $0 - 1$ laws can be applied to show, for example, that E_0 is not smooth. (This is a variation of the usual argument for the existence of a non-measurable set.)

The following remarkable theorem is due to Harrington, Kechris and Louveau [HKL90]:

Theorem 52 *Let E be a Borel equivalence relation on a standard Borel space X. The either:*

- *E is smooth.*

  or

- *E_0 is reducible to E.*

So either E is completely classifiable using numerical invariants or E is at least as complicated as the Vitali equivalence relation.

We now define another equivalence relation "Eq^+". The field of this relation is the countable product of the Baire space with itself, which we will denote $(^\omega\omega)^\omega$. We note that this space is homeomorphic with the Baire space, and if X is any Polish space (e.g., the unit circle in $\mathbb{C}$) then $(^\omega\omega)^\omega$ is Borel isomorphic to the space X^ω. We will view elements of $(^\omega\omega)^\omega$ as sequences $\langle x_n : n \in \omega \rangle$.

Definition 53 *Let $\langle x_n : n \in \omega \rangle$ and $\langle y_m : m \in \omega \rangle$ be elements of $(^\omega\omega)^\omega$. Then $\langle x_n : n \in \omega \rangle Eq^+ \langle y_n : n \in \omega \rangle$ iff:*

$$\forall n \exists m (x_n = y_m) \wedge \forall m \exists n (y_m = x_n)$$

In other words, two sequences $\langle x_n \rangle$ and $\langle y_m \rangle$ are equivalent iff each x_n occurs as a y_m, and each y_m as an x_n Note that the form of the definition shows immediately that the equivalence relation is Borel.

The following proposition is somewhat surprising:

Proposition 54 *If E is a Borel equivalence relation on X in which each class is countable. Then E is Borel reducible to Eq^+.*

To see this: view Eq^+ as an equivalence relation on X^ω. By the Luzin-Novikov uniformization theorem there is a Borel map from X to X^ω, defined by $x \mapsto \langle x_n : n \in \omega \rangle$ such that for all $x \in X$, $\langle x_n : n \in \omega \rangle$ is an enumeration of the E equivalence class of x. Then clearly xEy if and only if $\langle x_n : n \in \omega \rangle Eq^+ \langle y_n : n \in \omega \rangle$.

Definition 54 generalizes to any equivalence relation: if E is a Borel equivalence relation then so is E^+, where $\langle x_n : n \in \omega \rangle E^+ \langle y_m : m \in \omega \rangle$ iff $\langle x_n \rangle$ and $\langle y_m \rangle$ have the same set of E equivalence classes.

Another operation on equivalence relations E is to take the countable product of E, E^ω, where E^ω is the product equivalence relation on X^ω.

The "+ process" can be iterated transfinitely. Hjorth, Kechris and Louveau have characterized which Borel equivalence relations are reducible to some equivalence relation in this hierarchy, see [HKL98].

We finish this section with a classical example. We remind the reader that the linear operators on a Hilbert space H are a standard Borel space. (See the remarks preceeding Proposition 47.) The normal operators are a Borel subset of the linear operators, and hence it makes sense to discuss Borel subsets of the normal operators.

Example 55 *Let H be a separable Hilbert space, and $\mathcal{N}$ be the collection of normal operators on H with the weak operator topology. Let $I_\mathcal{N}$ be the relation of unitary equivalence on $\mathcal{N}$. Let E be the relation of measure equivalence on the space of measures $\mathcal{M}$ on $[0,1]$, where this space is given the weak* topology.*

Then the spectral theorem says that there is a Borel map S from $\mathcal{N}$ to $(\mathcal{M})^\omega$ such that $N_1 I_N N_2$ iff $S(N_1)E^+S(N_2)$ iff $S(N_1)E^\omega S(N_2)$ (see [Con85]). In particular, unitary equivalence is a Borel equivalence relation.

To completely explicate this example we show:

Example 56 *Let $\mathcal{M}$ be the space of measures on $[0,1]$ with the weak*-topology. Then the relation of measure equivalence is a Borel equivalence relation.*

⊢   Note that $\mu << \nu$ iff μ is absolutely continuous with respect to ν, in the sense that for all $\epsilon > 0$ there is a $\delta > 0$ such that for all open sets O if $\nu(0) < \epsilon$, then $\mu(O) < \delta$. This is easily seen to be a $\underline{\Pi}^0_3$ condition.   ⊣

5.1   K-Automorphisms

The Vitali relation comes up naturally in Ergodic theory as the following theorem of Ornstein and Shields shows:

Theorem 57 *(Ornstein, Shields) [OS73] For any fixed entropy $e > 0$ there is a Borel function mapping $K :^\omega 2 \to \{K\text{-automorphisms of entropy } e\} \subset \mathcal{H}$ that reduces E_0 to the conjugacy relation.*

Feldman [Fel74] shows that this could be done for 0-entropy transformations as well. In that same paper, Feldman remarks that this theorem shows that the isomorphism relation on the K-automorphism cannot be smooth. In particular there can be no complete numerical invariant on the K-automorphisms. However, as we shall see, this does not rule out a classification of the K-automorphisms along the lines of the Halmos-von Neumann classification of the transformations with discrete spectrum.

At the time of this writing (October 1999) the author does not know the answers to the following questions:

Questions: Consider the Borel set $\mathbb{K}$ of K-automorphisms as a Borel space with the structure inherited from $\mathcal{H}$. Let E be the equivalence relation of conjugacy.

1. Is $E \restriction \mathbb{K}$ a Borel subset of $\mathbb{K} \times \mathbb{K}$?

2. Is $E \restriction \mathbb{K}$ reducible to Eq^+?

3. Is $E \restriction \mathbb{K}$ reducible to E_0?

As we will see a positive answer to any of these questions implies a positive answer to the questions above it.

5.2   Discrete Spectrum Transformations

In this section we show that there are reductions of the conjugacy relation between ergodic discrete spectrum transformation and Eq^+ on $(^\omega 2)^\omega$ and vice versa. This characterizes the complexity of the equivalence relation of conjugacy on the ergodic discrete spectrum transformations. (This was proved by Foreman and Louveau in the fall of 1995.)

We begin with a summary of the Halmos-von Neumann theory. This summary is taken with only very minor modifications from Walters' book [Wal82]. An important consequence of this theory is that for ergodic discrete spectrum transformations on X, unitary equivalence of the associated unitary transformation on $L^2(X)$ implies isomorphism.

As noted earlier every measure preserving transformation T on a measure space X induces a natural unitary transformation $U_T : L^2(X) \to L^2(X)$ given by $U_T(f) = f \circ T$. (We will frequently write $T(f)$ for $U_T(f)$.) We will write (f, g) for the inner product of f and g in $L^2(X)$.

Throughout the remainder of this paper we will use $\mathbb{T}$ to denote the unit circle as a subset of $\mathbb{C}$.

The next lemma describes the eigenvalues of the operator U_T:

Lemma 58 *Let T be a measure preserving transformation.*

1. *If λ is an eigenvalue of U_T, then $|\lambda|=1$*

2. *T is ergodic iff all of the eigenvalues of T each have multiplicity 1. Moreover, if T is ergodic then the eigenvalues form a subgroup of the unit circle, $\mathbb{T}$.*

3. *If T is ergodic each eigenfunction has constant magnitude and if f, g are eigenfunctions then either f is a scalar multiple of g or $f \perp g$.*

To see this: For 1: Suppose that f is an eigenfunction with eigenvalue λ. Then $(f, f) = (U_T(f), U_T(f)) = (\lambda f, \lambda f) = |\lambda|^2(f, f)$. Hence $|\lambda|^2 = 1$

Now suppose that T is ergodic. Since each eigenvalue has norm 1, for all eigenfunctions $f, \{x : m \le |f(x)| \le M\}$ is a T-invariant set. Hence has measure zero or one. Thus the function $|f|$ is constant almost everywhere.

If f, g are eigenfunctions with eigenvalues λ, η then $\frac{f}{g} \in L^2(X)$, and $U_T(\frac{f}{g}) = \frac{f \circ T}{g \circ T} = \frac{\lambda f}{\eta g}$. Hence the collection of eigenvalues form a subgroup of the unit circle. To see that they have multiplicity one, suppose that f, g are eigenfunctions for λ and f is not a scalar multiple of g. Then $\frac{f}{g}$ is a non-constant invariant function for U_T contradicting ergodicity. If the eigenvalue 1 has multiplicity 1, then every bounded invariant measurable function must be constant almost everywhere. This clearly implies ergodicity.

For 3: we must see that if f, g are eigenfunctions for distinct eigenvalues λ, μ then $f \perp g$. Otherwise: $(f, g) = (U_T(f), U_T(g)) = \lambda\bar{\mu}(f, g)$. But then $\lambda\bar{\mu} = 1$ and hence $\lambda = \mu$.

Definition 59 *An ergodic measure preserving system (X, B, μ, T) is said to have* discrete spectrum *iff there is a basis for $L^2(X)$ consisting of eigenfunctions for U_T.*

Example 60 *[Hal60] Let $\alpha \in \mathbb{T}$ be an irrational rotation (e.g., not a root of unity.) Define $T : \mathbb{T} \to \mathbb{T}$ by $T(\xi) = \alpha\xi$. Then T is ergodic and has discrete spectrum.*

To see this note that if we let $e_n(\xi) = \xi^n$, then $\{e_n : n \in \mathbb{Z}\}$ form a basis for $L^2(\mathbb{T})$ and each is an eigenfunction, since $T(e_n) = \alpha^n e_n$. Each eigenvalue is of the form α^n and has multiplicity one.

We are now ready to state the Halmos-von Neumann theorem [HvN42]:

Theorem 61 *If (X, B, μ, T) and (Y, C, ν, S) are ergodic discrete spectrum measure preserving systems with the same groups of eigenvalues, then $(X, B, \mu, T) \cong (Y, C, \nu, S)$. Moreover, given a countable subgroup K of the unit circle there is an ergodic discrete spectrum transformation with K as its group of eigenvalues.*

⊢   Let U be a unitary operator from $L^2(Y)$ to $L^2(X)$. Then U is said to *intertwine* with measure preserving transformations S and T iff the following diagram commutes:

$$
\begin{array}{ccc}
L^2(Y) & \xrightarrow{S} & L^2(Y) \\
\downarrow U & & \downarrow U \\
L^2(X) & \xrightarrow{T} & L^2(X)
\end{array}
$$

If there is an intertwining operator that is induced by a measure preserving transformation, then the S and T are conjugate. By Theorem 15, to show the isomorphism in the Halmos-von Neumann theorem, suffices to find a unitary operator $W : L^2(X) \to L^2(Y)$ that is multiplicative on $L^\infty(X)$ and intertwines with T and S. Clearly such an operator must take an eigenfunction for T with value λ to an eigenfunction for S with value λ.

The following lemma is quoted from [Wal82], from where this presentation is heavily based. We use the same notation and proofs.

Lemma 62 *Let H be an abelian group and K a divisible subgroup. Then there is a homomorphism ϕ of H onto K that is the identity on K. (Such a ϕ will be called a retraction.)*

⊢ Use Zorn's lemma to find a maximal subgroup $H' \subset H$ and homomorphism ρ. If $x \in H \backslash H'$ break into cases according to whether a power of x lies in H' or not, to show that ρ can be extended to the subgroup of H generated by H' together with $\{x\}$. ⊣

For each eigenvalue λ choose eigenfunctions $f_\lambda \in L^2(X)$ and $g_\lambda \in L^2(Y)$ that have constant absolute value 1. Define $r(\lambda, \mu) \in \mathbb{T}$ by setting $f_\lambda f_\mu = r(\lambda, \mu) f_{\lambda\mu}$.

Let $H = \mathbb{T}^X$. We can view $\mathbb{T}$ as the subgroup of H given by the constant functions and apply the lemma to get a retraction ϕ. Define $f_\lambda^* = \overline{\phi(f_\lambda)} f_\lambda$. Then f^* is an eigenvector for λ and $\{f_\lambda^*\}$ are a basis for $L^2(X)$. We have:

$$
\begin{aligned}
f_\lambda^* f_\mu^* &= \overline{\phi(f_\lambda)\phi(f_\mu)} f_\lambda f_\mu \\
&= \overline{\phi(f_\lambda f_\mu)} r(\lambda, \mu) f_{\lambda\mu} \\
&= \overline{\phi(r(\lambda, \mu) f_{\lambda,\mu})} r(\lambda, \mu) f_{\lambda\mu} \\
&= \overline{r(\lambda, \mu)} r(\lambda, \mu) \phi(f_{\lambda\mu}) f_{\lambda\mu} \\
&= \overline{\phi(f_{\lambda\mu})} f_{\lambda\mu} = f_{\lambda\mu}^*.
\end{aligned}
$$

Thus without loss of generality we can assume that for all λ, μ we have $f_\lambda f_\mu = f_{\lambda\mu}$ and similarly for the g_λ's.

Hence if we define $W(f_\lambda) = g_\lambda$, then W is a multiplicative map from the orthonormal bases $\{f_\lambda\}$ onto the orthonormal basis $\{g_\lambda\}$. It can be extended to finite linear combinations of the f_λ's and finally to all of L^2. Thus this W is the desired multiplicative intertwining operator witnessing the isomorphism between T and S.

We now show the second assertion of the Halmos-von Neumann theorem, that every countable subgroup of the unit circle is the group of eigenvalues of an ergodic, discrete spectrum measure preserving system.

For this we will use the duality theory for abelian groups. See [HR79] or [Wal82] for information on this. We very briefly summarize it in the following paragraphs.

Let G be a locally compact abelian group. Define

$$\widehat{G} = \{\chi | \chi : G \to \mathbb{T} \text{ is a continuous homomorphism}\}.$$

Then $\widehat{G}$ is an abelian group with a natural topology.

Then G has a countable base for its topology iff $\widehat{G}$ does. Moreover G is compact iff $\widehat{G}$ is discrete. Further for compact G the map sending $a \to \widehat{\widehat{a}}$ where $\widehat{\widehat{a}}(\gamma) = \gamma(a)$ is an isomorphism of G with $\widehat{\widehat{G}}$ as topological groups. (In particular, if H is discrete, then $\widehat{\widehat{H}} \cong H$.) Finally, the characters of G form a basis for $L^2(G)$.

Example 63 $\widehat{\mathbb{T}} = \mathbb{Z}$. *This is a restatement of the fact that the exponential functions are exactly the characters of the unit circle to itself. Explicitly: let $e_n(\xi) = \xi^n$. (In other notation these are the functions $e^{2\pi i n\theta}$.) Then $\{e_n : n \in \mathbb{Z}\}$ form a basis for $L^2(\mathbb{T})$, and each e_n is a character of the unit circle.*

Suppose now that G is a compact abelian group and $a \in G$. Suppose that $T : G \to G$ is given by $T(g) = ag$. Standard theory of compact groups says that there is a unique translation invariant probability measure on G, called Haar measure. Clearly this measure is preserved by this T.

The next theorem generalizes Example 60.

Theorem 64 *T is ergodic iff the powers of a are dense in G. If T is ergodic, then T has discrete spectrum. Every eigenfunction of T is a scalar multiple of a character and the eigenvalues of T are $\{\gamma(a) : \gamma \in \widehat{G}\}$.*

⊢ To see this, we first show that every character is an eigenfunction. In fact this is immediate: suppose that χ is a character. Then $(\chi \circ T)(g) = \chi(ag) = \chi(a)\chi(g)$, hence χ is an eigenfunction with eigenvalue $\chi(a)$. If the powers of a are dense in G, then χ is completely determined by $\langle \chi(a^n) : n \in \mathbb{Z} \rangle = \langle \chi(a)^n : n \in \mathbb{Z} \rangle$. This latter set is determined by $\chi(a)$. Since the characters form a basis, each eigenvalue has multiplicity 1.

On the other hand, if the powers of a are not dense, let $K = cl\{a^n : n \in \mathbb{Z}\}$. Since K is a closed subgroup there are distinct characters χ_1, χ_2 whose kernel contains K. Hence the eigenvalue 1 does not have multiplicity one. Thus we have showed that T is ergodic just in case the powers of a are dense.

Since the characters of G form a basis for $L^2(X)$, T has discrete spectrum and every eigenfunction is a scalar multiple of a character. Moreover the eigenvalues are exactly the eigenvalues of the characters:

$$\{\chi(a) : \chi \in \widehat{G}\}.$$

To finish the Halmos-von Neumann theorem, we must see that every countable subgroup K of $\mathbb{T}$ is the group of eigenvalues of an ergodic discrete spectrum translation.

Let $G = \widehat{K}$. Then G is a compact, separable abelian group. Consider the character $a : K \to \mathbb{T}$ given by the identity map (obviously a character), and let $T : G \to G$ be translation by a.

Claim: T is ergodic and K is the group of eigenvalues of T.

By duality, K is naturally isomorphic to $\widehat{\widehat{K}}$ by the map $k \to \tilde{k}$ where $\tilde{k}(\chi) = \chi(k)$, and the functions $\{\tilde{k} : k \in K\}$ form a basis for $L^2(G)$.

Suppose that f is an invariant function for T. Expand f in its Fourier series with respect to this basis:

$$f(g) \sim \sum b_j \tilde{k}_j(g)$$

So:

$$
\begin{aligned}
f \circ T &\sim \sum b_j \tilde{k}_j(ag) \\
&= \sum b_j \tilde{k}_j(a) \tilde{k}_j(g) \\
&= \sum b_j k_j \tilde{k}_j(g)
\end{aligned}
$$

Since $f = f \circ T$, by the uniqueness of the Fourier expansion:

$$\sum b_j \tilde{k}_j(g) = \sum b_j k_j \tilde{k}_j(g)$$

and thus for all j, $b_j = b_j k_j$. The only way for this to happen is if either $b_j = 0$ or $k_j = 1$. Hence only the constant term in the expansion of f has non-zero coefficient. So f is a constant function.

To finish, we compute the eigenvalues. By our previous remarks the eigenvalues are exactly $\{\tilde{k}(a) : k \in K\}$. But $\tilde{k}(a) = a(k) = k$. Hence the group of eigenvalues is just K. (We note that there is an equally elementary argument that shows the stronger result that the transformation is uniquely ergodic. We argue as above because it foreshadows later similar arguments.)

This completes our summary of the Halmos-von Neumann theorem.   ⊣

Theorem 65 *(Foreman-Louveau, 1995)*

Let $X = {}^\omega 2$, and $\mathcal{D} \subset \mathcal{H}$ be the collection of ergodic discrete spectrum transformations. Then there are functions Φ and Ψ such that:

1. $\Phi : \mathcal{D} \to X^\omega$ *is a Borel reduction of the conjugacy equivalence relation on $\mathcal{D}$ to Eq^+ on X^ω.*

2. $\Psi : X^\omega \to \mathcal{D}$ *is a reduction of Eq^+ to the conjugacy relation on $\mathcal{D}$.*

To prove this we first note that X is Borel isomorphic with the unit circle $\mathbb{T}$ in $\mathbb{C}$, so it suffices to define $\Phi : \mathcal{D} \to \mathbb{T}^\omega$.

By the Halmos-von Neumann theorem if we define $\Phi(T)$ to be an enumeration of the eigenvalues of T, then Φ is a reduction of the conjugacy relation to Eq^+. It remains to see that there is such a Φ that is Borel.

We first note that the collection of transformations that have discrete spectrum is a Borel subset of $\mathcal{H}$. This will be proved in considerable more generality in Section 9, "An extended example", so we defer it until that time. (For all $\alpha < \omega_1$, every ergodic measure preserving transformation has a maximal measure distal factor of ordinal height less than or equal α. We will see that for each ordinal α the map sending T to the maximal distal factor of norm less than or equal α is a Borel map. Moreover the collection of measure distal transformations of norm less than or equal α forms a Borel set. The discrete spectrum transformations are exactly those measure distal transformations of norm 0.)

By Proposition 47, $\{(L, \lambda) : \lambda$ is an eigenvalue of $L\}$ is a Borel subset of $\mathcal{L} \times \mathbb{C}$.

Since $\mathcal{D}$ is a Borel subset of $\mathcal{L}$, the collection $\{(L, \lambda) : \lambda$ is an eigenvalue of L and $L = U_T$, where T has discrete spectrum $\}$ is Borel. Hence, again, by the Lusin-Novikov uniformization theorem, there is a Borel set $B \subset \mathcal{L} \times \mathbb{C}^\omega$ such that for all $T \in \mathcal{D}, (B)_{U_T}$ is an enumeration of the eigenvalues of U_T, which proves the claim that the discrete spectrum transformations can be reduced to Eq^+ on $\mathbb{T}^\omega$.

To show the other reduction let $P \subset \mathbb{T}$ be a perfect set of algebraically independent complex numbers. (Such sets exist by a Baire Category argument.) Then P is homeomorphic to X, and hence we can view the domain of Ψ as P^ω.

There is a Borel function $gen : P^\omega \to \mathbb{T}^\omega$ with the property that for all $\vec{p}, gen(\vec{p})$ is a 1-1 enumeration of the subgroup of $\mathbb{T}$ generated by $\vec{p}$.

Fix a basis $\{e_i : i \in \omega\}$ for $L^2([0,1])$. Define $\Psi : P^\omega \to \mathcal{U}(L^2([0,1]))$ by letting $\Psi(\vec{p})$ be the operator taking $e_i \to \lambda_i e_i$, where λ_i is the ith element of the enumeration given by $gen(\vec{p})$.

Given $\vec{p}$, if $T_{\vec{p}} : \widehat{gen(\vec{p})} \to \widehat{gen(\vec{p})}$ is given (as in the proof of the Halmos-von Neumann theorem) by translation by the identity character, and $U_{\vec{p}}$ is the associated unitary operator on $L^2(\widehat{gen(\vec{p})})$, then there is a unitary isomorphism between $U_{\vec{p}}$ and $\Psi(\vec{p})$, that is multiplicative on elements of $L^\infty(\widehat{gen(\vec{p})})$. Hence by Theorem 15 $\Psi(\vec{p})$ is induced by a measure preserving transformation whose action is isomorphic to $U_{\vec{p}}$.

We finish by observing that if $\vec{p}$ and $\vec{q}$ are elements of P^ω, the $\vec{p}Eq^+\vec{q}$ iff $\{p_n : n \in \omega\}$ generates the same subgroup of $\mathbb{T}$ as $\{q_n : n \in \omega\}$. The latter is the case iff $\Psi(\vec{p})$ is conjugate to $\Psi(\vec{q})$ and we have shown the other direction of the theorem.

6   The Structure of $\underset{\sim}{\Pi}^1_1$-sets

In this section we develop some of the structure of $\underset{\sim}{\Pi}^1_1$-sets, and the tools for showing that a $\underset{\sim}{\Pi}^1_1$-set is not Borel. One of the most important ideas is that of a $\underset{\sim}{\Pi}^1_1$-norm. Roughly speaking these are uniform stratifications of non-Borel $\underset{\sim}{\Pi}^1_1$-sets into Borel sets. An example in Ergodic Theory of such a norm is the

Furstenberg norm on the measure distal transformations.

We begin by giving a canonical example of a $\prod_1^1$-set that is not Borel. We will consider the graphs of linear orderings of the natural numbers. If $<$ is a linear ordering of the natural numbers, then we can canonically view $< \subset \omega \times \omega$ by considering its graph. Its characteristic function $\chi_< : \omega \times \omega \to \{0, 1\}$ can then be viewed as an element of $2^{\omega \times \omega}$. We view $\chi_<$ as the "code" for $<$.

We let:

$$\mathcal{LO} = \{\chi : \chi \text{ codes a linear ordering}\}$$

and

$$\mathcal{WO} = \{\chi : \chi \text{ codes a well ordering}\}.$$

We now check that $\mathcal{LO}$ is a Borel set. Namely:

$$\chi \in \mathcal{LO}$$

iff

$$
(\forall m, n, p) \quad [(\chi(n.m) = 1 \to \chi(m, n) = 0) \land \\
((\chi(n, m) = 1 \land \chi(m, p) = 1) \to \chi(n, p) = 1) \land \\
(\chi(n, n) = 0) \land \\
(n = m \lor \chi(n, m) = 1 \lor \chi(m, n) = 1)].
$$

Similarly we can see that $\mathcal{WO}$ is a $\prod_1^1$-set:

$$\chi \in \mathcal{WO}$$

iff

- $\chi \in \mathcal{LO}$

  and

- $(\forall A \subset N)(A = \emptyset \lor \exists m \in A \forall n \in A(n \neq m \to \chi(m, n) = 1)).$

As we will see presently, $\mathcal{WO}$ is the quintessential example of a non-Borel $\prod_1^1$-set.

To make this notion precise, we again define the notion of a reduction: Let $A \subset X$ and $B \subset Y$, then $f : X \to Y$ is a *reduction* of A to B iff for all $x \in X$,

$$x \in A \text{ iff } y \in B.$$

Note that this is the same notion as Definition 48 with respect to the equivalence relation induced by the partitions $\{A, X\backslash A\}$ and $\{B, Y\backslash B\}$ with the additional proviso that the elements of A are sent to elements of B.

We will say that A is *Borel* or *continuously* reducible to B according to whether there is a Borel or continuous reduction of A to B.

Recall that "pointclass" is jargon for any collection of subsets of Polish spaces. The pointclasses we will be interested in are the classes of Borel, analytic and co-analytic sets. If Γ is a point class, we let $\check{\Gamma} = \{B : X\backslash B \in \Gamma, X$ a Polish space.$\}$.

Remark 66 *In the following we will be only working with the* boldface *pointclasses* Γ. *These are characterized by the property that if $A \subset X \times Y$ is in Γ then for all $x \in X$, we have that $A_x \subset Y$ is in Γ. We will have this as an unstated hypothesis for the rest of this section. These classes include the classes we are most concerned about.*

Definition 67 $C \subset Y$ is complete *for a pointclass* Γ *iff for all X and $A \subset X, A \in \Gamma$ there is a Borel reduction of A to C.*

Kechris [Kec97] showed that for the two classes of analytic and co-analytic sets, C is complete with respect to Borel reductions just in case C is complete with respect to continuous reductions.

The notion of *completeness* is thus a universal notion of being maximally complicated. Clearly any two complete sets are mutually reducible, and it is easy to show that a complete set for either $\Gamma = \Sigma_1^1$ or $\Gamma = \Pi_1^1$ cannot be Borel.

Remark 68 *It is a consequence of the existence of a measurable cardinal that every non-Borel co-analytic set is complete, and similarly for analytic sets. This fact is not a consequence of ZFC, as it fails in Gödel's constructible universe L. The techniques to show this fall outside the scope of this paper.*

The following result is fundamental:

Theorem 69 *(Lusin-Sierpinski [Kec95])* $WO \subset {}^{\omega}\omega$ *is a* Π_1^1*-complete set.*

We now give an example of a complete co-analytic set. For this we use the very useful notion of a "norm".

Definition 70 *Let $A \subset X$. A* norm *on A is a function:*

$$\phi : X \to OR \cup \{\infty\}$$

such that for all $x \in X, x \in A$ iff $\phi(x) \in OR$.

Here "∞" is just used for an arbitrary point not in the ordinals, which we view as larger than all of the ordinals. We remark that this causes no foundational problems, since for any given ϕ, there is an ordinal α such that the range of ϕ is a subset of $\alpha \cup \{\infty\}$.

If ϕ is a norm on a set $A \subset X$, then ϕ induces pre-orderings $<_\phi$ and $\leq_\phi$ where:

$$x <_\phi y \text{ iff } \phi(x) < \phi(y)$$

and

$$x \leq_\phi y \text{ iff } \phi(x) \leq \phi(y).$$

(For these purposes we take the point ∞ to be larger than every ordinal.)

These pre-orderings have the following property: for any given non-empty set $I \subset X$, there is an $x \in I$ such that for all $y \in I, y \not<_\phi x$ and for all $y \in I, x \leq_\phi y$. For this reason we will often refer to them as *pre-wellorderings*

Definition 71 *Let X be a Polish space. If $A \in \Gamma$, then ϕ is a Γ-norm iff both $<_\phi$ and $\leq_\phi$ are Γ-subsets of $X \times X$.*

Given any ϕ, there is a ϕ' that induces the same pre-orderings, and is such that the ordinals in the range of ϕ' are a transitive set. Moreover, ϕ' is completely determined by $<_\phi, \leq_\phi$. Because of this we will view the Γ-norm as being *comprised* of the pair $\{<_\phi, \leq_\phi\}$.

In fact, an equivalent definition (see [Kec95]) is the existence of two relations $\leq^\Gamma \subset X \times X$, $\leq^{\check{\Gamma}} \subset X \times X$ that lie in Γ and $\check{\Gamma}$ respectively such that for $y \in A$,

$$\phi(x) \leq \phi(y) \text{ iff } x \leq^\Gamma y \text{ iff } x \leq^{\check{\Gamma}} y.$$

This definition is subtle in that it does *not* require that $\leq^\Gamma = \leq^{\check{\Gamma}}$, only that they agree on $\{x : \phi(x) \leq \phi(y)\}$ for all $y \in A$. This has an immediate consequence that for all $y \in A, \{x : \phi(x) \leq \phi(y)\}$ is in $\Gamma \cap \check{\Gamma}$.

Let ϕ be a norm that maps onto an initial segment of the ordinals. We will call an ordinal α the *length* of the norm ϕ iff ϕ maps onto α. Note that given a Γ-norm $\{<, \leq\}$, there is a unique ϕ inducing this norm that maps onto an initial segment of the ordinals, and we will define the length of the norm to be the length of ϕ.

Example 72 *Define a function $\phi : {}^\omega 2 \to \omega_1 \cup \{\infty\}$ by setting $\phi(\chi) = \alpha \in \omega_1$ iff $\chi \in WO$ and $<_\chi$ has length α as a well-ordering; and $\phi(\chi) = \infty$ otherwise. Then ϕ is a $\underline{\Pi}_1^1$-norm, which we will call the canonical norm on WO.*

⊢   To see this, we note that for $\chi \in \mathcal{WO}$ and all $\eta \in 2^{\omega \times \omega}$, we have the following equivalent statements:

1. $\phi(\eta) \leq \phi(\chi)$.

2. η is a well ordering and there is no isomorphism from the ordering $<_\chi$ to a proper initial segment of the ordering $<_\eta$. We write this as a $\amalg_1^1$-definition:

$$(\forall A \subset \omega)(A = \emptyset \vee (\exists n \in A)(\forall m \in A)(n \neq m \to \eta(n, m) = 1)) \text{ and}$$

$$(\forall f : \omega \to \omega)[\{(\exists k)(\forall m)\eta(f(m), k) = 1\} \to \{(\exists m, n)(\chi(m, n) = 1 \wedge \eta(f(m), f(n)) = 0)\}]$$

3. There is either an order preserving map from $<_\eta$ into $<_\chi$ or $<_\chi$ is not well-ordered. To see that this is Σ_1^1 we note that it can be written in the following way:

$$(\exists f : \omega \to \omega)[(\forall n, m)(\eta(n, m) = 1 \to \chi(f(n), f(m)) = 1)$$

$$\vee \; \forall n \forall m(n < m \to \chi(f(m), f(n)) = 1)]$$

Thus we have defined two relations $\leq^{\Sigma_1^1}$ and $\leq^{\amalg_1^1}$ whose graphs agree below a given $\chi \in \mathcal{WO}$.   ⊣

The following results follow from the completeness of $\mathcal{WO}$ and can be found in [Kec95], [Mos80].

Theorem 73 *Let X be a perfect Polish space, and A a $\amalg_1^1$ set. Then there is a $\amalg_1^1$ norm on A of length less than or equal ω_1. Moreover, A is not Borel iff some $\amalg_1^1$-norm on A has length ω_1 iff every $\amalg_1^1$-norm on A has length ω_1.*

The following result is of considerable significance of its own:

Theorem 74 *(Boundedness Theorem) Let A be a $\amalg_1^1$-set and ϕ a $\amalg_1^1$ norm on A with values in ω_1. Suppose that $B \subset A$ is a Σ_1^1-set. Then there is a countable ordinal α such that $\{\phi(x) : x \in B\} \subset \alpha$.*

7   The Conjugacy Action on Polish Groups

In this section we give an example of an equivalence relation that is Σ_1^1, but not Borel. We will use this example to give a Polish group acting on itself by conjugacy in a way that the orbit equivalence relation is not Borel. We will discuss this in connection with unitary equivalence and isomorphism of measure preserving systems. (For more information on the descriptive set theory of Polish Group actions, see the book of Becker and Kechris [BK96].)

7.1   A Non-Borel Equivalence Relation

An example of a natural non-Borel equivalence relation is that of isomorphism of countable linear orderings. More precisely, for $\chi_1, \chi_2 \in \mathcal{LO}$ we let $\chi_1 I \chi_2$ iff $<_{\chi_1}$ is isomorphic to $<_{\chi_2}$ as a linear ordering with domain ω.

We note that I is an analytic equivalence relation:

$$\chi_1 I \chi_2$$

$$\text{iff}$$

$$(\exists \text{ a bijection} f : \omega \rightarrow \omega)(\forall m, n)(\chi_1(m,n) = 1 \leftrightarrow \chi_2(f(m), f(n)) = 1).$$

Theorem 75 *(Folk) [Kec95] I is not Borel.*

⊢   Let $\leq^{WO}$ be the canonical $\underset{\sim}{\Pi}_1^1$-norm on $\mathcal{WO}$ with associated function $\phi : 2^{\omega \times \omega} \rightarrow \omega_1 \cup \{\infty\}$. From the definition of a norm, for each countable ordinal $\xi, L_\xi = \{\chi \in \mathcal{WO} : \phi(\chi) < \xi\}$ is a Borel set.

We claim that the Borel ranks of these sets are unbounded in ω_1. Using the $\underset{\sim}{\Pi}_1^1$-completeness of $\mathcal{WO}$ and the Boundedness Theorem (74), every Borel set A is continuously reducible to L_ξ for some $\xi \in \omega_1$. If the Borel ranks of the L_ξ were bounded by δ, then every Borel set would have rank bounded by δ, a contradiction.

Suppose now that I were Borel. Then it would have to have rank δ for some $\delta \in \omega_1$. For all $\xi \in \mathcal{WO}$, we can associate ξ_n coding the ordering of ξ on $\{m : m <_\xi n\}$ in a (say) $\underset{\sim}{\Sigma}_4^0$ way. ($\xi_n(m, l) = 1$ iff $k_m <_\xi k_l$ where $\langle k_m : m \in \omega \rangle$ is an enumeration of $\{j : j <_\xi n\}$ in the usual ordering of ω.)

Then $L_\xi = \{\chi : \chi <^{WO} \xi\} = \{\chi : (\exists n)(\chi I \xi_n)\}$. Hence $L_\xi \in \underset{\sim}{\Sigma}_{\delta+5}^0$. This contradicts the fact that the L_ξ's have unbounded rank.   ⊣

We note that this proof works equally well on the collection of codes for linear orderings that have a least and greatest element and where every element of the linear ordering has an immediate successor. This is a Borel subset of $\mathcal{LO}$. We will call this class $\mathcal{ALO}$.

7.2   A non-Borel Conjugacy Equivalence Relation

The equivalence relation induced by conjugacy on a Polish group is a $\underset{\sim}{\Sigma}_1^1$ equivalence relation. In many cases however it turns out to be Borel. (We give examples of such relations at the end of this section.)

In this section we give an example of a Polish group acting on itself in a way that the associated conjugacy relation is analytic, but not Borel. This example is due to the author who believes it to be the original such example.

Put the discrete topology on $\mathbb{Q}' = \mathbb{Q} \cap [0,1]$, and let $\mathbb{Q}'^{\mathbb{Q}'}$ be the product of countably many copies of $\mathbb{Q}'$ indexed by elements of $\mathbb{Q}'$. Put the product topology on this space. This space is a Polish space homeomorphic to the Baire space. We will view it as the space of all functions from $\mathbb{Q}'$ to $\mathbb{Q}'$ together with the topology of pointwise convergence.

Let G be the group of automorphisms of the structure $\mathfrak{A} = \langle \mathbb{Q}', < \rangle$ which we will call $Aut(\mathbb{Q}')$. So G consists of those functions f from $\mathbb{Q}'$ to $\mathbb{Q}'$ with the property that:

1. $\forall q \in \mathbb{Q}' \exists r \in \mathbb{Q}'(f(r) = q)$

2. $\forall q, r \in \mathbb{Q}'(f(r) = f(q) \rightarrow q = r)$

3. $\forall q, r \in \mathbb{Q}'(q < r \leftrightarrow f(q) < f(r))$

Thus $Aut(\mathbb{Q}')$ is a $\underset{\sim}{\Pi}^0_2$ subset of the Polish space $\mathbb{Q}'^{\mathbb{Q}'}$, and hence is a Polish Group.

Let E be the conjugacy equivalence relation on $Aut(\mathbb{Q}')$, so gEh iff $\exists k \in Aut(\mathbb{Q}')$ such that $h = kgk^{-1}$.

Theorem 76 *(Foreman) E is an analytic, non-Borel equivalence relation.*

⊢   We will define a Borel reduction Φ from $\mathcal{ALO}$ to $Aut(\mathbb{Q}')$ so that $\chi_1 I \chi_2$ iff $\Phi(\chi_1)$ is conjugate to $\Phi(\chi_2)$. Since I is not Borel, it follows that E is not Borel.

Call an element $f \in Aut(\mathbb{Q}')$ *closed* iff whenever $f(q) \neq q$ there is a greatest fixed point of f less than q and a least fixed point of f greater than q, and, further, $\{f^n(q) : n \in \mathbb{Z}\}$ is dense between these two fixed points.

We first have a lemma that is a criterion for conjugacy:

Lemma 77 *Let $f, g \in Aut(\mathbb{Q}')$ be closed. Then f and g are conjugate iff the collection of fixed points of f is isomorphic to the collection of fixed points of g.*

⊢   Clearly conjugacy implies that the collections of fixed points are isomorphic as linear orderings. To see the other direction, let F and G be the collection of fixed points of f and g. Let $\psi_0 : F \to G$ be an isomorphism. We must extend ψ_0 to the collection of non-fixed points.

The closure of f and g implies that every non-fixed point q lies between successive fixed points. For each pair x_0, x_1 of successive fixed points of f, a "back and forth argument" builds the isomorphism $\psi \upharpoonright (x_0, x_1)$ between the interval $(x_0, x_1) \cap \mathbb{Q}'$ and $(\psi_0(x_0), \psi_0(x_1)) \cap \mathbb{Q}'$, so that for all $q \in (x_0, x_1) \cap \mathbb{Q}'$

the orbit $\{f^n(q) : n \in \mathbb{Z}\}$ is sent isomorphically to the orbit $\{g^n(\psi(q)) : n \in \mathbb{Z}\}$. ⊣

To prove Theorem 76 we define Φ so that for all $\chi \in \mathcal{ALO}, \Phi(\chi) \in Aut(\mathbb{Q}')$ is closed and the fixed points of $\Phi(\chi)$ are isomorphic to the ordering given by χ.

Given χ, we must embed the ordering χ into the fixed points of $\Phi(\chi)$ in such a way that there are no "gaps" in the range of the embedding. This is a again done with a back and forth argument:

For a fixed χ we build a function $e = e_\chi : \omega \to \mathbb{Q}'$ by specifying its graph $\{(n, q) : e(n) = q\}$. At a given stage l in the construction we will have a collection of ordered pairs $\{(n_i, q_i) : i < i_l\}$ that define a partial function from ω to $\mathbb{Q}'$ that preserves the ordering given by χ.

Enumerate $\mathbb{Q}'$ as $\langle q^j : j \in \omega \rangle$.

At stages 0 and 1 in the construction, we send the least element of ω according to the ordering $<_\chi$ to 0 and the greatest element to 1.

At odd stages $2k + 1$, consider the least natural number m such that m is not already in the domain of the function we are defining, and choose the least j such that adding the pair (m, q^j) to $\{(n_i, q_i) : i < i_{2k}\}$ is order preserving.

At even stages $2k$, consider the least $j \geq k - 1$ such that q^j is not in the range of the partial function built at stage $2k - 1$. Then q^j lies between two successive elements of $\{q_i : i < i_{2k}\}$, say q_{i_0} and q_{i_1}.

- If n_{i_0} and n_{i_1} are successive elements of ω according to the ordering $<_\chi$, then we add no new element to the collection of ordered pairs.

- If n_{i_0} and n_{i_1} are not successive elements of ω according to χ, then we choose the least n (in the usual ordering) in ω that lies between n_{i_0} and n_{i_1} in the χ-ordering and add the pair (n, q^j) to the collection of ordered pairs $\{(n_i, q_i) : i < i_{2k-1}\}$.

The resulting function e with graph $\{(n_i, q_i) : i \in \omega\}$ is easily checked to be order preserving and to have the property that every element of $\mathbb{Q}'$ not in the range of e is between two successive elements of the range of e.

We note that the map $\chi \mapsto e_\chi$ is a Borel map from $\mathcal{ALO}$ to the space of functions from ω into $\mathbb{Q}'$, $\mathbb{Q}'^\omega$.

Yet another back and forth argument shows that given a function $e \in \mathbb{Q}'^\omega$, such that every $q \in \mathbb{Q}'$ lies between two successive elements of the range of e, we can build a closed element $\Psi(e)$ of $Aut(\mathbb{Q}')$ that has the range of e as its collection of fixed points. This map, from a Borel subset of $\mathbb{Q}'^\omega$ to $Aut(\mathbb{Q}')$, is also Borel. Finally the map $\chi \mapsto \Psi(e_\chi)$ is Borel, thus proving the theorem. ⊣

Note that very similar arguments show the same result for $Aut(\mathbb{Q})$ and $Aut(\mathbb{Q} \cap [0, 1))$, etc.

Remark 78 *The conjugacy relation on the group of unitary operators on a separable Hilbert space and on the group of permutations of the natural numbers both turn out to be Borel. To see that the conjugacy relation on the group of permutations of the natural numbers is Borel, note that two permutations ϕ and ψ are conjugate in the group iff they have the same cardinalities of orbits with the same multiplicity. This is a Borel statement. Unitary equivalence is Example 55.*

7.3   Unitary Equivalence vs. Conjugacy

As shown by Examples 55 and 56, the conjugacy relation on the normal operators on a separable Hilbert space is a Borel equivalence relation. In some sense this is the heart of spectral theorem, from the point of view of descriptive set theory: it gives a way of associating invariants to normal operators so that a concrete countable amount of "information" describes whether they are conjugate.

One of the main goals of this paper is to raise the question about whether this is possible for measure preserving transformations. This question is discussed further at the end of this section, but for the benefit of the logicians, we give an example showing that there are non-conjugate, ergodic measure preserving transformations whose associated unitary operators are conjugate. (See [Hal60] for a discussion of this phenomenon.)

We again follow [Wal82] quite closely for the language and proof used in this example. We say that two measure preserving transformations are *spectrally isomorphic* iff the associated linear operators are unitarily equivalent.

Note that Theorem 15 says that two transformations are conjugate just in case they are spectrally isomorphic by an intertwining unitary transformation that preserves multiplication on the bounded functions from X to the complex numbers. This gives a positive condition for verification of conjugacy.

We now give a sufficient condition for spectral isomorphism:

Definition 79 *T has* countable Lebesgue spectrum *iff there are $\{f_i\}_{i=0}^{\infty}$ with $f_0 \equiv 1$ such that $\{f_0\} \cup \{U_T^n(f_i) : n \in \mathbb{Z}, i \geq 1\}$ is an orthonormal basis.*

The following proposition is immediate:

Proposition 80 *Suppose that S, T are measure preserving transformation that have countable Lebesgue spectrum. Then S, T are spectrally isomorphic.*

The following lemma then gives the desired examples:

Lemma 81 *If (X, B, μ, T) is a Bernoulli shift on a finite alphabet, then T has countable Lebesgue spectrum.*

⊢   To see this: If T is the shift on $X = n^{\mathbb{Z}}$ then for all $\vec{i} = (i_0, \ldots, i_k) \subset \mathbb{Z}$ we can define $g_{\vec{i}} \in L^2(X)$ by setting

$$g_{\vec{i}}(s) = e^{2\pi i (s_{i_0} + s_{i_1} \cdots + s_{i_k})}$$

(where $s = \langle s_i : i \in \mathbb{Z} \rangle$.) These form a basis witnessing countable Lebesgue spectrum.   ⊣

But now it is clear that there are spectrally isomorphic, non-conjugate measure preserving systems. For example the Bernoulli shift given by a fair two headed coin has different entropy from the Bernoulli shift given by a fair three headed coin, and hence the corresponding measure preserving transformations are not conjugate. However Lemma 81 says that they are spectrally isomorphic.

Remarks: It is an interesting question as to whether the conjugacy relation on various classes of measure preserving transformation is Borel. For Bernoulli transformations, Theorem 38 has a positive answer as an easy consequence. The Halmos-von Neumann theorem (together with the remarks around Theorem 65) imply that the conjugacy relation on the discrete spectrum transformations is Borel.

For K-automorphisms, the situation remains quite open: the Ornstein-Shields Theorem (Theorem 57) shows that E_0 can be reduced to conjugacy on the K-automorphisms, but this is a relatively weak statement: E_0 is reducible to $=^+$ (Proposition 54), which is the exact complexity of the conjugacy relation on the discrete spectrum transformations. Hence, for all we know, there are invariants analogous to the Halmos-von Neumann theory that completely classify the K-automorphisms.

For measure distal transformations the situation is more complicated, as we shall see. Halmos [Hal60] showed that all skew products of $\mathbb{T}^2$ of the form $T_k(\xi_0, \xi_1) = (\alpha\xi_0, k\xi_0\xi_1)$, for a fixed irrational α, are spectrally isomorphic. It is an exercise, using the methods of Section 8, to show (for example) that T_1 and T_2 are not isomorphic. Hence even for measure distal transformations of height 2, spectral isomorphism does not imply conjugacy.

Hence the situation is more complicated. Briefly summarized, the current state of knowledge is: For each countable ordinal α there is a Borel map from $\mathcal{H}$ to $\mathcal{H}$ that calculates the maximal distal factor of a transformation that has height less than or equal α. (In particular, setting $\alpha = 0$ it calculates the Kronecker factor.) Moreover, for a given countable ordinal α, the measure distal transformations that have height α form a Borel set.

We do not know whether the conjugacy relation on the measure transformations of height α is Borel (even for $\alpha = 2$), but we do know this for a large class (what Zimmer called the *normal* discrete spectrum transformations). Roughly speaking, this class is the collection of distal transformations achieved by iterating skew products of compact groups. (The non-normal ones include skew products into homogeneous spaces of compact groups.)

Author's Note: Subsequent to the Luminy meeting, Hjorth showed that the conjugacy relation on the collection of *all* measure preserving transformations is Σ_1^1 and not Borel. However the collection of transformations witnessing the result were highly non-ergodic. Indeed the ergodic components of the transformation were discrete spectrum. The complexity of Hjorth's transformations is due to the complexity of the ergodic decomposition rather than inherent complexity of a class of ergodic measure preserving transformations. Thus, unfortunately, it says relatively little about the problem of classifying ergodic measure preserving transformations.

Hjorth, however, defined a class of equivalence relations that he calls *turbulent* which is exactly characterized by a certain non-classifiability property. This property is analogous to the category $0 - 1$ laws used to show that E_0 cannot be smooth. Hjorth showed that a turbulent equivalence relation can be reduced to the relation of conjugacy on the ergodic transformations. We refer the reader to Hjorth's forthcoming book for details.

8   Furstenberg's Structure Theorem

In this section we review the Furstenberg structure theorem for ergodic measure preserving transformations. Furstenberg developed this structure theorem, in part, to show deep results in combinatorial number theory. This theory and its generalizations and extensions continue to produce remarkable results (see [PS96] for a survey). We refer the reader to other sources for applications of the theory in this direction.

The exposition is heavily dependent on Furstenberg's book [Fur81], in particular, we largely adopt his notation and many of his arguments. This section may be viewed as a partial summary of the appropriate sections of his book.

We now state the obligatory prerequisites.

Theorem 82 *(The Ergodic Theorem) Suppose that* (X, B, μ, T) *is a measure preserving system and* $f \in L^1(\mu)$. *For almost all* x, *the averages*

$$\frac{1}{n} \sum_{k=0}^{n-1} f(T^k(x))$$

converge to a T-*invariant function* $F(x)$. *Moreover if* $1 \leq p < \infty$, *and* $f \in L^p(\mu)$ *then the convergence also occurs in the* p-*norm.*

This theorem is often summarized by saying that for almost all x the time averages exist and, since $\int f d\mu = \int F d\mu$, the average of the time averages is equal to the space average of f.

Since convergence in norm implies convergence in the weak topology, the ergodic theorem implies:

Theorem 83 *Suppose that (X, B, μ, T) is a measure preserving system and that $g \in L^2(X)$. Then there is a T-invariant function $G \in L^2(X)$ such that for all $f \in L^2(X)$ the averages:*

$$\frac{1}{n} \sum_{k=0}^{n-1} \int f(x)g(T^k(x))d\mu$$

converge to $\int f(x)G(x)d\mu$.

This is the form of the ergodic theorem used in the proof of the Furstenberg structure theorem and follows from a rather "soft" fact about contractions on Hilbert Spaces:

Theorem 84 *Let $U : \mathbb{H} \to \mathbb{H}$ be a linear operator on the Hilbert Space $\mathbb{H}$ and suppose that for all vectors $h, \| U(h) \| \leq \| h \|$. Let M be the subspace of invariant vectors for U and P_I be the projection onto M. Then for all h, l:*

$$(l, \frac{1}{n} \sum_{k=0}^{n-1} U^k(h)) \to (l, P_I(h)).$$

In other words P_I is the weak limit of the averages $\frac{1}{n} \sum_{k=0}^{n-1} U^k(h)$.

Definition 85 *We repeat the definition of weakly mixing and introduce the orthogonal notion of a compact transformation. Let (X, B, μ, T) be a measure preserving system.*

- *T is weakly mixing iff for all $U, V \in B$, we have*

$$\lim_D \mu(T^n(U) \cap V) = \mu(U)\mu(V)$$

  ($\lim_D$ means limit with respect to the density filter.)

- *T is compact iff there is a dense set of $f \in L^2(X)$ such that*

$$cl_{\| \ \|}\{T^n(f)\}_{n \in \omega}$$

  is compact.

Recall by Example 29 that every Bernoulli shift is weakly mixing (in fact strongly mixing).

Example 86 *Let $\alpha \in \mathbb{T}$ be a non-root of unity. (i.e., $\alpha = e^{2\pi i \theta}$ for some irrational θ.) Let $T : \mathbb{T} \to \mathbb{T}$ be given by $T(\xi) = \xi \alpha$. Then T is ergodic (by Theorem 64) and is a compact transformation.*

⊢ To see the assertion in this example, note that the action of translation is continuous on the subcollection of continuous functions with domain $\mathbb{T}$. Hence the orbit of any continuous function in pre-compact. Since the continuous functions are dense in $L^2(X)$ the assertion follows.   ⊣

8.1 A Characterization of the Weakly Mixing Transformations

The next theorem is a structural characterization of the weakly mixing transformations.

Theorem 87 *Let T be a measure preserving transformation on X. Then T is weakly mixing iff $T \times T$ is ergodic on the space $X \times X$.*

As a byproduct of the proof we will see that T is weakly mixing iff its product with any ergodic transformation is ergodic. The proof of this theorem we give here is taken directly from Furstenberg's book ([Fur81]).

⊢ To start with, we note that for bounded sequences of real numbers a_n, b

$$\lim_D a_n = b$$

iff

$$\frac{1}{n} \sum_{k=0}^{n-1} (a_n - b)^2 \to_n 0$$

iff

$$\frac{1}{n} \sum_{k=0}^{n-1} |a_n - b| \to_n 0.$$

Suppose that T is weakly mixing. First show that for all real valued $f, g \in L^2(X)$,

$$\frac{1}{n} \sum_{k=0}^{n-1} \{ \int f(x) g(T^k(x)) d\mu - \int f(x) \int g(x) d\mu \}^2 \to 0$$

For characteristic functions of measurable sets this is just the definition of weak mixing together with the equivalences above. For general functions you first approximate with simple functions and follow the usual procedures.

This implies that for complex $f, g \in L^2(X)$,

$$\frac{1}{n} \sum_{k=0}^{n-1} \{\int f(x)\bar{g}(T^k(x))d\mu - \int f(x) \int \bar{g}(x)d\mu\}^2 \to 0$$

(here $^-$ refers to complex conjugation.)

Now let (Y, C, ν, S) be ergodic. We will show that the measure preserving system $(X \times Y, B \otimes C, \mu \otimes \nu, T \times S)$ is ergodic.

Since the ergodic theorem shows that averages of the $T \times S$ iterates of every function $h(x, y) \in L^2(X \times Y)$ converge a.e., it suffices to show that for all $g(x, y), h(x, y) \in L^2(X \times Y)$

$$\frac{1}{n} \sum_{k=0}^{n-1} \int g(x, y)\bar{h}(T^k(x), S^k(y))d(\mu \otimes \nu) \to$$

$$\int g(x, y)d(\mu \otimes \nu) \int \bar{h}(x, y)d(\mu \otimes \nu),$$

since this shows that

$$\frac{1}{n} \sum_{k=0}^{n-1} h(T^k(x), S^k(y)) \to \int h(x, y)$$

weakly.

Since linear combinations of the form $\sum_i c_i f_i(x)g_i(y)$ are dense in $L^2(X \times Y)$, it suffices to show that for all $f_1(x), f_2(x) \in L^2(X), g_1(y), g_2(y) \in L^2(Y)$

$$\frac{1}{n} \sum_{k=0}^{n-1} \int f_1(x)f_2(T^k(x))d\mu \int g_1(y)g_2(S^k(y))d\nu$$

has limit

$$\int f_1(x)d\mu \int f_2(x)d\mu \int g_1(y)d\nu \int g_2(y)d\nu$$

Since every function is a sum of a constant function and a mean 0 function we can treat these cases separately:

Case 1 f_1 is a constant:

Then $\int f_1(x)f_2(T^n(x))d\mu = \int f_1 d\mu \int f_2 d\mu$, and by the weak ergodic theorem $\frac{1}{n}\sum_{k=0}^{n-1}\int g_1(y)g_2(S^k(y))d\nu$ converges to $\int g_1(y)d\nu \int g_2(y)d\nu$.

<u>Case 2</u> $\int f_1(x)d\mu = 0$:
Then, by the Cauchy-Schwartz inequality:

$$\{\frac{1}{n}\sum_{k=0}^{n-1}\int f_1(x)f_2(T^k(x))d\mu \int g_1(y)g_2(S^k(y))d\nu\}^2$$

is less than or equal to:

$$\frac{1}{n}\sum_{k=0}^{n-1}\{\int f_1(x)f_2(T^k(x))d\mu\}^2 \times \frac{1}{n}\sum_{k=0}^{n-1}\{\int g_1(x)g_2(S^k(x))d\mu\}^2.$$

Since f_1 has mean zero we see:

$$\frac{1}{n}\sum_{k=0}^{n-1}\{\int f_1(x)f_2(T^k(x))d\mu - 0\}^2 \to 0.$$

Hence the first multiplicand goes to 0 and thus the product also goes to 0.

We now show the converse. Suppose that $T \times T$ is ergodic on $X \times X$. Let $A, B \subset X$ be measurable and let $f = \chi_A, g = \chi_B$. Consider $f(x)f(x')$ and $g(x)g(x')$ as members of $L^2(X \times X)$.

By the weak ergodic theorem for $T \times T$:

$$\frac{1}{n}\sum_{k=0}^{n-1}(\int f(x)g(T^k(x))d\mu)^2 = \frac{1}{n}\sum_{k=0}^{n-1}\int f(x)f(x')g(T^k(x))g(T^k(x'))d(\mu \times \mu)$$

converges to

$$(\int f(x)d\mu)^2(\int g(x)d\mu)^2.$$

Let $\alpha_k = \int f(x)g(T^k(x))d\mu$ and $\alpha = \int f(x)d\mu \int g(x)d\mu$. Then we have shown that $\frac{1}{n}\sum_{k=0}^{n-1}\alpha_k \to \alpha$ and $\frac{1}{n}\sum_{k=0}^{n-1}\alpha_k^2 \to \alpha^2$. Hence we have that:

$$\frac{1}{n}\sum_{k=0}^{n-1}(\alpha_k - \alpha)^2 \to 0.$$

Hence the α_k converge to α in density, and weak mixing follows.   ⊣

We summarize:

- T is ergodic iff

$$\frac{1}{n}\sum_{k=0}^{n-1}\int f(x)g(T^k(x)) - \int f\int g \rightarrow_n 0$$

- T is weakly mixing iff

$$\frac{1}{n}\sum_{k=0}^{n-1}\{\int f(x)g(T^k(x)) - \int f\int g\}^2 \rightarrow_n 0$$

8.2 Height One Transformations

The Furstenberg theory (and its parallel developed by Zimmer) exhausts a measure preserving transformation by atomic extensions of a particularly simple sort, followed by a "relatively" weak mixing extension. In the case of height one transformations these relativized notions simplify exactly to the definition of weakly mixing and compact. In this section we develop this theory for height one extensions of the trivial flow; i.e., weakly mixing and compact transformations. The height one case contains many of the ideas of the general case without some of the complicating technicalities. We begin with two illustrative theorems.

Theorem 88 *Let (X, B, μ, T) be a an ergodic measure preserving system. Then*

1. (X, B, μ, T) is compact iff X has discrete spectrum.

2. (X, B, μ, T) is weakly mixing iff the only eigenvalue for U_T is 1.

Thus the notions of compact and weakly mixing extensions are, in some sense orthogonal. These two results have many ideas in common, so we'll do the easy directions of both theorems first.

⊢ Suppose that (X, B, μ, T) is a discrete spectrum transformation. Since linear combinations of the eigenfunctions are dense, it suffices to see that if $g = \sum_{i<N}\alpha_i f_i$ is a finite sum of eigenfunctions, then $\{U_T^n(g) : n \in \mathbb{Z}\}$ is precompact. However $U_T^n(g) = \sum_{i<N}\lambda_i^n\alpha_i f_i$ (where λ_i is the eigenvalue associated to f_i.) Hence the closure of the orbit of g is isomorphic to a closed subgroup of $\mathbb{T}^N$.

Suppose that (X, B, μ, T) is weakly mixing, and that f is an eigenfunction with an eigenvalue $\lambda \neq 1$. Since $\int f d\mu = \int (f \circ T)d\mu = \int \lambda f d\mu$ and $|\lambda| = 1$ we know that $\int f d\mu = 0$.

Since T is weakly mixing:

$$\frac{1}{n}\sum_{k=0}^{n-1} |\int f\overline{U_T^k(f)} - \int f d\mu \int \bar{f} d\mu| \to 0.$$

Hence $\frac{1}{n}\sum_{k=0}^{n-1} |\int f\overline{U_T^k(f)}d\mu| \to 0$. But this is the same as

$$\frac{1}{n}\sum_{k=0}^{n-1} |\lambda|^n \int f\bar{f} d\mu \to 0.$$

Hence $f\bar{f} = 0$ a.e., and so $f \equiv 0$.

To see the converse of 1. and 2. we need a means of generating eigenfunctions. For this we will use the Hilbert-Schmidt operators. The following fact can be found in any textbook on functional analysis:

Fact 89 *Suppose that X, Y are Lebesgue spaces and $H(x, y) \in L^2(X \times Y)$. Then the operator $L_H : L^2(Y) \to L^2(X)$ given by*

$$L_H(g) = \int H(x, y)g(y)d\nu(y)$$

is a compact operator. Moreover, if $H(x, y) = \overline{H(y, x)}$, then L_H is Hermitian.

We will call the operator L_H the *Hilbert-Schmidt operator* with kernel H.

Lemma 90 *Let (X, B, μ, T) be an ergodic measure preserving transformation. Let $T \times T$ be the product transformation on $X \times X$. Then:*

- *If $H \in L^2(X \times X)$ is a non-constant invariant function, then the range of L_H has a basis of eigenfunctions, one of which is non-constant.*

- *If f is non-zero and $o(f) = \{U_T^n(f) : n \in \mathbb{Z}\}$ is precompact, let $\tilde{f}(x) = f(x)\overline{f(x')} \in L^2(X \times X)$. Then $P_I(\tilde{f}) \neq 0$, where P_I is the projection onto the invariant functions.*

- *If $\{f : o(f) \text{ is precompact}\}$ is dense, then the set of images of Hilbert-Schmidt operators from bounded invariant H is also dense.*

Note that the first assertion of this lemma combined with Theorem 87 implies that a measure preserving transformation with no non-trivial eigenvalues is weakly mixing.

Also, the first assertion combined with the third assertion immediately give that a compact transformation has discrete spectrum.

To see the <u>first assertion</u>, suppose that $H(x, x') \in L^2(X \times X)$ is invariant and non-constant. Note that we can assume that H is real valued (and even bounded).

Hence the Hilbert-Schmidt operator L_H is compact and Hermitian. By the spectral theorem, $L^2(X)$ is a direct sum of eigenspaces for L_H that are finite dimensional (except perhaps for the eigenvalue 0). If we can show that the range of L_H contains a non-constant function (so in particular L_H is not the 0 operator) and that the eigenspaces of L_H are invariant under U_T, then we have found a finite dimensional subspace of $L^2(X)$ that is invariant under U_T. Since U_T is unitary, it is diagonalizable on this invariant subspace. In particular the subspace has a basis consisting of eigenfunctions for U_T.

To see that the eigenspaces of L_H are invariant under U_T, it suffices to show that L_H intertwines with U_T:

$$U_T L_H = L_H U_T.$$

Since then, if f is an eigenfunction of U_T with value λ, we have

$$
\begin{aligned}
\lambda U_T(f) &= U_T(\lambda f) \\
&= U_T(L_H(f)) \\
&= L_H(U_T(f)).
\end{aligned}
$$

Hence $U_T(f)$ is also an eigenfunction of L_H.

To see $U_T L_H = L_H U_T$, let $f \in L^2(X)$. Then:

$$
\begin{aligned}
U_T(L_H(f))(x) &= \int H(Tx, y)f(y)d\mu(y) \\
&= \int H(Tx, Ty)f(Ty)d\mu(y) \\
&= \int H(x, y)f(Ty)d\mu(y) \\
&= L_H(U_T(f))(x).
\end{aligned}
$$

It remains to show that there is a non-constant function in the range of L_H. Let $\{e_i(x) : i \in \omega\} \cup \{1\}$ be a basis for $L^2(X)$. Then $\{e_i(x)e_j(x') : i, j \in \omega\} \cup \{1\}$ is a basis for $L^2(X \times X)$. (Where we denote pairs from $X \times X$ as (x, x').) Since H is non-constant, there is some (i, j) such that $(H, e_i(x)e_j(x')) \neq 0$. We claim that $(e_i, L_H(\overline{e_j})) \neq 0$ and hence $L_H(\overline{e_j})$ is non-constant. Indeed,

$$
\begin{aligned}
(e_i, L_H(\overline{e_j})) &= \int e_i(x) \overline{\int H(x, x') \overline{e_j(x')} d\mu(x')} d\mu(x) \\
&= \int \int \overline{H(x, x')} e_i(x) e_j(x') d\mu(x) d\mu(x') \\
&= \overline{(H(x, x'), e_i(x) e_j(x'))} \\
&\neq 0.
\end{aligned}
$$

To see the <u>second assertion</u> suppose that $f \neq 0$ and $o(f)$ is precompact. We want to see that the projection of $\tilde{f} = f(x)\overline{f(x')}$ to the invariant functions is not zero.

By Theorem 84 it suffices to see that

$$
\frac{1}{n} \sum_{k=0}^{n-1} U_T^k(\tilde{f})
$$

does not converge to 0.

Let $g_1, \ldots, g_m$ be ϵ-dense in $o(f)$. If $P_I(f) = 0$ then for all $1 \leq j \leq m$, we have $(\overline{g_j}(x) g_j(x'), P_I(\tilde{f})) = 0$.

By Theorem 84:

$$
(\overline{g_j}(x) g_j(x'), \frac{1}{n} \sum_{k=0}^{n-1} U_T^k(\tilde{f})) =
$$

$$
\frac{1}{n} \sum_{k=0}^{n-1} \int \overline{g_j}(x) g_j(x') f(T^k(x)) \overline{f(T^k(x'))} (d\mu \times d\mu) \to 0.
$$

So

$$
\frac{1}{n} \sum_{k=0}^{n-1} |\int \overline{g_j(x)} f(T^k(x)) d\mu|^2 \to 0.
$$

Combining these for the various j, we get

$$
\frac{1}{n} \sum_{k=0}^{n-1} \{\sum_{j=1}^{m} |\int \overline{g_j(x)} f(T^k(x)) d\mu|^2\} \to 0.
$$

Hence for any ϵ, there is a k for all j,

$$
|\int \overline{g_j(x)} f(T^k(x)) d\mu|^2 < \epsilon
$$

Choose a j such that $\| U_T^k(f) - g_j \|^2 < \epsilon$. Then:

$$\| U_T^k(f) - g_j \|^2 = (U_T^k(f) - g_j, U_T^k(f) - g_j)$$
$$= \| U_T^k(f) \|^2 - 2Re(U_T^k(f), g_j)$$
$$+ \| g_j \|^2 .$$

This implies that

$$\| U_T^k(f) - g_j \|^2 + |2Re \int \overline{g_j} f(T^k(x)) d\mu| > \| U_T^k(f) \|^2 .$$

Hence

$$2\epsilon > \| U_T^k(f) \|^2 .$$

Since ϵ was arbitrary this implies that $f \equiv 0$.

Note that this argument gives more: If f is non-constant, then $P_I(f)$ is non-constant. This follows because we can take f to have mean zero. Then the mean of $P_I(f)$ is the same as the mean of f. Since $P_I(f)$ is not the zero function, it must be non-constant.

To see the <u>third assertion</u> we need to see that if the precompact orbits are dense then the images of the Hilbert-Schmidt operators are also dense.

If not, let $0 \neq f \in L^2(X)$ be orthogonal to the ranges of all L_H, for H bounded and invariant, and be such that $o(f)$ is precompact. Define $\tilde{f}$ as before.

Let $H(x, x') = P_I(\tilde{f})$. By the previous claim, $P_I(\tilde{f}) \neq 0$. Let H_M be equal to H, where $|H| \leq M$, and 0 otherwise. Then H_M is bounded and invariant. Since $f \perp L_{H_M}(f)$ we have:

$$\int \overline{f(x)} \int H_M(x, x') f(x') d\mu(x') d\mu(x) = 0.$$

Hence

$$\int \int \overline{f(x)} f(x') H_M(x, x') d\mu(x) d\mu(x') = 0.$$

In other words: $(\tilde{f}, H_M)_{L^2(X \times X)} = 0$. Since H_M is invariant under T, this implies that $(U_T^k(\tilde{f}), H_M)_{L^2(X \times X)} = 0$, for all k. But then we see that $(\frac{1}{n} \sum_{k=0}^{n-1} U_T^k(f), H_M) = 0$, for all n. Since $\frac{1}{n} \sum_{k=0}^{n-1} U_T^k(f)$ converges to $P_I(\tilde{f}) = H$, we see that $H \perp H_M$, for all M. But $(H, H_M) = \| H_M \|^2$, hence $H_M = 0$ for all M, a contradiction.

This proves Lemma 90 and completes the proof of Theorem 88. ⊣

8.3   The Relativized Theory

In this section we show how Lemma 90 and Theorem 88 are generalized to
the context of extensions of non-trivial measure preserving systems.

First we begin by considering three equivalent ways of viewing a factor
(Y, C, ν, S) of a measure preserving system (X, B, μ, T). The first is via the
definition. We describe two additional ways.

Suppose that (Y, C, ν, S) is a factor of (X, B, μ, T) and that $\pi : X \to Y$
is the function witnessing this. Then there is an injective homeomorphism of
the σ-algebra C into the σ-algebra B given by

$$\iota(U) = \pi^{-1}(U).$$

The image of this map is an invariant sub-σ-algebra D of B.

Conversely, let (X, B, μ, T) be a measure preserving system and D be a
T-invariant sub-σ-algebra of B. Put an equivalence relation on X by setting
$x_1 \sim x_2$ iff for all $U \in D$, $x_1 \in U$ iff $x_2 \in U$. The collection Y of $\sim$-equivalence
classes together with the σ-algebra induced by D (modulo measure zero sets)
forms a standard Borel space. Since D is invariant under T, the action of
T on Y is well-defined. Further, μ induces a measure ν on D. Thus we
get a measure preserving system (Y, C, ν, S) that is a factor of (X, B, μ, T).
We have shown that there is a canonical correspondence between invariant
sub-σ-algebras of (X, B, μ, T) and factors.

The third way of viewing factors is via the associated L^2. If (Y, C, ν, S)
is a factor of (X, B, μ, T), with witness π, then the map $f \mapsto f \circ \pi$ yields a
linear isometric injection of $L^2(Y)$ into $L^2(X)$, associating with (Y, C, ν, S) a
closed invariant subspace of $L^2(X)$.

The subspaces associated with factors can be identified by the following
lemma:

Lemma 91 *(See [Zim76a] or [BF96]) Suppose that $\mathcal{A} \subset L^2(X)$ is a closed
subspace of $L^2(X)$, such that $\mathcal{A}$ contains the constant functions and is closed
under the following operations:*

1. *$f \mapsto \overline{f}$, where $\overline{f}$ is the complex conjugate of f,*

2. *multiplication by elements of $L^\infty(X)$,*

3. *$f \mapsto f \wedge M$ for positive real valued, where $0 < M < \infty$ ($f \wedge M$ is the
truncation of f at M).*

*Then there is a factor (Y, C, ν, S) of (X, B, μ, T) such that $\mathcal{A}$ is the subspace
of $L^2(X)$ associated with $L^2(Y)$.*

Viewing factors as certain subspaces of $L^2(X)$ is the mechanism used to
generate factors with certain properties. We give an example:

Example 92 *Let (X, B, μ, T) be a measure preserving system and suppose that $K \subset \mathbb{T}$ is the collection of eigenvalues. Let $\mathcal{A}$ be the closed subspace of $L^2(X)$ spanned by the eigenvectors corresponding to elements of K. Then $\mathcal{A} = L^2(Y)$ for a factor (Y, C, ν, S). Moreover, the factor has discrete spectrum and is the maximal discrete spectrum factor of (X, B, μ, T).*

⊢ As we showed in Theorem 61 we can take a multiplicative subcollection of the eigenfunctions that span the eigenfunctions. By Lemma 91, this subspace of $L^2(X)$ corresponds to a factor. ⊣

In order to give the relativized versions of weak mixing and compactness we need some measure theoretic facts.

Conditional Expectation and Disintegrations

Let (X, B, μ) be a Lebesgue space and let C be a sub-σ-algebra of B. An application of the Radon-Nikodym theorem shows that for every B-measurable function f there is a C-measurable function $E(f|C)$ such that for all $W \in C$, $\int_W f d\mu = \int_W E(f|C) d\mu$. The function $E(f|C)$ is called the *conditional expectation* of f with respect to C. (In the case of finite measure spaces this is simply Bayes' Law.)

The conditional expectation satisfies the following properties:

- The map $f \to E(f|C)$ is a positive bounded linear functional from $L^1(X, B, \mu)$ to $L^1(X, C, \mu)$.

- If g is C-measurable, then $E(fg|C) = gE(f|C)$

As we remarked previously, there is a canonical correspondence between factors of a measure preserving system (X, B, μ, T) and invariant sub-σ-algebras of B. In an abuse of notation, if (Y, C, ν, S) is a factor of (X, B, μ, T), we will write $E(f|Y)$ for the conditional expectation with respect to the subalgebra corresponding to the factor Y.

An important tool for analyzing factor maps is the *disintegration of measures*. Suppose that $\pi : X \to Y$ is measurable and $\mu(\pi^{-1}(W)) = \nu(W)$ for all $W \in C$. Suppose that (X, B, μ) is a Lebesgue space.

Theorem 93 *There is a measurable map from Y to $\mathcal{M}(X)$ $(y \mapsto \mu_y)$ such that*

- *For all $f \in L^1(X)$, $f \in L^1(\mu_y)$, and $E(f|Y)(y) = \int f d\mu_y$ for a.e. y.*

- *$\int \{\int f d\mu_y\} d\nu(y) = \int f d\mu$ for all $f \in L^1(X)$.*

Moreover the map $y \mapsto \mu_y$ is essentially unique.

Note that this theorem says that for separable space there is always a form of Fubini's theorem. Namely, the measures μ_y can be shown to concentrate on the fibers $\pi^{-1}(y)$, and then the second equation says that the integral of f is the same as horizontally integrating the μ_y integrals of the vertical sections.

Definition 94 *The map $y \mapsto \mu_y$ in Theorem 93 will be called the* disintegration *of μ with respect to ν.*

Relative Products

Let (Y, C, ν, S) be a factor of (X, B, μ, T). Let $\langle \mu_y \rangle$ be the disintegration of μ with respect to ν. The *relative product* (or "fiber product") of X with itself over Y, $X \times_Y X$, is the space $(X \times X, B \otimes B, \mu \times_Y \mu, T \times T)$, where $\mu \times_Y \mu$ is the measure $\int (\mu_y \times \mu_y) d\nu(y)$.

More generally, if (Y, C, ν, S) is a factor of (X, B, μ, T) and also of (Z, D, λ, T'), we can define the relative product of X and Z over Y, $X \times_Y Z$. Let $\langle \lambda_y \rangle$ be the disintegration of λ with respect to ν. Then $X \times_Y Z$ is the space $(X \times Z, B \otimes D, \mu \times_Y \lambda, T \times T')$, where $\mu \times_Y \lambda = \int (\mu_y \times \lambda_y) d\nu(y)$.

If π and π' are the factor maps for X and Z, then the measure $\mu \times_Y \lambda$ concentrates on $\{(x, z) : \pi(x) = \pi'(z)\}$. Hence we can view this set as the underlying set in the relative product.

It is routine to check that $X \times_Y Z$ is a measure preserving system and that various "universal" properties hold, such as the fact that whenever (W, D, η, T^*) is a measure preserving system and $\pi_1 : W \to X$ and $\pi_2 : W \to Z$ are factor maps that agree on the projections to Y, then there is an essentially unique factor map $\rho : W \to X \times_Y Z$ that commutes with π_1 and π_2.

Relativized Weak Mixing and Compactness

We now relativize the definitions of *ergodic, weakly mixing* and *compact*. Note that by Theorem 87 these definitions are equivalent to the usual definitions when (Y, C, ν, S) is taken to be the trivial measure preserving system.

Definition 95 *Suppose that (Y, C, ν, S) is a factor of (X, B, μ, T).*

- *X is a* relatively ergodic *extension of Y iff every T invariant function in $L^2(X)$ lies in $L^2(Y)$.*

- *X is a* relatively weakly mixing *extension of Y iff $X \times_Y X$ is a relatively ergodic extension of Y. (Compare with Theorem 87.)*

- *X is a* relatively compact *extension of Y iff for a dense set of $f \in L^2(X)$, for all $\epsilon > 0$, there are $g_1, \ldots, g_k \in L^2(X)$ such that for almost* *all y and all n there is an i, such that $\| U_T^n(f) - g_i \|_{\mu(y)} < \epsilon$. Such a function will be called* relatively compact *over Y. (Compare with Definition 85.)*

The latter definition says that the orbit closure of f in $L^2(X)$ is compact in each $L^2(\mu_y)$, uniformly in y.

We give a fairly representative example of a compact extension.

Example 96 *Let (Y, C, ν, S) be a measure preserving transformation and K a compact group (with Haar measure). Let $\phi : Y \to K$ be a measurable function. Define the transformation $T : Y \times K \to Y \times K$ by setting*

$$T(y, k) = (Sy, \phi(y)k).$$

Then T is measure preserving, and (X, B, μ, T) is a compact extension of (Y, C, ν, S). An example of a compact extension of the circle is the transformation on the torus defined by:

$$T(\xi_0, \xi_1) = (\alpha\xi_0, \xi_0\xi_1)$$

where α is an irrational rotation.

The Furstenberg Structure Theorem

Let (X, B, μ, T) be a measure preserving system with factor (Y, C, ν, S). Let $H(x, x')$ be a $T \times_Y T$ invariant function on $X \times_Y X$. Then we get a generalized Hilbert-Schmidt operator $L_H : L^2(X) \to L^2(X)$ by setting:

$$L_H(f)(x) = \int H(x, x')f(x')d\mu_{\pi(x)}(x').$$

The general situation of factors of (X, B, μ, T) that extend (Y, C, ν, S) can be summed up by the following. (Note the analogy to Theorem 88 and Lemma 90.)

- The invariant functions in $X \times_Y X$ are spanned by the weak limits of the averages $\frac{1}{n} \sum_{k=0}^{n-1} U_T^k(T \times_Y T)(g)$ of elements g in $L^2(X \times_Y X)$ of the form $g(x, x') = f(x)\overline{f(x')}$.

- The elements of the ranges of L_H, as H ranges over the bounded invariant functions in $L^2(X \times_Y X)$, form a basis for the relatively compact functions over Y. Moreover the subspace generated by these functions is a direct integral (over Y) of finite dimensional subspaces of $L^2(\mu_y)$.

- The subspace of $L^2(X)$ generated by the relatively compact functions satisfies Lemma 91, and hence corresponds to a factor of X. This factor is the maximal compact factor of X extending Y.

From these we see:

Theorem 97 *If (X, B, μ, T) is an extension of (Y, C, ν, S) that is not relatively weakly mixing, then there is a factor (Z, D, λ, T^*) of (X, B, μ, T) that is a proper compact extension of (Y, C, ν, S).*

This theorem allows us to to exhaust the measure preserving system (X, B, μ, T) by starting at the trivial flow and taking compact extensions until we reach a factor (Y, C, ν, S) such that (X, B, μ, T) is a relatively weak mixing extension of (Y, C, ν, S).

Since this process may not finish in a finite number of steps we must say what to do at limit ordinals. (In general all countable ordinals are needed, see [BF96].)

For this it is convenient to consider factors as subspaces of $L^2(X)$. Given a system of factors $\{Y_i : i \in I\}$ (where I is a directed set and the factor maps commute) with $\mathcal{A}_i$ the subspace of $L^2(X)$ corresponding to $L^2(Y_i)$, the closure of $\bigcup_{i \in I} \mathcal{A}_i$ in $L^2(X)$ satisfies the conditions of Lemma 91, and hence corresponds canonically to a factor of (X, B, μ, T). This factor is isomorphic to the inverse limit of the system $\{Y_i : i \in I\}$.

We define a transfinite sequence of subspaces of $L^2(X)$ corresponding to factors:

Let $L^2(X_0)$ be the constant functions. This corresponds to the trivial flow on a one element set with the atomic probability measure.

If we are given X_α for an ordinal α, we let $X_{\alpha+1}$ be the extension of X_α determined by taking the subspace of $L^2(X)$ generated by the relatively compact functions over X_α.

At limit stages β, we let $\mathcal{A}_\alpha$ be the subspace of $L^2(X)$ corresponding to $L^2(X_\alpha)$, and X_β the factor of X determined by the closure of $\bigcup_{\alpha < \beta} \mathcal{A}_\alpha$.

Since $L^2(X)$ is a separable space, this process must stop at some countable stage (depending on the transformation T). If θ is least such that $X_\theta = X_{\theta+1}$, then it follows that X is a relatively weak mixing extension of X_θ. (Were there invariant functions $H \in L^2(X \times_{X_\theta} X)$ not in $L^2(X_\theta)$, then there would have to be some $f \in L^2(X)$ such that $L_H(f) \notin L^2(X_\theta)$, contradicting $X_\theta = X_{\theta+1}$).

We have outlined the proof of Furstenberg's Structure theorem for measure preserving transformations.

Theorem 98 *(Furstenberg Structure Theorem). Suppose that (X, B, μ, T) is an ergodic measure preserving system. Then there is a countable ordinal θ and a system of factors $\langle X_\alpha : \alpha \leq \theta \rangle$ such that:*

- X_0 is the trivial measure preserving transformation and X is a relatively weak mixing extension of X_θ.

- $X_{\alpha+1}$ is a compact extension of X_α.

- For α a limit ordinal, $X_\alpha = \varprojlim \{X_\beta : \beta < \alpha\}$.

We can now give the definition of *measure distal* using the notation of the theorem above.

Definition 99 *An ergodic measure preserving system* (X, B, μ, T) *is measure distal (or in [Zim76a], generalized discrete spectrum) iff there is a sequence of compact extensions* $\langle X_\alpha : \alpha \leq \theta \rangle$ *such that* $X = X_\theta$.

If (X, B, μ, T) is a measure distal system, then we will call a sequence satisfying the conclusion of the Furstenberg Structure Theorem with $X = X_\theta$ an *approximating tower*.

8.4   Skew Products

In this section we describe in more detail a class of extensions generalizing those in Example 96. Let $\mathbb{K}$ be a compact group and $\mathbb{H}$ a closed subgroup. Then the homogeneous space $\mathbb{K}/\mathbb{H}$ carries a projection of Haar measure and $\mathbb{K}$ acts on the coset space by measure preserving transformations. Let $T : Y \to Y$ be a measure preserving transformation and suppose that $\phi : Y \to \mathbb{K}$ is a measurable function. Then there is a measurable transformation $T_\phi : Y \times \mathbb{K}/\mathbb{H} \to Y \times \mathbb{K}/\mathbb{H}$ defined by

$$T_\phi(y, [k]) = (T(y), [\phi(y)k]).$$

For our purposes (i.e., $\mathbb{Z}$-actions) we will refer to ϕ as a "cocycle", and the transformation T_ϕ as the compact extension induced by ϕ on $\mathbb{K}/\mathbb{H}$. Most of the theory works for arbitrary locally compact groups but the definition of cocycle is more complicated.

Rather surprisingly, this is the general form of compact extensions:

Theorem 100 *Suppose that X is an ergodic compact extension of Y. Then there is a compact group P, a cocycle from Y to P and a homogeneous space P/Q such that X is isomorphic to the compact extension of Y induced by ϕ on P/Q.*

This can be seen along the following lines (see [BMZ97] for details):

- From the definition of a compact extension, we can write $L^2(X)$ as a direct integral over Y of Hilbert space direct sums of finite dimensional spaces $(H_n)_y$, that are invariant under the action of T. In the language of disintegrations, $L^2(\mu_y) = \sum_n (H_n)_y$ and the action of U_T takes $(H_n)_{Sy}$ to $(H_n)_y$.

- It is possible to find "global bases" for the spaces $H_n = \int (H_n)_y d\nu(y)$, i.e., functions $\{f_n^m : m < \dim (H_n)\}$ in $L^2(X)$ in such a way that for almost all y, $\{f_n^m\}$ is an orthonormal basis for $(H_n)_y$. Since T acts by mapping $(H_n)_{Sy}$ to $(H_n)_y$, this gives a cocycle into the unitary group U_l, where $l = \dim (H_n)$. Then there is a natural factor map from X to the induced action of this cocycle on S^l (the l-dimensional sphere). Moreover, since S^l is a homogeneous space of U_l, this factor is a transformation of the form given above.

- There is a natural inverse limit system structure on these factors and taking the limit one gets a representation of X as a homogeneous space of a subgroup $P \subset \prod U_l$ via a map from Y to P.

Definition 101 *Let (Y, T) be a measure preserving transformation. Suppose that ϕ and ψ are cocycles of Y into a compact group $\mathbb{K}$. Then ϕ and ψ are cohomologous or cocycle equivalent iff there is a map $\beta : Y \to \mathbb{K}$ such that*

$$\phi(y)\beta(y) = \beta(Ty)\psi(y).$$

This is an equivalence relation, and tracing diagrams it is easy to see that if ϕ and ψ are equivalent cocycles, then they induce isomorphic transformations on $Y \times \mathbb{K}$. If they induce ergodic transformations then there is a converse to this proved by Zimmer. We outline the statement of the theorem.

A cocycle ϕ from Y into $\mathbb{K}$ induces a cocycle from Y into the group of unitary operators on $L^2(\mathbb{K})$. Since $\mathbb{K}$ is compact, the Peter-Weyl theorem gives $L^2(\mathbb{K})$ as a direct sum of finite dimensional spaces determined by the irreducible subrepresentations of the left regular representation of $\mathbb{K}$ into the group of unitary operators. Composing with ϕ we get a collection of irreducible cocycles into finitary unitary groups. Let $\{\phi_n : n \in \omega\}$ be the collection of irreducible representations associated with ϕ. Then Zimmer [Zim76a] proved the following theorem:

Theorem 102 *Let (Y, T) be an ergodic measure preserving transformation. Suppose that $\phi : Y \to \mathbb{K}$ and $\psi : Y \to \mathbb{H}$ are two cocycles from Y into compact groups, inducing ergodic transformations. If the set of cocycle equivalence classes of $\{\phi_n : n \in \omega\}$ is equal to the set of cocycle equivalence classes of $\{\psi_n : n \in \omega\}$, then the transformations $T_\phi : Y \times \mathbb{K} \to Y \times \mathbb{K}$ and $T_\psi : Y \times \mathbb{H} \to Y \times \mathbb{H}$ are isomorphic.*

Thus the "+" operation applied to the cocycle equivalence relation gives canonical invariants for ergodic compact group extensions, analogous to the invariants in the Halmos-von Neumann theorem.

Definition 103 *Let (X,T) be an ergodic measure distal system. The (X,T) is normal iff there is a sequence of compact group extensions $\langle X_\alpha : \alpha \leq \theta \rangle$ such that $X = X_\theta$.*

Using Theorem 102 one can see that for any given countable ordinal α, the isomorphism relation on the collection of normal transformations of "norm α" is Borel.

9 An Extended Example

In this section we give an example of a natural subset of the group $\mathcal{H}$ of measure preserving transformations which is not Borel. It is thus distinguished from any class of transformations that *is* Borel.

For each measure distal transformation T we can associate the least ordinal that is the length of an approximating tower. We will call this ordinal the *Furstenberg norm* of T.

Theorem 104 *(Beleznay-Foreman) The class of measure distal transformations is a complete $\coprod_1^1$ set. Moreover the Furstenberg norm is a $\coprod_1^1$-norm.*

The following corollary follows immediately from the completeness of the measure distal transformations, the fact that the Furstenberg norm is a $\coprod_1^1$-norm, and the theory of $\coprod_1^1$-sets:

Corollary 105 *For each countable ordinal α, there is a measure distal transformation of Furstenberg norm α.*

Thus every countable ordinal occurs as the least ordinal of an approximating tower to some measure distal transformation. The set of measure distal transformations of norm bounded by α is a Borel set. It follows easily that the collection of transformations of norm exactly α is a Borel set.

In particular, the measure preserving transformations that are distal of norm zero are exactly the discrete spectrum transformations, and thus the discrete spectrum transformations form a Borel set. (A computation shows that it is a $\coprod_3^0$-set.) This pays the remaining debt for the proof of Theorem 65.

While this is beyond the scope of this paper, we note a connection between this result and the ergodic proof of the Szemeredi theorem.

Given a set of positive lower Banach density $A \subset \mathbb{Z}$, we let χ_A be its characteristic function. Let X be the closure of the orbit of χ_A in $2^{\mathbb{Z}}$ under the shift map. A diagonal procedure gives an ergodic invariant measure. The ergodic Szemeredi theorem is then proved by studying the recurrence properties of this system.

By our previous results, the collection of measure preserving transformations of entropy zero is a Borel set. Moreover, by the remarks following Proposition 37, there is a Borel function from the zero entropy transformations into the space of measures on $2^{\mathbb{Z}}$ that assigns to each ergodic zero entropy T a shift invariant measure that is uniquely ergodic and isomorphic to T. Choosing a generic point x in $2^{\mathbb{Z}}$ with respect to μ, and letting $A = \{n : x(n) = 1\}$ we see:

Corollary 106 *For every countable ordinal α there is a set A of positive lower Banach density such that there is a unique ergodic measure on the orbit closure of A and with respect to this measure the shift map is distal and has Furstenberg norm α.*

As a corollary of (the proof of) this theorem we also get the following result:

Corollary 107 *Fix a countable ordinal α. Then there is a Borel function that associates to each ergodic transformation T the largest distal factor of T of norm less than or equal to α.*

We now attempt to exposit some of the main ideas of this theorem and its corollaries.

The issues faced in the proof are:

- The collection of measure distal flows is a $\coprod_1^1$-set.

- The Furstenberg norm is a $\coprod_1^1$-norm.

- The collection of measure distal flows is a complete $\coprod_1^1$-set.

9.1   Some Methods for Computation

We saw in the section on the Furstenberg Structure Theorem that if (Y, C, ν, S) is a factor of (X, B, μ, T), then there is a maximal compact extension of (Y, C, ν, S) inside (X, B, μ, T). (Note that this factor exists by the remarks preceding Theorem 98.) The following lemma is a step towards characterizing this concretely.

Lemma 108 *Let (Y, C, ν, S) be a compact factor of (Y', C', ν', S') which in turn is a factor of (X, B, μ, T). Then (Y', C', ν', S') is the maximal compact factor of (Y, C, ν, S) in (X, B, μ, T) iff $X \times_Y X$ is a relatively weak mixing extension of $Y' \times_Y Y'$.*

Note that one direction of this is easy: a basis for the compact functions is of the form $L_H(f)$ for $H \in L^2(X \times_Y X)$ invariant under $T \times_Y T$. By the fact that $X \times_Y X$ is a relatively weak mixing extension of $Y' \times_Y Y'$, $H \in L^2(Y' \times_Y Y')$, and hence any $L_H(f) \in L^2(Y')$.

The other direction involves computing L_H for an invariant H that lies in $\{L^2(X \times_Y X) \backslash L^2(Y' \times_Y Y')\}$ with respect to an appropriate basis and showing that the result is not in $L^2(Y')$.

We saw in the section on the Furstenberg Structure Theorem that a basis for the invariant elements of $L^2(X \times_Y X)$ can be obtained by weak limits of averages of a basis for $L^2(X \times_Y X)$.

Suppose that $Y' \cong Y \times Z'$, where Z' is a Lebesgue space and $\nu'(y)$ is the disintegration of ν' over ν, and suppose that $X \cong Y' \times Z$ is similar.

Let $\mathcal{H}$ be a basis for $L^2(Y)$ and $\mathcal{G}_1$ be a basis for $L^2(Z')$ and $\mathcal{G}_2$ be a basis $L^2(Z)$. Then $\mathcal{F} = \{f(y)g_1(z')g_2(z) : f \in \mathcal{H}, g_i \in \mathcal{G}_i\}$ can be canonically identified with a basis for $L^2(X)$ and

$$\mathcal{D} = \{f(y)g_1(z_0')g_2(z_0)g_1'(z_1')g_2'(z_1) : f \in \mathcal{H}, g_i, g_i' \in \mathcal{G}_i\}$$

can be identified with a basis for $L^2(X \times_Y X)$.

Let $T^* = T \times_Y T$.

Lemma 109 *Y' is the maximal compact extension of Y in X iff for every $f, \phi_0 \in \mathcal{D}, \phi_1 \in \mathcal{G}_1, \phi_2 \in \mathcal{G}_2$ and $\psi \in \mathcal{F}$ with $\phi_2 \neq 0$:*

$$\frac{1}{n} \sum_{k=0}^{n-1} \int_{Z' \times Z} \int_{Z' \times Z} \int_Y U_{T^*}^m (f) \overline{\phi_0(y)\phi_1(z')\phi_2(z)} \psi(y, z', z) \to 0.$$

Note that quantifying over $\mathcal{D}, \mathcal{G}_1, \mathcal{G}_2, \mathcal{F}$ is quantifying over a countable set.

The following lemma yields a universal property of the maximal compact extensions:

Lemma 110 *Suppose that (Z, D, λ, S') is a factor of (Y, C, ν, S) and that (Y, C, ν, S) is a factor of (X, B, μ, T). Suppose that $f \in L^2(X)$ is compact over (Z, D, λ, S') and $f \notin L^2(Y)$. Then f is compact over (Y, C, ν, S).*

In particular, it shows that if Z' is the maximal compact extension of Z and Y is any extension of Z, then there is a dense collection of functions in $L^2(Z')$ that are compact over Y. Thus Z' is a factor of the maximal compact extension Y' of Y.

The next proposition relates some of these ideas to their complexity:

Proposition 111 *Let X be an extension of Y and suppose that $M \supset L^2(Y)$ is a closed subspace of $L^2(X)$. Then the following statements are Borel (in M, X, Y):*

1. *There is a T^*-invariant $H \in L^2(X \times_Y X)$ and $f \in L^2(X)$ such that $L_H(f) \notin M$.*

2. *M is spanned by $\{L_H(f) : H \in L^2(X \times_Y X) \text{ is } T^*\text{-invariant}, f \in L^2(X)\}$.*

⊢   Let $\mathcal{F}$ be a basis for $L^2(X)$. Then (as above) the first statement is equivalent to the statement that there are $f, g, h \in \mathcal{F}$ such that if H is the weak limit of $\frac{1}{n} \sum_{k=0}^{n-1} U_{T^*}^k (f \otimes g)$, then $L_H(h) \notin M$. To show that this is a Borel condition, we use Theorem 7, e.g., we show that this condition is both Σ_1^1 and Π_1^1.

We note that by the von Neumann ergodic theorem the weak limit always exists and is unique. Hence the statement is equivalent to both of the following statements (individually):

- There are $f, g, h \in \mathcal{F}$ and there is an $H \in L^2(X \times_Y X)$ such that H is the weak limit of $\frac{1}{n} \sum_{k=0}^{n-1} U_{T^*}^k (f \otimes g)$ and $L_H(h) \notin M$.

- There are $f, g, h \in \mathcal{F}$ such that for all $H \in L^2(X \times_Y X)$, if H is the weak limit of $\frac{1}{n} \sum_{k=0}^{n-1} U_{T^*}^k (f \otimes g)$, then $L_H(h) \notin M$.

The first form is Σ_1^1, the second Π_1^1.

Similarly if $\mathcal{G}$ is a basis for M, then the second statement is equivalent to the statement that every element of $\mathcal{G}$ can be approximated arbitrarily closely by rational combinations of some $L_H(f)$'s. It follows as in the previous statement that this is Borel.   ⊣

9.2   Maximal Towers

Recall that the collection of acceptable linear orderings, $\mathcal{ALO}$, is the subset of $\mathcal{LO}$ consisting of those linear orderings with a least and greatest element that are such that every element (except the last) has an immediate successor. For convenience we will assume that the least element is always 0 and if $i \in I$ we will denote its immediate successor by $i + 1$ and the maximal element will called i_{max}.

We generalize our notion of approximating tower to any sequence of factors. Let I be an acceptable linear ordering and $\langle X_i : i \in I \rangle$ a sequence of factors with a commuting family of factor maps making X_i a factor of X_j if $i < j$. Then $\langle X_i : i \in I \rangle$ is an *approximating tower* iff the following three conditions hold:

- X_0 is the trivial flow, $X_{i_{max}} = X$,

- For all $i \neq i_{max}, X_{i+1}$ is a compact extension of X_i,

- If $i \in I$ is not a successor element, then X_i is the inverse limit of X_j for $j < i$.

We will call an approximating tower $\langle X_\alpha : \alpha \leq \theta \rangle$ *maximal* iff for all $\alpha < \theta$, $X_{\alpha+1}$ is the maximal compact extension of X_α in X.

Lemma 112 *Let $I \in \mathcal{ALO}$ and $T \in \mathcal{H}$. Then the statement "$\langle X_i : i \in I \rangle$ is a maximal approximating tower" is a Borel condition (in I, T and $\langle X_i : i \in I \rangle$).*

⊢ We show that the equivalent statements in $L^2(X)$ are Borel. We again fix a dense $\mathcal{F} \subset L^2(X)$. We verify that each of the requirements in the definition of being a maximal approximating tower are Borel.

To say that X_0 is the trivial flow is to say that $L^2(X_0) = \mathbb{C}$. To say that $X_{i_{max}}$ is X is to say that every element of the canonical basis for $L^2(X)$ lies in $L^2(X_{i_{max}})$.

To say that if $i \neq i_{max}$, X_{i+1} is the maximal compact extension of X_i, we say that if $i \neq i_{max}$, then $L^2(X_{i+1})$ contains all of the appropriate $L_H(f)$'s and that it is spanned by such functions. By Proposition 111 this is Borel.

The third condition on being an approximating tower is that if i is a limit point in the ordering I, then X_i is the inverse limit of X_j for $j <_I i$. This is equivalent to the statement that $\bigcup_{j <_I i} L^2(X_j)$ is dense in $L^2(X_i)$.   ⊣

Lemma 113 *Suppose that X is measure distal. Let $\langle Y_\beta : \beta \leq \gamma \rangle$ be a sequence of compact extensions. Then X is not a (non-trivial) weak mixing extension of Y_γ. Further there is a unique maximal approximating tower and the length of this tower gives the Furstenberg norm of X.*

⊢ Let $\langle X_\alpha : \alpha \leq \theta \rangle$ be an approximating tower of X, and, towards a contradiction, suppose that X is a non-trivial weak mixing extension of Y_γ. Let α be minimum such that there is a function $f \in L^2(X_{\alpha+1}) \backslash L^2(Y_\gamma)$. (Such a function exists since $L^2(X)$ is not a subset of $L^2(Y_\gamma)$.) Then $L^2(X_\alpha) \subset L^2(Y_\gamma)$, and so, by Lemma 110, f is compact over $L^2(Y_\gamma)$. This contradicts the statement that X is a relatively weakly mixing extension of Y_γ.

In particular this shows that the system of factors obtained by taking the maximal compact extension at each successor stage eventually leads to X and hence, if X is measure distal, a maximal approximating tower exists.

We show that there is a unique maximal approximating tower. Note that if $\langle X_\alpha : \alpha \leq \theta \rangle$ is maximal, and $\langle X'_\beta : \beta \leq \theta' \rangle$ is an arbitrary approximating tower (with $\theta, \theta' \in OR$), then by Lemma 110, for all $\beta \leq \theta, X'_\beta$ is a factor of X_β. If both towers are maximal, then by symmetry, $X_\beta = X'_\beta$ and $\theta = \theta'$.

A similar argument shows that if $\langle Y_\beta : \beta \leq \gamma \rangle$ is an approximating tower for X, then for all $f \in L^2(X)$, if $f \in L^2(Y_\beta)$, then $f \in L^2(X_\beta)$. Hence θ is the minimal length of an approximating tower.   ⊣

We note that the proof of this theorem also shows that:

Proposition 114 X *is measure distal iff for every proper factor Y, there is a compact function in $L^2(X)$ over Y*

The next lemma is essential for one direction of our reduction of $\mathcal{WO}$ to the distal transformations.

Lemma 115 *If (X, B, μ, T) is measure distal and $\langle X'_i : i \in I \rangle$ is a maximal approximating tower on an acceptable linear ordering I, then I is a well-ordering.*

⊢   Let $\langle X_\alpha : \alpha < \theta \rangle$ be the canonical maximal approximating tower (with $\theta \in OR$). Let I^* be the maximal well-ordered initial segment of I. Then I^* is isomorphic to some ordinal α_0, and the tower $\langle X'_i : i \in I^* \rangle$ is level by level identical to $\langle X_\beta : \beta < \alpha_0 \rangle$.

If I^* is not all of I, we show by induction on $\alpha \geq \alpha_0$, that X_α is a factor of X'_i for all $i \in I \backslash I^*$.

Suppose that we know this for α, we show it for $\alpha + 1$. Since $i \in I \backslash I^*$, there is a $j < i, j \in I \backslash I^*$. By the induction hypothesis, X_α is a factor of X'_j. Since X'_{j+1} is the maximal extension of X'_j, by Lemma 110, $X_{\alpha+1}$ is a factor of X'_{j+1}, which in turn is a factor of X'_i. Hence $X_{\alpha+1}$ is a factor of X_i.

Suppose now that α is a limit and for all $\beta < \alpha, X_\beta$ is a factor of X'_i. Then for all $\beta < \alpha, L^2(X_\beta) \subset L^2(X'_i)$. Hence $L^2(X_\alpha) \subset L^2(X'_i)$ and thus X_α is a factor of X'_i.   ⊣

9.3   The Measure Distal Flows are $\underset{\sim}{\Pi}^1_1$

We now show:

Theorem 116 *The collection of measure distal transformations is a $\underset{\sim}{\Pi}^1_1$ subset of $\mathcal{H}$.*

Using Proposition 114 one can check:

Proposition 117 *Fix a dense subset $\mathcal{F}$ of $L^2(X)$. The following are equivalent:*

1. X is measure distal.

2. For all proper subspaces $M \subsetneq L^2(X)$ satisfying Lemma 91 (i.e., invariant, and conjugacy, multiplication and truncation closed) there are $f, g, h \in \mathcal{F}$ such that if H is the weak limit of

$$\frac{1}{n} \sum_{k=0}^{n-1} U_{T^*}^k (f \otimes g),$$

then $L_H(h) \notin M$.

We now indicate that the second condition is a $\coprod_1^1$ condition. We first note that the form of the condition is:

$$(\forall \text{subspaces } M)(A(M) \Rightarrow B(M))$$

Hence it suffices to show that the quantifier "$\forall$subspaces M" is a quantifier over a Polish space and the conditions $A(M)$ and $B(M)$ are Borel conditions on M.

The fact that $B(M)$ is a Borel condition is the content of Proposition 111. We can quantify over elements of $L^2(X)$ by quantifying over convergent sequences of elements of $\mathcal{F}$. Thus, subspaces of $L^2(X)$ are determined by a countable collection of sequences of elements of $\mathcal{F}$. Thus the quantifier "for all subspaces" is a universal quantifier over sequences of sequences of elements of a countable set, and hence a universal quantifier over a Polish space.

To say that a subspace generated by a countable collection of convergent sequences of elements of $\mathcal{F}$ is proper is to state that there is some n such that all rational combinations of the convergent sequences are at least $\frac{1}{n}$ away from an element of the standard basis $\{e_i\}$. This is clearly Borel.

To say that M is invariant, and conjugacy, multiplication and truncation closed is also Borel. For example, we show that the condition of being conjugacy closed is Borel. It clearly suffices to show that if f is one of the generating elements of M, then $\overline{f} \in M$. For this we say that for every element of the generating set for M, and all n, there is a rational combination of elements of the generating set that is within $\frac{1}{n}$ of $\overline{f}$. Since M is presented to us by way of its generating set, and this is countable, all of the quantifiers are over a countable set and this condition is Borel. Hence, the condition $A(M)$ is a Borel condition on sequences of sequences of elements of $\mathcal{F}$.

9.4 The Furstenberg Norm

Theorem 118 *The Furstenberg norm is a $\coprod_1^1$-norm.*

⊢   Let $o(Y)$ be the length of a maximal approximating tower for Y, if Y is measure distal, and ∞ otherwise. We want to show that o is a $\prod_1^1$ norm. For this we must see that "$o(X) \leq o(Y)$" and "$o(X) < o(Y)$" are both $\prod_1^1$ conditions.

The first follows by noting that $o(X) \leq o(Y)$ iff X is distal and for all maximal approximating towers $\mathcal{T}$ for X and $\mathcal{T}'$ for Y there is no order preserving embedding from the length of $\mathcal{T}'$ to a proper initial segment of the length of $\mathcal{T}$.

We have seen that the statement that X is distal is $\prod_1^1$. If I is an element of $\mathcal{ALO}$, then by Lemma 112, the statement "$\mathcal{T} = \langle X_i : i \in I \rangle$ is a maximal approximating tower" is Borel in the sequence $\mathcal{T}$. Similarly for $\mathcal{T}' = \langle Y_j : j \in J \rangle$. To say that there is no embedding from the length of I to the length of J is the universal statement that for all $f : I \to J$, either f is not order preserving or f is cofinal in J. This latter is $\prod_1^1$.

Hence we see that the form of this statement is:

$(\forall \mathcal{T}, \mathcal{T}')(\mathcal{T}, \mathcal{T}'$ are maximal approximating towers for X and $Y \Rightarrow$

$(\forall f)(f$ is not an order preserving map from I to $J))$.

This is of the form:

$$\forall \mathcal{T}, \mathcal{T}'(A(\mathcal{T}, \mathcal{T}') \Rightarrow B(I, J)).$$

Since A is Borel by Lemma 112 and B is $\prod_1^1$, the whole statement is $\prod_1^1$.

Since $o(X) < o(Y)$ iff X is measure distal, and whenever I and J are the lengths of maximal approximating towers to X and Y respectively, there is no order preserving embedding of J into I, a similar argument shows that this relation is $\prod_1^1$.

Corollary 107 is now proved very similarly to the Lemma 112.

9.5   The Measure Distal Flows are Complete $\underset{\sim}{\prod_1^1}$.

To show this we must find a complete $\underset{\sim}{\prod_1^1}$ set $W \subset {}^\omega \omega$, and a Borel function $R : {}^\omega \omega \to \mathcal{H}$ such that for all $x \in {}^\omega \omega, x \in W$ iff $R(x)$ is measure distal. It is quite natural here that we will take $W = \mathcal{WO}$, the collection of well orderings. We saw in Section 6, "The Structure of $\prod_1^1$-sets", that this is a complete co-analytic set.

We will do more: if I is a well-ordering isomorphic to an ordinal α, we will arrange that $R(I)$ is measure distal of norm α.

In our discussion we will be somewhat ambivalent about the way we view a linear ordering I. For most of the discussion we will view I as a set I with a linear ordering $<_I$. We will refer to elements of I as i and refer to $i <_I j$ when we discuss the ordering itself.

On the other hand, our official definition of $\mathcal{ALO}$ is as a collection of functions $I : \omega \times \omega \to \{0, 1\}$. When we are taking this view we will identify I as an ordering of ω. In particular, this allows us to view I as a limit of finite partial orderings $I \upharpoonright N = \{(n, m) : n <_I m \text{ and } n, m < N\}$. With this view, $I \upharpoonright (N + 1)$ adds a new element to the ordering $I \upharpoonright N$, by sticking the point $\{N\}$ into the ordering somewhere.

Given an acceptable linear ordering I, the reduction R will construct a transformation on the infinite dimensional torus, $\mathbb{T}^\omega$. This will be constructed continuously in $I \upharpoonright N$.

If I is an acceptable linear ordering, then we will order $\mathbb{T}^\omega$ as $\mathbb{T}^I$ by ordering the index set by I. It will turn out from the form of the transformations $R(I)$, that if we take $X_i = \mathbb{T}^{<i}$ to be $\mathbb{T}^A$, where $A = \{j : j <_I i\}$, then X_i is a factor of $X = \mathbb{T}^I$. Moreover, it will be the case that X_{i+1} will be the skew product of X_i with $\mathbb{T}$ by a measurable map.

It follows from these remarks that if $i \in I$ is a limit, then X_I is the inverse limit of $\{X_j : j <_I i\}$, and that X_{i+1} is a compact extension of X_i, i.e., that the sequence of factors $\langle X_i : i \in I \rangle$ is an approximating tower for X. The work will be to show that this approximating tower is always maximal.

Lemma 119 *Suppose that for all $I \in \mathcal{ALO}, R(I)$ is a measure preserving transformation on $X = T^I$, so that $\langle X_i : i \in I \rangle$ is a maximal approximating tower. Then R is a reduction of WO to the measure distal transformations.*

To see this we note that if I is a well-ordering, then, by Lemma 113, $R(I)$ is distal of norm the ordinal of I. If I is not a well-ordering, then again Lemma 115 shows that $R(I)$ is not measure distal.

We must construct transformations of arbitrary ordinal height. We will focus on how to do this, and afterwards remark that we have made a construction that depends continuously on I, and hence is a reduction.

Finite Norm Distal Flows

We begin by describing the classical distal flows on the torus. These have all finite norms, and we will have to generalize them in a non-trivial way to get infinite norm transformations. We will switch to additive notation and identify the unit circle with addition on the unit interval modulo one. We will write $\oplus$ for this operation, when it is important to distinguish it from ordinary addition.

These are of the form $T_{\vec{k}} : \mathbb{T}^n \to \mathbb{T}^n$, where T is defined by:

$$T_{\vec{k}}(\xi_0, \xi_1, \ldots, \xi_{n-1}) =$$

$$(\alpha \oplus \xi_0, k_0\xi_0 \oplus \xi_1, \ldots, k_{n-2}\xi_{n-2} \oplus \xi_{n-1}),$$

where α is irrational and $\vec{k} \in \{\mathbb{Z}\backslash 0\}^{n-2}$.

Lemma 120 $T_{\vec{k}}$ *is ergodic, and for all* $m < n$, $\mathbb{T}^{m+1}$ *is the maximal compact extension of* $\mathbb{T}^m$ *in* $\mathbb{T}^n$.

We start by showing that these flows are ergodic. We will give the style of argument from our paper, which is a variation on the standard proof.

Let $e_k(\theta) = exp(2\pi i k\theta)$, for $\theta \in [0,1]$ and $k \in \mathbb{Z}$. The functions e_k are a basis for $L^2(\mathbb{T})$ and products of these functions form a basis for $L^2(\mathbb{T}^n)$. We show by induction on n, that for all $\vec{k} \in (\mathbb{Z}\backslash\{0\})^n$, $T_{\vec{k}}$ is ergodic.

For $n = 1$, suppose that $f \in L^2(X)$ is an invariant function. Expand f by its Fourier series $f(\theta) \sim \sum b_i e_i(\theta)$. Then:

$$\begin{aligned} f = U_T(f) \quad &\sim \quad \sum b_i e_i(\theta + \alpha) \\ &= \sum e_i(\alpha) b_i e_i(\theta) \\ &= \sum b_i e_i(\theta). \end{aligned}$$

The latter equality holds, since $f = U_T(f)$.

But this shows that for all $i, b_i = b_i e_i(\alpha)$. Since α is irrational, $e_i(\alpha) \neq 1$ unless $i = 0$. Hence for all $i \neq 0, b_i = 0$. But this shows that f is constant.

Now suppose that all skew products of $\mathbb{T}^{n-1}$ with non-zero integer coefficients are ergodic. Suppose that $k_{n-2} \neq 0$.

Let f be an invariant function and expand f with respect to the last coordinate:

$$f \sim \sum_i g_i(\xi_0, \ldots, \xi_{n-2}) e_i(\xi_{n-1}),$$

where $g_i \in L^2(\mathbb{T}^{n-1})$.

Applying T we see

$$\begin{aligned} U_T(f) \quad &\sim \quad \sum g_i(T(\xi_0, \ldots, \xi_{n-2})) e_i(k_{n-2}\xi_{n-2} + \xi_{n-1}) \\ &= \sum_{i \neq 0} g_i(T(\vec{\xi})) e_{ik_{n-2}}(\xi_{n-2}) e_i(\xi_{n-1}) \\ &\quad + g_0(T(\vec{\xi})) \\ &= \sum g_i(\vec{\xi}) e_i(\xi_{n-1}). \end{aligned}$$

For $i = 0, g_0(\vec{\xi}) = g_0(T(\vec{\xi}))$, hence g_0 is an invariant function of $n - 1$ variables, and hence is constant.

For all $i \neq 0$, we have

$$g_i(\vec{\xi}) = g_i(T(\vec{\xi})) e_{ik_{n-2}}(\xi_{n-2}). \tag{$\dagger$}$$

Let $\vec{\xi}' = (\xi_0, \ldots, \xi_{n-3})$ (unless $n = 2$ in which case g_i is a function of one variable). Note that $T(\vec{\xi})_{n-2} = \xi_{n-2} + F(\vec{\xi}')$. Expand g_i in terms of ξ_{n-2}:

$$
\begin{aligned}
g_i &\sim \sum_j h_j(\vec{\xi}')e_j(\xi_{n-2}) \\
&= \left\{ \sum_j h_j(T(\vec{\xi}'))e_j(T(\xi)_{n-2}) \right\} e_{ik_{n-2}}(\xi_{n-2}) \\
&= \left\{ \sum_j h_j(T(\vec{\xi})) \left[e_j(F(\xi'))e_j(\xi_{n-2}) \right] \right\} e_{ik_{n-2}}(\xi_{n-2}) \\
&= \sum_j h_j(T(\vec{\xi}'))e_j(F(\xi'))e_{j+ik_{n-2}}(\xi_{n-2}).
\end{aligned}
$$

Then $h_j(\xi) = h_{j-(ik_{n-2})}(T(\vec{\xi}'))e_j(F(\xi'))$, in particular $\| h_j \| = \| h_{j-(ik_{n-2})} \|$. But we must have $\| h_j \| \to 0$, by the Riemann-Lebesgue lemma. The only way this can happen is if $\| h_j \| = 0$, for all j.

The equation (†) is called a *cocycle equation*.

Remark We must also calculate norms. For this, by Lemma 108 it suffices to show that $\mathbb{T}^n \times_{\mathbb{T}^m} \mathbb{T}^n$ is a relative ergodic extension of $\mathbb{T}^{m+1} \times_{\mathbb{T}^m} \mathbb{T}^{m+1}$. Failure of relative ergodicity also gives a form of a cocycle equation for an invariant function on $\mathbb{T}^n \times_{\mathbb{T}^m} \mathbb{T}^n$. From this and induction on n, one can find an invariant function whose expansion just mentions one power $e^{2\pi i k z_{n-1}}$ for each copy of $\mathbb{T}^n$ (where z_{n-1} is the last coordinate of an element of $\mathbb{T}^n$.)

Such an invariant function would have to be a weak limit of averages, as in Lemma 109. That these averages tend to zero, if there is non-trivial dependence on z_n, is then established directly by integrating.

Skew Products with Fractional Coefficients

If we simply generalized this construction to norm ω it would indeed give us a measure distal transformation of norm ω. However, if we want a measure distal transformation of norm $\omega + 1$, we must take a skew product of $\mathbb{T}^\omega$ with $\mathbb{T}$ by a measurable function ϕ. In order that the last coordinate not be swallowed by a maximal compact extension at a finite stage, we must make sure that ϕ depends non-trivially on infinitely many coordinates in $\mathbb{T}^\omega$. For this reason, we must allow fractional coefficients in the definition of the skew product. Even in finite dimensions this is somewhat problematical.

Let $\vec{t} = \langle t_i^j : j < i \le n \rangle$ be numbers between 0 and 1. Define the transformation $T_{\vec{t}} : \mathbb{T}^n \to \mathbb{T}^n$ by the formula:

$$
(T_{\vec{t}}(\vec{\xi}))_i = \xi_i \oplus \sum_{j<i} t_i^j \xi_j.
$$

Theorem 121 *For all n there is a set of measure one $B \subset \mathbb{R}^n$ such that for all choices $\vec{t}$ from $B \cup \mathbb{Q}^n$:*

- *$T_{\vec{t}}$ is ergodic.*

- *$\mathbb{T}^{i+1}$ is the maximal compact extension of $\mathbb{T}^i$.*

This theorem is proved using two lemmas. The first shows that the analogue of equation (†) is equivalent to the non-ergodicity of $T_{\vec{t}}$. (The function "exp" is the exponential function.)

Lemma 122 *Let $T_{\vec{t}} : \mathbb{T}^{n+1} \to \mathbb{T}^{n+1}$ be as above. Suppose that $T_{\vec{t}} \upharpoonright \mathbb{T}^n$ is ergodic. Then $T_{\vec{t}}$ is ergodic iff there is no measurable, non-zero $c_k : \mathbb{T}^n \to \mathbb{C}$ satisfying:*

$$c_k(x_0, \ldots, x_{n-1}) =$$

$$c_k(T_{\vec{t}}(x_0, \ldots, x_{n-1}))\, exp(2\pi i k \sum_{j=0}^{n-1} t_n^j x_j).$$

Note that if there is a c_k as in this lemma, then $|c_k|$ is an invariant function on $\mathbb{T}^n$, and hence is constant almost everywhere. Since c_k is non-zero, we have that c_k is almost never equal to zero.

The next lemma says that the c_k's satisfying the cocycle equation are invariant under certain arithmetic operations.

Lemma 123 *If c_k and d_k are witnesses to the equations above for $\vec{s}$ and $\vec{t}$, then $c_k d_k$ and c_k/d_k are witnesses for $\vec{s} + \vec{t}$ and $\vec{s} - \vec{t}$ respectively.*

Suppose now that there is either a rational counterexample or a positive measure set of counterexamples to Theorem 121. Then Lemma 123 gives a counterexample where the last coefficient vector $\vec{t}_n$ consists of integers. (We can add and subtract counterexamples. The closure of a positive measure set under addition contains an integer vector.) Now an argument similar to the one for purely integer coefficients works.

The assertion that $\mathbb{T}^{i+1}$ is the maximal compact extension of $\mathbb{T}^i$ is proved using more complicated cocycle arguments in somewhat the same spirit.

Approximating Infinite Dimensional Transformations

For an acceptable linear ordering I, we will build a transformation

$$T = T_I : \mathbb{T}^I \to \mathbb{T}^I.$$

For $j \in I, \vec{x} \in \mathbb{T}^I$, we will write $(\vec{x})_j$ for the jth coordinate of $\vec{x}$. The form of T will be:

1. $T(\vec{x})_0 = \vec{x}_0 \oplus \alpha$, for some fixed irrational α,

2. $T(\vec{x})_j = (\vec{x})_j \oplus \sum_{i <_I j} t^i_j(\vec{x})_i$,

for some sequences $\langle t^i_j : i <_I j \rangle$ of rational numbers.

The sequences t^i_j will be essentially arbitrary as long as they tend to zero fast enough (but this must be shown).

We will be given the linear ordering I as increasing finite pieces $I \restriction N$. At stage N, we will have defined a measure preserving map $T_N : \mathbb{T}^N \to \mathbb{T}^N$ corresponding to the finite linear ordering I gives of the first N elements of ω. At stage $N + 1$, a new element of I will be inserted into this ordering and some more t^i_j's must be defined. (To be precise, if $i = N$ is the new element added to the ordering, then we must define t^N_j for all j with $j <_I N$ and the t^j_N for $N <_I j$.)

By Theorem 121, each T_N is ergodic and has well behaved maximal compact extensions. The transformation T_I is built as a limit of T_N's, and it must be shown that these properties are preserved by the limit.

This is done by showing that if the t^i_j's decay fast enough, then the criterion given by Proposition 24 is preserved in the limit.

Note that if j is the I-successor of i, then there is a large enough N such that for all $N' > N$, j is the $I \restriction N'$-successor of i. Hence $T^{<j}$ is the maximal compact extension of $T^{<i}$ for $T_{I \restriction N'}$. A condition equivalent to the relative ergodicity hypothesis in Lemma 108 is preserved in the limits, provided again that the t^i_j go to zero fast enough. This condition is technical and the reader is referred to [BF96].

10 Topologically Distal Transformations

We now briefly discuss Theorem 46.

In [Fur63] Furstenberg showed that there is a nice structure theorem for distal transformations of compact metric spaces. This was extended by Ellis [Ell78], Veech, and others to more general contexts such as the case of compact Hausdorff spaces. This area is surveyed by Glasner in his paper in this volume; [Gla00]. We refer the reader to that paper for details.

In these structure theorems, there is the analogous notion of a *factor*, where the maps must be continuous. (A slight disanalogy is that we expect the maps to be defined everywhere, instead of just on a comeager set.) The structure theorems are of the same form as the Furstenberg structure theorem for measure distal flows (indeed the structure theorem for the topological case was proved first and the latter result was modeled on the result for topologically distal transformations.)

In these results, the notion of a compact extension is replaced by the notion of an isometric, or equicontinuous extension. Very briefly, the idea of disintegrating $L^2(X)$ over Y, and requiring the transformation $T : X \to X$ to act on the fibers in a way to preserve some finite dimensional $L^2(Y)$-modules, is replaced by the requirement that X be homeomorphic to $Y \times Z$, where Z is compact Hausdorff, and T sends fibers over Y to fibers over Y equicontinuously.

The structure theorem then is:

Theorem 124 *Let X be a compact Hausdorff space and $T : X \to X$ be a minimal distal transformation. Then there is an ordinal θ and a commuting sequence of factors $\langle X_\alpha : \alpha < \theta \rangle$ such that:*

- *X_0 is the trivial transformation and X is the inverse limit of $\langle X_\alpha : \alpha < \theta \rangle$.*

- *If $\alpha < \theta$, then $X_{\alpha+1}$ is an isometric extension of X_α.*

- *If α is a limit ordinal, then X_α is the inverse limit of $\langle X_\beta : \beta < \alpha \rangle$.*

The theory contains many analogous results to the theory in the measure case outlined above, such as a result implying the existence of maximal equicontinuous extensions. In particular, there is an analogous notion of an approximating tower. (Glasner calls these I-systems in [Gla00].)

The actual theorem that has Theorem 46 as a consequence is:

Theorem 125 *Let $\mathcal{K}_0$ be the collection of compact subsets of the square of the Hilbert Cube. Then*

1. *$\{T : T$ is a distal homeomorphism $\}$ is a complete $\coprod_1^1$-set.*

2. *The function $T \mapsto o(T)$, where $o(T)$ is the least ordinal of an approximating tower for T, maps from the collection of distal flows into the countable ordinals and is a $\coprod_1^1$-norm.*

This theorem has several corollaries, including some consequences for the theory of certain function spaces (see, e.g., [Kna67]).

In addition a Löwenheim-Skolem argument gives the following theorem:

Theorem 126 *Let (T, X) be the universal minimal distal transformation for compact Hausdorff spaces. Then the maximal approximating tower for (T, X) has length ω_1, the least uncountable ordinal.*

An easy corollary of the proof of this theorem is the fact that every minimal flow from a separable group action on a compact Hausdorff space is an inverse limit of flows on compact metric spaces.

In Glasner's article, a structure theorem for arbitrary minimal transformations on compact Hausdorff is described, in terms of a transfinite procedure. It differs from the one for distal transformations (among other ways) in that the transfinite tower associated with a minimal transformation does not necessarily consist of factors of the original transformation. We note that a similar theorem to 126 holds for the length of these transfinite towers as well. Namely they have length at most ω_1 and there are examples of every countable length, as well as ω_1.

The basic outline of the proof of these theorems is similar to the one given above for Theorem 104. Since the notion of maximal isometric extension is a topological one, the techniques of construction must be different.

Definition 127 *Let F be a Polish space, and $\mathcal{D}$ be a countable collection of dense open subsets of F. A point $f \in F$ is said to be $\mathcal{D}$-generic iff $f \in \bigcap \mathcal{D}$. A property is said to be generic iff there is a set $\mathcal{D}$ such that it holds for all $\mathcal{D}$-generic points f.*

Let X be compact Hausdorff, and suppose that $T : X \to X$ is a homeomorphism. If $\mathbb{K}$ is a group and $f : X \to \mathbb{K}$ is a continuous function, we can form the skew product $T_f : X \times \mathbb{K} \to X \times \mathbb{K}$ by setting

$$T_f(x, k) = (T(x), f(x)k).$$

We will call the resulting skew product $X \times_f \mathbb{K}$. If T is distal and $\mathbb{K}$ is compact, this is a minimal distal homeomorphism and $X \times_f \mathbb{K}$ is an equicontinuous extension of X.

A typical result proved in [BF95] is:

Lemma 128 *Let X, Y be compact metric spaces and $\mathbb{K}$ a compact group. Suppose that $T : X \to X$ is distal and Y is a factor of (T, X) with maximal compact extension Y' inside X. Then for generic $f : X \to \mathbb{K}$, Y' is still the maximal compact extension of Y inside $X \times_f \mathbb{K}$.*

This Lemma allows transfinite constructions by generic skew products. In particular it implies that for a sufficiently generic f the distal tower of $X \times_f \mathbb{K}$ extends the distal tower for X by one step.

11 Some Questions

In this section we list some questions that are of interest to the author. The first few questions deal with "classification" of measure preserving systems:

1. Is the isomorphism relation on the ergodic measure preserving transformations Borel?

2. Is the isomorphism relation on the K-automorphisms a Borel relation? Is it reducible to the relation $=^{+}$? Is it reducible to E_0?

3. Is the isomorphism relation on the measure distal transformations the restriction of a Borel subset of $\mathcal{H} \times \mathcal{H}$ to {measure distal} × {measure distal}?

4. Let $\mathcal{S}$ be the collection of transformations that are isomorphic to smooth transformations that preserve a measure on a compact smooth manifold (in the same measure class as Lebesgue measure). How complicated is the isomorphism relation on $\mathcal{S}$?

5. What is the complexity of $\mathcal{S}$? Is it Borel?

6. Are there examples of distal elements of $\mathcal{S}$ that have infinite Furstenberg norm?

   Note that a negative answer to either of the questions above would give an example of a finite entropy (or even a measure distal in the second case) transformation that is not isomorphic to an element of $\mathcal{S}$. This is also true of an affirmative answer to the next question:

7. Is there a Borel function $M : \mathcal{S} \to \mathcal{H}$ such that for all $T \in \mathcal{S}, M(T)$ is the maximal distal factor of T? Perhaps there is a way of calculating "separating sieves" in a concrete way for transformations on compact manifolds.

References

[Add59] John W. Addison. Separation principles in the hierarchies of classical and effective descriptive set theory. *Fund. Math.*, 46:123–135, 1959.

[BF95] Ferenc Beleznay and Matthew Foreman. The collection of distal flows is not Borel. *Amer. J. Math.*, 117(1):203–239, 1995.

[BF96] Ferenc Beleznay and Matthew Foreman. The complexity of the collection of measure-distal transformations. *Ergodic Theory and Dynam. Systems*, 16(5):929–962, 1996.

[BK96] Howard Becker and Alexander S. Kechris. *The descriptive set theory of Polish group actions.* London Math Soc. Lecture Note Series, Vol. 232. Cambridge University Press, Cambridge, 1996.

[BMZ97] Vitaly Bergelson, Randall McCutcheon, and Qing Zhang. A Roth theorem for amenable groups. *Amer. J. Math.*, 119(6):1173–1211, 1997.

[Con85] John B. Conway. *A course in functional analysis.* Springer-Verlag, New York, 1985.

[DGS76] Manfred Denker, Christian Grillenberger, and Karl Sigmund. *Ergodic theory on compact spaces.* Lecture Notes in Mathematics, Vol. 527. Springer-Verlag, Berlin, 1976.

[DJK] Randall Dougherty, Steven Jackson and Alexander S. Kechris. The structure of hyperfinite Borel equivalence relations. *Trans. Amer. Math. Soc.*, 341(1):193-225, 1994.

[Ell78] Robert Ellis. The Furstenberg structure theorem. *Pacific J. Math.*, 76(2):345–349, 1978.

[End77] Herbert B. Enderton. *Elements of set theory.* Academic Press, New York, 1977.

[Fel74] Jacob Feldman. Borel structures and invariants for measurable transformations. *Proc. Amer. Math. Soc.*, 46:383–394, 1974.

[Fur63] Hillel Furstenberg. The structure of distal flows. *Amer. J. Math.*, 85:477–515, 1963.

[Fur81] Hillel Furstenberg. *Recurrence in ergodic theory and combinatorial number theory.* M. B. Porter Lectures, Princeton University Press, Princeton, N.J., 1981.

[Gla00] Eli Glasner. *Structure theory as a tool in topological dynamics.* In Matthew Foreman, Alexander S. Kechris, Alain Louveau, and Benjamin Weiss, editors, This Volume. Cambridge University Press, 2000.

[Hal60] Paul R. Halmos. *Lectures on ergodic theory*. Chelsea Publishing Co., New York, 1960.

[HKL90] Leo A. Harrington, Alexander S. Kechris, and Alain Louveau. A Glimm-Effros dichotomy for Borel equivalence relations. *J. Amer. Math. Soc.*, 3(4):903–928, 1990.

[HKL98] Greg Hjorth, Alexander S. Kechris, and Alain Louveau. Borel equivalence relations induced by actions of the symmetric group. *Ann. Pure Appl. Logic*, 92(1):63–112, 1998.

[HR79] Edwin Hewitt and Kenneth A. Ross. *Abstract harmonic analysis. Vol. I.* Second edition, Springer-Verlag, Berlin, 1979.

[HvN42] Paul R. Halmos and John von Neumann. Operator methods in classical mechanics. II. *Ann. of Math. (2)*, 43:332–350, 1942.

[JW96] Winfried Just and Martin Weese. *Discovering modern set theory I, The basics.* American Mathematical Society, Providence, RI, 1996.

[Kec95] Alexander S. Kechris. *Classical descriptive set theory*. Springer-Verlag, New York, 1995.

[Kec97] Alexander S. Kechris. On the concept of Π_1^1-completeness. *Proc. Amer. Math. Soc.*, 125(6):1811–1814, 1997.

[Kna67] Anthony W. Knapp. Distal functions on groups. *Trans. Amer. Math. Soc.*, 128:1–40, 1967.

[Kri70] Wolfgang Krieger. On entropy and generators of measure-preserving transformations. *Trans. Amer. Math. Soc.*, 149:453–464, 1970.

[Lév79] Azriel Lévy. *Basic set theory*. Springer-Verlag, Berlin, 1979.

[Mos80] Yiannis N. Moschovakis. *Descriptive set theory*. North-Holland Publishing Co., Amsterdam, 1980.

[Orn74] Donald S. Ornstein. *Ergodic theory, randomness, and dynamical systems.* James K. Whittemore Lectures in Mathematics given at Yale University, Yale Mathematical Monographs, No. 5. Yale University Press, New Haven, Conn., 1974.

[OS73] Donald S. Ornstein and Paul C. Shields. An uncountable family of K-automorphisms. *Advances in Math.*, 10:63–88, 1973.

[OW91] Donald S. Ornstein and Benjamin Weiss. Statistical properties of chaotic systems, *Bull. Amer. Math. Soc. (N.S.).* 24:11–116, 1991. With an appendix by David Fried.

[Par68] William Parry. Zero entropy of distal and related transformations. In *Topological Dynamics (Symposium, Colorado State Univ., Ft. Collins, Colo., 1967)*, pages 383–389. Benjamin, New York, 1968.

[Par69] William Parry. *Entropy and generators in ergodic theory.* W. A. Benjamin, Inc., New York-Amsterdam, 1969.

[PS96] Mark Pollicott and Klaus Schmidt, editors. *Ergodic theory of $\mathbf{Z}^d$ actions,* Cambridge University Press, Cambridge, 1996.

[Roh48] Vladimir A. Rohlin. A "general" measure-preserving transformation is not mixing. *Doklady Akad. Nauk SSSR (N.S.),* 60:349–351, 1948.

[Shi73] Paul Shields. *The theory of Bernoulli shifts.* Chicago Lectures in Mathematics. The University of Chicago Press, Chicago, Ill.-London, 1973.

[Smo71] Meir Smorodinsky. *Ergodic theory, entropy.* Lecture Notes in Mathematics, Vol. 214. Springer-Verlag, Berlin, 1971.

[Sus17] Mikhail Ya. Suslin. Sur une définition des ensembles B sans nombres transfinis. *Compte Rendus Acad. Sciences, Paris,* 164:88–91, 1917.

[Wal82] Peter Walters. *An introduction to ergodic theory.* Springer-Verlag, New York, 1982.

[WeiBA] Benjamin Weiss. Unpublished notes.

[Zim76a] Robert J. Zimmer. Extensions of ergodic group actions. *Illinois J. Math.,* 20(3):373–409, 1976.

[Zim76b] Robert J. Zimmer. Ergodic actions with generalized discrete spectrum. *Illinois J. Math.,* 20(4):555–588, 1976.

Structure Theory as a Tool in Topological Dynamics

Eli Glasner
Department of Mathematics
Tel Aviv University
Ramat Aviv, Israel

0   Introduction

Consider the following two fixed point theorems:

Theorem 0.1 *Let E be a locally convex topological vector space, $Q \subset E$ a compact convex subset and $\gamma \mapsto T_\gamma$ a representation of the group Γ as a group of affine homeomorphisms of Q. If the system (Q, Γ) is distal then there exists a Γ-fixed point in Q.*

Theorem 0.2 *Let E be a locally convex topological vector space, $Q \subset E$ a compact convex subset and $\gamma \mapsto T_\gamma$ a representation of the group Γ as a group of affine homeomorphisms of Q. Let X be the closure of the set $\mathrm{ext}(Q)$ of extreme points of Q. Clearly X is a closed Γ-invariant set. If the system (X, Γ) is distal, then there exists a Γ-fixed point in Q.*

Here is a proof of the first theorem:

Proof of theorem 0.1. By Zorn's lemma there exists a compact convex Γ-invariant subset Q_0 of Q which is minimal with respect to these properties. If x and y are points in Q_0, then so is $z = 1/2(x+y)$ and we have, by minimality of Q_0,

$$\overline{\mathrm{co}}(\Gamma z) = Q_0,$$

where $\overline{\mathrm{co}}$ stands for "convex closed hull". This equality implies $\overline{\Gamma z} \supset \mathrm{ext}(Q_0)$ and we deduce the existence of a net $\gamma_i \in \Gamma$ and a point $w \in \mathrm{ext}(Q_0)$ with $\lim \gamma_i z = w$. We can assume that the limits $\lim \gamma_i x = w_1$ and $\lim \gamma_i y = w_2$ exist as well, so that $w = 1/2(w_1 + w_2)$. Hence $w = w_1 = w_2$; i.e., the points x and y are proximal and our assumption that the Γ action is distal, implies $x = y$. Thus Q_0 consists of a single point which is Γ-fixed. □

This straightforward geometric proof is just what one expects. It is therefore surprising to find out that the only proof known for the second theorem involves a considerable amount of dynamics. In fact it requires Furstenberg's structure theorem for distal systems (for the proof see section 4 below).

Our purpose in these notes is to describe the structure theory of minimal dynamical systems with emphasis put on its few known applications (one of which is the above fixed point theorem). The notes present an elaborate version of a short series of lectures given during the Descriptive Set Theory and Dynamical Systems joint workshop held at Luminy, in June 1996. They are intended primarily for students and workers in other fields of mathematics who wish to learn quickly the elements of the abstract theory of minimal dynamical systems. The basic ideas of this theory as well as many of its theorems are due to R. Ellis. The first four sections present a sketch of this theory, where a considerable part of the material as well as some basic examples are presented in a series of exercises. The more difficult of these are followed by an indication of a solution (within double brackets [[...]] (as in

[Vr])) or a convenient reference. I have neither tried to systematically trace the historical development of the subject nor to attribute precisely results or ideas to their original authors. This information can be traced using the reference list.

In the last section I describe a more recent application of the general structure theorem ([G,5]). The result (theorem 5.1) shows that for a minimal $\mathbb{Z}$-system (X, T), up to a proximal extension, the asymptotic behaviour of the transformation $\tau = T \times T^2 \times \cdots T^{n+1}$ on points of the diagonal $\Delta_{n+1} \subset X^{n+1}$ is determined by the corresponding action of τ on the canonical rank-n PI-factor of X. This result arises as an analogue in topological dynamics of Hillel Furstenberg's notion of characteristic factors in ergodic theory ([F,4]). The latter is the main tool in Furstenberg's proof of the multiple recurrence theorem ([F,2]) (a direct consequence of which is the celebrated Szemerédi theorem). At the end of the section we also show how to derive the topological multiple recurrence theorem and van der Waerden's theorem from $\beta\mathbb{Z}$ theory. I thank Robert Ellis for the elegant proof he provided for proposition 5.4.

1   Definitions and some basic examples

The objects we study are **topological dynamical systems**, or just **systems**, (X, Γ) where X is a compact metric space and Γ a group represented on X as a group of self homeomorphisms. For simplicity we take Γ to be infinite countable and discrete. (The term "flow" is often used in the literature instead of "dynamical system"; here we prefer the latter and thus avoid the confusion with the "$\mathbb{R}$-system" meaning of "flow"). We emphasize the fact that, unless stated explicitly otherwise, our dynamical systems are assumed to be metrizable. However we will have to deal with non-metrizable dynamical systems when we introduce the "universal" dynamical systems.

We let γx denote the image of $x \in X$ under the homeomorphism corresponding to the element $\gamma \in \Gamma$. We let Γx or $\mathcal{O}_\Gamma x$ be the Γ orbit of x; i.e., the set $\{\gamma x : \gamma \in \Gamma\}$. $\bar{\mathcal{O}}_\Gamma x$ will denote the orbit closure of x. If (X, Γ) is a system and Y a closed Γ-invariant subset, then we say that (Y, Γ) is a **subsystem** of (X, Γ) . For dynamical systems (X, Γ) and (Y, Γ), their **product system** $(X \times Y, \Gamma)$ is defined by the diagonal action: $\gamma(x, y) = (\gamma x, \gamma y)$.

Mostly our examples will be of dynamical systems with acting group $\mathbb{Z}$. In this case we let T represent the generator 1 of $\mathbb{Z}$ and we write (X, T) instead of $(X, \mathbb{Z})$. The system (X, Γ) is called **topologically transitive** if for every nonempty open sets U, V in X there exists $\gamma \in \Gamma$ with $\gamma U \cap V \neq \emptyset$. This is easily seen to be equivalent to the existence of a point $x \in X$ with $\bar{\mathcal{O}}_\Gamma x = X$. It is often convenient when working with topologically transitive systems (X, Γ) to distinguish a point x_0 with dense orbit. We call such a system

(X, x_0, Γ) a **pointed system**. When (X, x_0, Γ) and (Y, y_0, Γ) are pointed systems, then their **join** $(X, x_0, \Gamma) \vee (Y, y_0, \Gamma)$ is by definition the subsystem $\bar{\mathcal{O}}_\Gamma(x_0, y_0) \subset X \times Y$. The join construction applies to any number of topologically transitive pointed systems.

The dynamical system (X, Γ) is called **minimal** if $\bar{\mathcal{O}}_\Gamma x = X$ for every $x \in X$. A point x in a system (X, Γ) is called **almost periodic** if the subsystem $\bar{\mathcal{O}}_\Gamma x$ is minimal. We say that the system (X, Γ) is **semisimple** (or **pointwise almost periodic**) if it is the union of its minimal subsystems. If this union is dense in X (i.e., the almost periodic points are dense in X) we say that (X, Γ) is **Bronstein**.

When (X, Γ) and (Y, Γ) are dynamical systems and $\pi : X \to Y$ a continuous onto map which intertwines the Γ actions (i.e., $\pi \gamma x = \gamma \pi x$ for all $x \in X$, $\gamma \in \Gamma$), we say that the system (Y, Γ) is a **factor** of the system (X, Γ), or that (X, Γ) is an **extension** of (Y, Γ) and denote this by $(X, \Gamma) \xrightarrow{\pi} (Y, \Gamma)$. The map π is called a **homomorphism** of dynamical systems or an **extension** or a **factor map**. Such a factor map defines an invariant closed equivalence relation on X (an **icer**):

$$R_\pi = \{(x, x') : \pi(x) = \pi(x')\}.$$

Conversely, an icer defines a factor system $(X, \Gamma) \xrightarrow{\pi} (Y, \Gamma)$ with $\pi(x)$ denoting the equivalence class of x.

Let (X, Γ) be a Γ-system; a self homeomorphism ϕ of X is called an **automorphism** of (X, Γ) if $\phi \gamma x = \gamma \phi x$ for all $x \in X$ and $\gamma \in \Gamma$. We let $\mathrm{Aut}(X, \Gamma)$ denote the group of automorphisms of (X, Γ). With the topology of uniform convergence of homeomorphisms and their inverses $\mathrm{Aut}(X, \Gamma)$ is a Polish topological group.

If K is a compact subgroup of $\mathrm{Aut}(X, \Gamma)$ then the map $x \mapsto Kx$ defines a factor map $(X, \Gamma) \xrightarrow{\pi} (Y, \Gamma)$ with $Y = X/K$ and

$$R_\pi = \{(x, kx) : x \in X, \ k \in K\}.$$

Such an extension is called a **group extension**.

An extension $(X, \Gamma) \xrightarrow{\pi} (Y, \Gamma)$ is called **isometric extension** if there exists a continuous function $d : R_\pi \to \mathbb{R}$ such that for every $y \in Y$ the function d restricted to $\pi^{-1}(y)$ is a metric and for every pair $(x, x') \in R_\pi$, and $\gamma \in \Gamma, d(\gamma x, \gamma x') = d(x, x')$. A basic result shows that an extension of minimal systems $(X, \Gamma) \xrightarrow{\pi} (Y, \Gamma)$ is an isometric extension iff there exists a commutative diagram:

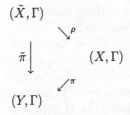

where $(\tilde{X}, \Gamma) \xrightarrow{\tilde{\pi}} (X, \Gamma)$ is a group extension with some compact group K and the map ρ is obtained by "dividing" $\tilde{X}$ by a closed subgroup H of K. Thus $Y = \tilde{X}/K$ and $X = \tilde{X}/H$.

Exercises:

(1) Show that when (X, Γ) is topologically transitive there exists in X a dense G_δ subset of points whose orbit is dense. (Recall the space X is assumed metrizable).

(2) The following are equivalent:

(a) (X, Γ) is minimal

(b) The empty set and X are the only subsystems of (X, Γ) .

(c) For every nonempty open set U in X the set $\Gamma U = \bigcup\{\gamma U : \gamma \in \Gamma\}$ is all of X.

(d) For every nonempty open set U in X there exists a finite subset $\{\gamma_1, \ldots, \gamma_k\}$ in Γ with $\bigcup_{j=1}^{k} \gamma_j U = X$.

(e) For every $x \in X$ and neighborhood U of x the set $N(x, U) = \{\gamma \in \Gamma : \gamma x \in U\}$ is (left) **syndetic** or (left) **cofinite** in Γ; i.e., there exists a finite set $F \subset \Gamma$ such that $FN(x, U) = \Gamma$.

(3) Use Zorn's lemma to show that every system has a minimal subsystem.

(4) Do the same without using Zorn's lemma.

(5) Let $\alpha = (\alpha_1, \ldots, \alpha_k)$ be an element of $\mathbb{T}^k$, the k-torus, with the set $\{1, \alpha_1, \alpha_2, \ldots, \alpha_k\}$ independent over $\mathbb{Q}$. Show that the $\mathbb{Z}$ system (X, T), where $X = \mathbb{T}^k$ and $Tx = x + \alpha$, is a minimal system (Kronecker's theorem).

(6) Let $\mathbb{A}$ be the dyadic group; i.e., $\mathbb{A} = \{0, 1\}^{\mathbb{N}}$ and addition is performed modulo 2 with carry to the right. We let $\mathbf{1}$ be the sequence $(1, 0, 0, \ldots)$ and equip $\mathbb{A}$ with the product topology. Then the system (X, T) with $X = \mathbb{A}$ and $Tx = x + \mathbf{1}$ is a minimal system called the **dyadic adding machine**. More generally for any infinite sequence $\mathbf{a} = (a_0, a_1, a_2, \ldots)$ of positive integers greater than one, let

$$\mathbb{A} = \prod_{n \geq 0} \{0, 1, \ldots, a_n - 1\}.$$

$\mathbb{A}$ becomes an abelian group—the **a-adic integers**—under the operation of addition mod a_n at the n-th coordinate with carry to the right. Denoting $\mathbf{1} = (1, 0, 0, \ldots)$, the transformation $Tx = x + \mathbf{1}$ defines a minimal dynamical system (X, T) —the **a-adic adding machine**.

(7) Let (Ω, T) be the $\mathbb{Z}$-system defined on $\Omega = \{0, 1, \ldots, s\}^{\mathbb{Z}}$ by left shift: $T\omega = \omega'$ where $\omega'_n = \omega_{n-1}$. Show that (Ω, T) is topologically transitive and find sequences in Ω whose orbit closure is infinite and minimal (i.e., nonperiodic almost periodic points). A dynamical system of the form (X, T), where X is a closed invariant subset of Ω, is called a **subshift**. More generally let (Ω, Γ) be the Γ-system defined on $\Omega = \{0, 1, \ldots, s\}^{\Gamma}$ by left shift: $(\gamma\omega)_{\gamma'} = \omega_{\gamma^{-1}\gamma'}$. We call the system (Ω, Γ) the **Bernoulli system** on Γ. Again this is a topologically transitive system.   ♠

A pair of points $(x_1, x_2) \in X \times X$ in a dynamical system (X, Γ) is called **proximal** if there exist a sequence $\gamma_i \in \Gamma$ and a point $z \in X$ such that

$$\lim \gamma_i x_1 = \lim \gamma_i x_2 = z.$$

We call the subset of $X \times X$ consisting of proximal pairs, the **proximal relation** of (X, Γ) and denote it by

$$P = P_X = \bigcap\{\Gamma V : V \text{ a neighborhood of the diagonal in } X \times X\}.$$

The proximal relation is a Γ-invariant symmetric and reflexive relation. Even for minimal systems it need be neither closed nor an equivalence relation. The system (X, Γ) is called **proximal** if every pair is proximal; i.e., $P = X \times X$. An extension $(X, \Gamma) \overset{\pi}{\to} (Y, \Gamma)$ is a **proximal extension** if $R_\pi \subset P$. We say that a pair of points (x, x') is **distal** if either $x = x'$ or $(x, x') \notin P$. The system (X, Γ) is **distal** if every pair in $X \times X$ is distal. An equivalent condition is: for every $x_1 \neq x_2$

$$\mathrm{Inf}_{\gamma \in \Gamma} d(\gamma x_1, \gamma x_2) > 0.$$

The extension $(X, \Gamma) \overset{\pi}{\to} (Y, \Gamma)$ is a **distal extension** when every pair in R_π is a distal pair. A point x in a system (X, Γ) is called **distal** if it is proximal only to itself. A minimal system is called **point distal** if it has at least one distal point. As we will see the existence of one distal point in a minimal system (X, Γ) implies that the set of distal points forms a dense G_δ subset of X. The only known proof of this fact is one that uses the structure theorem for point distal systems.

The system (X, Γ) is **equicontinuous** if the group Γ acts equicontinuously on X; i.e., $\forall \epsilon > 0, \exists \delta > 0$ such that $d(x_1, x_2) < \delta$ implies $d(\gamma x_1, \gamma x_2) < \epsilon$, for every $\gamma \in \Gamma$. Clearly every equicontinuous system is distal. A minimal equicontinuous system will be called **Kronecker**. There is a well known characterization of Kronecker systems: they are the systems of the form $(K/H, \Gamma)$ where K is a compact metrizable topological group, H a closed subgroup, $\tilde{\Gamma}$ a dense subgroup of K with a homomorphism onto $j : \Gamma \to \tilde{\Gamma}$, and the action of $\gamma \in \Gamma$ on the homogeneous space K/H is by left multiplication with $j(\gamma)$.

If Γ is abelian, then so is K and we can assume that $H = \{e\}$, the trivial subgroup consisting of the identity element of K. See exercises (5) and (6) above. For a non-abelian example, consider the natural action of a residually finite group Γ on its pro-finite completion. Or take Γ to be a countable dense subgroup of the Lie group $\mathbf{O}(n)$; we get a Kronecker action on a homogeneous space by letting Γ act on the sphere $S^{n-1} = \mathbf{O}(n)/\mathbf{O}(n-1)$.

The **regionally proximal relation** on X is defined by

$$Q_X = \bigcap \{ \overline{\Gamma V} : V \text{ a neighborhood of the diagonal in } X \times X \}.$$

It is easy to see that Q_X is trivial—i.e., equals the diagonal Δ—iff the system is equicontinuous. The relation Q_X is obviously Γ-invariant, closed, symmetric and reflexive. On a first glance there is no reason to think it is transitive and indeed there are examples of systems where it is not (see exercise 3.(12) below). However it turns out that, surprisingly, for a minimal system and abelian Γ, Q_X is a closed invariant equivalence relation, so that X/Q_X is the largest Kronecker factor of X.

When $(X, \Gamma) \xrightarrow{\pi} (Y, \Gamma)$ is an extension of minimal systems we let:

$$P_\pi = P \cap R_\pi = \bigcap \{ \Gamma V \cap R_\pi : V \text{ a neighborhood of the diagonal in } X \times X \}$$

and

$$Q_\pi = \bigcap \{ \overline{\Gamma V \cap R_\pi} : V \text{ a neighborhood of the diagonal in } X \times X \}.$$

Clearly the extension π is distal iff $P_\pi = \Delta$ and it is not hard to see that π is isometric iff $Q_\pi = \Delta$. Notice that in general we only have $Q_\pi \subset Q_X \cap R_\pi$.

The system (X, Γ) is called **weakly mixing** if the product system $(X \times X, \Gamma)$ is topologically transitive; i.e., for every four nonempty open sets $U_i, i = 1, 2, 3, 4$ there exists $\gamma \in \Gamma$ with $\gamma U_1 \cap U_2 \neq \emptyset$ and $\gamma U_3 \cap U_4 \neq \emptyset$. A well known result tells us that for abelian Γ a minimal system (X, Γ) is weakly mixing iff the system (X, Γ) admits no nontrivial Kronecker factors (so that by the above $Q_X = X \times X$). (See [KR]).

For Γ abelian this condition reduces to the following one. Every continuous **proper function** of (X, Γ) is trivial; i.e., if $f : X \to S^1$ ($S^1 = \{z \in \mathbb{C} : |z| = 1\}$) is continuous and for some character χ of Γ, called the **proper character**,

$$f(\gamma x) = \chi(\gamma) f(x) \qquad \forall x \in X, \gamma \in \Gamma,$$

then f is a constant and $\chi \equiv 1$.

An extension $(X, \Gamma) \xrightarrow{\pi} (Y, \Gamma)$ is called a **weakly mixing extension** if the system (R_π, Γ) is topologically transitive. It is easy to see that when an extension $(X, \Gamma) \xrightarrow{\pi} (Y, \Gamma)$ is weakly mixing then no nontrivial isometric extension of Y which is a factor of X exists. Unfortunately the converse does not always hold; i.e., the fact that $(X, \Gamma) \xrightarrow{\pi} (Y, \Gamma)$ is not weakly mixing does not always imply the existence of an intermediate nontrivial isometric extension (see, e.g., exercise (33) below). This fact makes structure theory in topological dynamics more complicated than the analogous theory in measurable dynamics. We will return to this question in the following sections.

Exercises:
(8) If P_X is closed it is already an icer.
(9) Show that a system (X, Γ) is equicontinuous iff it is **isometric**; i.e., there exists an equivalent Γ-invariant metric on X.
(10) Show that examples (5) and (6) are equicontinuous and (7) is weakly mixing.
(11) On $X = \mathbb{T}^2$ let $T(x, y) = (x + \alpha, y + x)$ for an irrational α. Show that (X, T) is minimal distal but not equicontinuous. The map $\pi(x, y) = x$ defines a factor map onto the largest Kronecker factor of (X, T) which is a group extension. [[The distality follows easily by considering separately pairs of points of the form $((x, y), (x, y'))$ and then pairs of the form $((x, y), (x', y'))$ with $x \neq x'$. For minimality show that if M is a *proper* minimal subset of $X \times X$ then $H = \{\beta \in \mathbb{T} : R_\beta M \cap M \neq \emptyset\}$ is the finite subgroup $H = \{0, 1/n, 2/n, \ldots, (n-1)/n\}$ for some positive integer n, where $R_\beta(x, y) = (x, y + \beta)$. Show that the set $\{(x, ny) : (x, y) \in M\}$ is a graph of a continuous function $f : \mathbb{T} \to \mathbb{T}$; then use the invariance of M to get $f(x + \alpha) - f(x) = nx$, which is impossible for $n \geq 1$. Finally for non-equicontinuity show that for a sequence $n_i \to \infty$ with $T^{n_i}((0, 0)) \to (0, 0)$, we have $T^{n_i}((0, 0), (1/(2n_i), 0)) \to ((0, 0), (0, 1/2)) \in Q$.]]
(12) Modify the transformation in the previous example to read $T(x, y) = (x + \alpha, y + 2x + \alpha)$. Again show that (X, T) is minimal distal. Check that $T^n(0, 0) = (n\alpha, n^2\alpha)$ and deduce that the sequence $\{n^2\alpha\}_{n=1}^\infty$ is dense in $\mathbb{T}$.
(13) Let $\mathbb{G}$ be the nilpotent group

$$\mathbb{G} = \left\{ \begin{pmatrix} 1 & q & y \\ 0 & 1 & x \\ 0 & 0 & 1 \end{pmatrix} : q \in \mathbb{Z}, \ y, x \in \mathbb{T} \right\}, \qquad H = \left\{ \begin{pmatrix} 1 & q & 0 \\ 0 & 1 & 0 \\ 0 & 0 & 1 \end{pmatrix} : q \in \mathbb{Z} \right\},$$

and

$$T = \begin{pmatrix} 1 & 1 & 0 \\ 0 & 1 & \alpha \\ 0 & 0 & 1 \end{pmatrix}.$$

Show that the system (X, T) from (11) is isomorphic to the **nil-system** $(\mathbb{G}/H, T)$, where T acts by multiplication on the left ([G,4]).

(14) Let $\mathbb{G} = SL(2, \mathbb{R})$, H a cocompact discrete subgroup and let

$$T_t = \begin{pmatrix} 1 & t \\ 0 & 1 \end{pmatrix}.$$

Then the $\mathbb{R}$-system $(\mathbb{G}/H, T_t)$—the **horocycle flow**—is a minimal weakly mixing flow (i.e., a $\mathbb{R}$-system). [[Minimality was first proved by G. A. Hedlund [H]. For the weak mixing use the commutation relation: $S_s T_t S_{s^{-1}} = T_{ts^2}$ for $t \in \mathbb{R}$, $s > 0$ and

$$S_s = \begin{pmatrix} s & 0 \\ 0 & s^{-1} \end{pmatrix},$$

to show that no nonconstant continuous proper function exists. In fact if f is a proper function with proper character $\chi_\theta(t) = e^{i\theta t}$, $\theta \notin \mathbb{Z}$, $t \in \mathbb{R}$, then $f_s(x) = f(S_s x)$ is a proper function with proper character $\chi_{\theta s^2}(t) = e^{i\theta s^2 t}$. Now show that whenever f, g are proper functions corresponding to two distinct proper values then $\|f - g\|_\infty > \sqrt{2}$. The separability of the Banach space $C(X)$ is now used to get a contradiction.]]

(15) If (X, Γ) is minimal and weakly mixing and Γ abelian then for every $x \in X$ the proximal cell: $P[x] = \{x' : (x, x') \in P\}$, is a dense G_δ subset of X ([F,3], theorem 9.12). Compare this result and its elementary proof (due to B. Weiss) to its relative version, theorem 3.2.(3) below, whose proof requires our whole machinery.

(16) A proximal minimal system is weakly mixing ([G,1], corollary II.2.2). ♠

The minimal system (X, Γ) is called **regular** if every almost periodic point in $X \times X$ is of the form $(x, \phi(x))$ for $\phi \in \mathrm{Aut}(X, \Gamma)$. We say that the minimal system (X, Γ) is **incontractible** if for every $n > 1$ the almost periodic points are dense in the product system X^n (i.e., X^n is Bronstein). The extension $(X, \Gamma) \xrightarrow{\pi} (Y, \Gamma)$ is **relatively incontractible** (or **RIC**) extension if it is open and for every $n > 1$ the almost periodic points are dense in the relation

$$R_\pi^n = \{(x_1, x_2, \ldots, x_n) : \pi(x_i) = \pi(x_j), \ 1 \le i, j \le n\}$$

(see exercise 3.(7) below). It is a **Bronstein** extension if $R_\pi = R_\pi^2$ is Bronstein, so that every RIC extension is also Bronstein. Let us call a minimal system (X, Γ) **prodal** if every topologically transitive Bronstein subsystem of $X \times X$ is necessarily minimal.

Exercises:

(17) For every automorphism ϕ of a system (X, Γ) and an almost periodic point $x \in X$ the pair $(x, \phi(x))$ is an almost periodic point. In particular when Γ is abelian and (X, Γ) is minimal, every pair $(x, \gamma x)$ is almost periodic. Similarly points of the form $(x, \gamma_1 x, \gamma_2 x, \ldots, \gamma_n x)$ are almost periodic points and we conclude that (X, Γ) is incontractible.

(18) For abelian Γ no nontrivial proximal and minimal system exists.

(19) Does there exist a minimal system for which the almost periodic points are dense in $X \times X$ but not dense in X^n for some $n > 2$ (i.e., X is not incontractible)? [[This is an open problem, perhaps not too hard.]]

(20) Show that the class of prodal systems is closed under factors and under distal and proximal extensions. In particular proximal and distal minimal systems are prodal. An incontractible weakly mixing prodal system is trivial.

(21) Proximal as well as Kronecker systems are regular; in fact it can be shown that for abelian Γ a (metrizable) prodal system is regular iff it is Kronecker ([G,3]). ♠

An extension $(X, \Gamma) \xrightarrow{\pi} (Y, \Gamma)$ is called an **almost 1-1 extension** if there exists a residual subset $X_0 \subset X$ with $\pi^{-1}(\pi(x)) = \{x\}$ for every $x \in X_0$.

Exercises:

(22) If $(X, \Gamma) \xrightarrow{\pi} (Y, \Gamma)$ is an extension of dynamical systems, and $\pi^{-1}(\pi(x_0)) = \{x_0\}$ for a point $x_0 \in X$ with dense orbit, then the set of points $Y_0 \subset Y$ whose inverse image under π is a singleton, forms a dense G_δ subset of Y. The set $X_0 = \pi^{-1}[Y_0]$ is dense G_δ in X. If moreover (Y, Γ) is minimal, then (X, Γ) is also minimal.

(23) For (X, Γ) topologically transitive an extension $(X, \Gamma) \xrightarrow{\pi} (Y, \Gamma)$ is an almost 1-1 extension iff X is the only closed subset of X mapped by π onto Y.

(24) An almost 1-1 extension of minimal systems is a proximal extension.

(25) There are extensions $(X, \Gamma) \xrightarrow{\pi} (Y, \Gamma)$ of minimal systems which are proximal but not almost 1-1 ([GW,1]).

(26) A minimal system which is an almost 1-1 extension of its maximal Kronecker factor is called **almost automorphic**. Show that every almost automorphic system is point distal.

(27) Let us define an element $\omega \in \Omega = \{0,1\}^\Gamma$ as follows. For a finite block $B \in \{0,1\}^n$ we let $\bar{B}$ be the dual block obtained by replacing each 0 with 1 and vice versa. Then inductively define blocks B_n by:

$$B_0 = 0, \ B_1 = 01, \ B_2 = 0110, ..., B_{n+1} = B_n \bar{B}_n.$$

Since for every n, B_n is the initial 2^n subblock of B_{n+1} we get this way an infinite block B_∞ and our ω is now defined by setting $\omega_0 \omega_1 ... = B_\infty$ and $\omega(n) = \omega(-n-1)$ for $n < 0$. We let $X = \bar{\mathcal{O}}_T \omega$. The sequence ω is the famous **Thue Morse sequence** and the system (X, T) is called the **Morse system**. Show that (X, T) is a minimal infinite subshift; see also [K].

(28) The Morse minimal system has the following structure: $X \xrightarrow{\rho} Y \xrightarrow{\sigma} Z$, where ρ is a group extension with group $K = \{f, e\}$ (where $f(\omega) = \bar{\omega}$ and e the identity map), and σ is an almost 1-1 extension of Z, the Kronecker factor of X. The latter coincides with the dyadic adding machine $\mathbb{A}$. Conclude that the Morse system is point distal ([G,2]).

(29) Define inductively blocks B_n by:

$$B_1 = 0010, \quad B_2 = 0010001010010, ..., B_{n+1} = B_n B_n 1 B_n.$$

We let ω be defined by setting for every n:

$$\omega_{-l_n}\omega_{-l_n+1} \cdots \omega_0\omega_1...\omega_{l_n-1} = B_n B_n,$$

where l_n is the length of B_n. We let $X = \bar{\mathcal{O}}_T\omega$. The sequence ω is called the **Chacon sequence** and the system (X,T) is the **Chacon system**. Show that (X,T) is a minimal infinite subshift. More difficult to show are the following facts. (X,T) is weakly mixing and regular (in fact every minimal set in $X \times X$ is a graph of some power of T). Moreover (X,T) is a **prime system**; i.e., it admits no nontrivial factors ([J],[P]).

(30) The last two examples are special cases of a general construction of minimal dynamical systems called **substitution systems**. For a finite alphabet $A = \{0, 1, \ldots, s\}$ a **substitution** is a function θ which assigns to each symbol of A a finite word on A; i.e., $\theta : A \to A^* = \cup_{n \geq 1} A^n$. When all these words have the same length, say d, we say that θ is of constant length d. Thus the Morse and the Chacon substitutions are defined on $A = \{0, 1\}$; the Morse substitution is of constant length, while Chacon's is not. Clearly θ induces maps, also denoted θ, from A^* into itself and from $A^{\mathbb{N}}$ into itself. In particular iterations of θ make sense. The substitution θ is called **primitive** if

(1) $\theta(0) = 0$.

(2) $|\theta^n(0)| \to \infty$, where $|w|$ denotes the length of the word w.

(3) There exists a positive integer k such that for every $r, t \in A$, r appears in $\theta^k(t)$.

It is easy to see that for a primitive θ the sequence $\theta^n(0)$ converges to a limit $\omega \in A^{\mathbb{N}}$ with $\theta(\omega) = \omega$. Define the subshift $X_\theta \subset A^{\mathbb{Z}}$ as the collection of all sequences $x \in A^{\mathbb{Z}}$ with the property that every finite word in x appears as a subword of ω. Show that the dynamical system (X_θ, T) is minimal. (See [Q].)

(31) For a primitive substitution θ of constant length, the minimal system (X_θ, T) is point distal ([Ma]).

(32) The subshift $X = \bar{\mathcal{O}}_T\omega$, where $\omega \in \{0, 1\}^{\mathbb{Z}}$ is the sequence

$$\omega(n) = \operatorname{sgn} \cos(2\pi n\alpha)$$

for $\alpha \notin \mathbb{Q}$, is a minimal subshift. Show that its maximal Kronecker factor is the rotation on $(\mathbb{T}, \alpha)$, where the extension $X \to \mathbb{T}$ is an almost 1-1 extension. Show that this extension is *not* a weakly mixing extension.

(33) Define a sequence $\omega \in \{0,1\}^{\mathbb{Z}}$ as follows. For every $n \in \mathbb{Z}$, $\omega(2n) = 0$, $\omega(4n+1) = 1$, $\omega(8n+3) = 0$, $\omega(16n+7) = 1$, and in general $\omega(2^k n + 2^{k-1} - 1)$ is zero or one according to whether k is even or odd. Let $X = \bar{\mathcal{O}}_T(\omega)$; show that the subshift (X, T) is a minimal almost 1-1 extension of the dyadic adding machine.

(34) In general a sequence $\omega \in A^{\mathbb{Z}}$ for an alphabet $A = \{0, 1, \ldots, s\}$ is a **Toeplitz sequence** if for every $n \in \mathbb{Z}$ there exists $p \geq 1$ such that for all $k \in \mathbb{Z}$, $\omega(n + kp) = \omega(n)$. A subshift $X \subset A^{\mathbb{Z}}$ is a **Toeplitz system** if X is the orbit closure of a Toeplitz sequence. Show that every Toeplitz system is minimal (see [W]).

(35) A minimal infinite subshift is Toeplitz iff it is an almost 1-1 extension of an adding machine ([MP]). ♠

2   The enveloping semigroup

In order to study the asymptotic behavior of a Γ-system (X, Γ) R. Ellis introduced in 1960 the **enveloping semigroup** $E = E(X, \Gamma)$. This is defined as the closure in X^X (with its compact, usually nonmetrizable, pointwise convergence topology) of the set Γ considered as a subset of X^X. (We identify here the elements of Γ with the corresponding homeomorphism of X. Since in most cases the action of Γ on X is **effective** (i.e., $\gamma x = x$, $\forall x \in X$ implies $\gamma = e$), this abuse of notation will cause no harm).

It follows directly from the definitions that, under composition of maps, E forms a compact semigroup in which the operations

$$p \mapsto pq \qquad \text{and} \qquad p \mapsto \gamma p$$

for $p, q \in E$, $\gamma \in \Gamma$, are continuous. Notice that this makes Γ act on E by left multiplication, so that (E, Γ) is a Γ-system (though usually nonmetrizable).

The elements of E may behave very badly as maps of X into itself; usually they are not even Borel measurable. However our main interest in E lies in its algebraic structure and its dynamical significance. A key lemma in the study of this algebraic structure is the following:

Lemma 2.1 *Let L be a compact Hausdorff semigroup in which all maps $p \mapsto pq$ are continuous. Then L contains an **idempotent**; i.e., an element v with $v^2 = v$.*

Proof. By Zorn's lemma, there exists a minimal compact subsemigroup $K \subset L$. For any $v \in K$, Kv is a compact subsemigroup of K whence $Kv = K$ and in particular for some $k \in K$, $kv = v$. Now the set $M = \{l \in K : lv = v\}$ is a nonempty closed subsemigroup of K, and again we deduce that $M = K$. In particular $vv = v$. $\square$

In the next series of exercises we state some useful properties of the enveloping semigroup E. Most of these are easy consequences of the definitions and lemma 2.1.

Exercises:
(1) A subset M of E is a minimal left ideal of the semigroup E iff it is a minimal subsystem of (E, Γ). In particular a left minimal ideal is closed. We will refer to it simply as **minimal ideal**. Minimal ideals M in E exist and for each such an ideal the set of idempotents in M, denoted by $J = J(M)$, is nonempty ([G,1] for exercises (1)–(3)).

(2) Let M be a minimal ideal and J its set of idempotents then:

(a) For $v \in J$ and $p \in M$, $pv = p$.

(b) For each $v \in J$, $vM = \{vp : p \in M\}$ is a subgroup of M with identity element v. For every $w \in J$ the map $p \mapsto wp$ is a group isomorphism of vM onto wM.

(c) $\{vM : v \in J\}$ is a partition of M. Thus if $p \in M$ then there exists a unique $v \in J$ such that $p \in vM$; we denote by p^{-1} the inverse of p in vM.

(3) Let K, L, and M be minimal ideals of E. Let v be an idempotent in M. Then there exists a unique idempotent v' in L such that $vv' = v'$ and $v'v = v$. (We write $v \sim v'$ and say that v' is **equivalent** to v.) If $v'' \in K$ is equivalent to v', then $v'' \sim v$. The map $p \mapsto pv'$ of M to L is an isomorphism of Γ-systems. ♠

Often one has to deal with more than one system at a time; e.g., we can be working simultaneously with two systems, their product, subsystems of the product, etc. Or given a dynamical system (X, Γ) we may have to work with associated systems like the actions induced on the space of probability measures $M(X)$ with its weak* topology, or the space of closed subsets 2^X, with its Hausdorff topology. It is then convenient to have one enveloping semigroup acting on all of the systems simultaneously. This can be easily done by considering the enveloping semigroup of the product of all the systems under consideration. In fact we will loose nothing and gain much in convenience as well as in added machinery if we work instead with a "universal" enveloping semigroup.

Let $\beta\Gamma$ denote the Stone-Čech compactification of Γ. Recall that $\{\bar{A} : A \subset \Gamma\}$ is a basis for the topology of $\beta\Gamma$ consisting of clopen sets. Moreover for $\gamma \in \Gamma$, $\gamma \in \bar{A}$ implies $\gamma \in A$. It will be convenient to consider $\beta\Gamma$ both as the space of ultrafilters on Γ and as the universal compactification of the discrete group Γ. The latter means that every map $f : \Gamma \to X$ where X is a compact space can be uniquely extended to a continuous map $\bar{f} : \beta\Gamma \to X$.

One can easily show, using this universal property, that $\beta\Gamma$ has a semigroup structure with the same properties as the enveloping semigroups discussed above. Moreover if (X, Γ) is a system and $x \in X$ then the map $\gamma \mapsto \gamma x$ can be extended uniquely to a map $p \mapsto px$ of $\beta\Gamma$ onto $\bar{\mathcal{O}}_\Gamma x$. This defines a (systems and semigroups) homomorphism of $\beta\Gamma$ onto $E(X, \Gamma)$. In fact it is convenient to forgo the use of $E(X, \Gamma)$ and use instead the universal enveloping semigroup $\beta\Gamma$. This will be our course from now on. Moreover we will fix once and for all a minimal ideal M in $\beta\Gamma$, let $J = J(M)$ and fix an idempotent u in J. We say that an idempotent is a **minimal idempotent** if it belongs to a minimal ideal and denote the set of minimal idempotents by $\hat{J}$. We will denote the group uM by G and use Greek letters for the elements of G.

Since right multiplication on $\beta\Gamma$ is continuous and since each $v \in J$ is a right identity in M, it follows that for each $\alpha \in G$ the map $p \mapsto p\alpha$ is a homeomorphism of M and in fact an automorphism of the (nonmetrizable) dynamical system (M, Γ). Thus every element of G defines an automorphism of M and it is easy to check that if we identify the elements of G with the corresponding automorphisms then actually $G = \text{Aut}(M, \Gamma)$. We will freely use both interpretations of G. Notice that every element p in M can be expressed uniquely as $p = v\alpha$ for $v \in J$ and $\alpha \in G$. We then have $p^{-1} = v\alpha^{-1}$.

Thus $\beta\Gamma$ is a universal enveloping semigroup, it "acts" on every system (X, Γ). If $p \in \beta\Gamma$ and $x \in X$, then px can be interpreted both as the image of x under the image of p in $E(X, \Gamma)$ or as the limit $\lim_p \gamma x$ where we now consider p as an ultrafilter (or a universal net) on Γ.

Associated with (X, Γ) we have the system $(2^X, \Gamma)$, on the space of closed subsets of X. To avoid confusion, the action of $\beta\Gamma$ on 2^X is denoted $p \circ A, p \in \beta\Gamma, A \in 2^X$. Thus in general $pA = \{px : x \in A\} \subset p \circ A = \{\lim_p \gamma_i x_i, x_i \in A\}$, where $\lim_p \gamma_i$ denotes a limit along the universal net p. Notice that the latter description of $p \circ A$ enables us to extend the definition of $p \circ A$ to arbitrary (not necessarily closed) subsets A of X. A minimal subsystem of $(2^X, \Gamma)$, is called a **quasifactor** of (X, Γ).

Exercises:

(4) The enveloping semigroup of the Bernoulli system (Ω, Γ) is isomorphic (as a Γ-system as well as a compact semigroup) to $\beta\Gamma$. [[Recall that $\{\bar{A} : A \subset \Gamma\}$ is a basis for the topology of $\beta\Gamma$ consisting of clopen sets. Next identify $\Omega = \{0, 1\}^\Gamma$ with the collection of subsets of Γ in the obvious way: $A \longleftrightarrow 1_A$. Now define an "action" of $\beta\Gamma$ on Ω by:

$$p * A = \{\gamma \in \Gamma : \gamma p \in \bar{A}\}.$$

(For $p \in \beta\Gamma$ the element γp is defined as the left translation of the ultrafilter p by γ.) It is easy to see that this action extends the action of Γ on Ω and defines an isomorphism of $\beta\Gamma$ onto $E(\Omega, \Gamma)$.]]

(5) Using the universal property show that $\beta\Gamma$ has a semigroup structure with the same properties as the enveloping semigroups ([G,1]).

(6) Let

$$(Z, z_0) = \bigvee \{(X, x_0) : (X, x_0) \text{ a pointed minimal system}\}$$

be the (nonminimal, nonmetrizable) join of all minimal pointed systems. Show that the Γ system Z is isomorphic to the enveloping semigroup $E(M)$ of the universal minimal system M. [[Given $p \in E(M)$ and a minimal system X, px is defined for every $x \in X$. Thus the map $\phi : E(M) \to Z$ defined by $\phi : p \mapsto pz_0$ is a homomorphism of the Γ systems. If $p, q \in E(M)$ are distinct elements then there exists $m \in M$ with $pm \neq qm$. Since (M, m) is a member of the family of pointed minimal systems, we have $pz_0 \neq qz_0$, hence ϕ is an isomorphism.]]

(7) The natural homomorphism $\psi : (\beta\mathbb{Z}, e) \to (E(M), id)$ is not 1-1, hence these two systems are not isomorphic. [[By the previous exercise (with $\Gamma = \mathbb{Z}$), identify $(E(M), id)$ with (Z, z_0). Let

$$\psi^* : C(Z) \to C_b(\mathbb{Z}) = l^\infty(\mathbb{Z}) \cong C(\beta\mathbb{Z}),$$

be the adjoint map from $C(Z)$, the algebra of real valued continuous functions on Z, into $l^\infty(\mathbb{Z})$. Let $\mathcal{A} = \psi^* C(\mathcal{Z})$ be its image in $l^\infty(\mathbb{Z})$. A subset A of $\mathbb{Z}$ is **small** if for every $k > 0$ there is an $N_k > 0$ such that in every interval of length N_k in $\mathbb{Z}$, there are k consecutive members of A^c. For a norm closed translation invariant subalgebra $\mathcal{D}$ of $l^\infty(\mathbb{Z})$ containing the constant functions, a subset A of $\mathbb{Z}$ is a $\mathcal{D}$-**interpolation set** if every bounded real valued function on A can be extended to a function in $\mathcal{D}$. By [GW,2] $\mathcal{A}$-interpolation sets are small subsets of $\mathbb{Z}$, while every subset of $\mathbb{Z}$ is an $l^\infty(\mathbb{Z})$-interpolation set. Thus $\mathcal{A} \neq l^\infty(\mathbb{Z})$, hence ψ is not 1-1. The universal property of $\beta\mathbb{Z}$ now shows that no isomorphism can exist between $\beta\mathbb{Z}$ and Z.]]

(8) Establish the following connections between the dynamical properties of the system (X, Γ) and the algebraic properties of $E(X, \Gamma)$.

(a) $\bar{\mathcal{O}}_\Gamma x = Ex$

(b) $\bar{\mathcal{O}}_\Gamma x$ is minimal (i.e., x is almost periodic) iff for every minimal ideal M in E, $\bar{\mathcal{O}}_\Gamma x = Mx$ iff in every minimal ideal there is an idempotent v such that $vx = x$. Thus JX is the set of almost periodic points of the system (X, Γ). Applying this to the product system we see that $J(X \times X)$ is the set of almost periodic points in $X \times X$. Every quasifactor of (X, Γ) has the form $\{p \circ A : p \in M\}$ for some $A \subset X$.

(c) The pair (x, x') is proximal iff $px = px'$ for some $p \in E$ iff there exists a minimal ideal M in E with $px = px'$ for every $p \in M$.

(d) If (X, Γ) is minimal, then

$$P[x] = \{x' \in X : (x, x') \in P\} = \{vx : v \in \hat{J}\}.$$

In particular $x \in X$ is a distal point iff $vx = x$ for every $v \in \hat{J}$.

(e) For $v \in \hat{J}$ every pair of points in vX is distal.

(f) A point distal system with a residual set of distal points (as we will see later the latter assumption is redundant) is prodal. [[Show first that every homomorphism of minimal systems is **semiopen**; i.e., the image of a nonempty open set has a nonempty interior. Now let $W \subset X \times X$ be topologically transitive and Bronstein; use the Bronstein property to show that the natural projections of W on X are semiopen. Let X_0 denote the residual set of distal points in X and show that $W \cap (X_0 \times X_0)$ is residual in W. Finally use topological transitivity and (b) and (d) above to show that W is minimal.]]

(g) The relation P_X is transitive iff E contains a unique minimal ideal. [[See [M,2] for a far reaching generalization of this result.]]

(h) (X, Γ) is distal iff E is a group.

(i) A distal system is semisimple.

(j) (X, Γ) is distal iff $X \times X$ is semisimple. Now conclude that a factor of a distal system is distal.

(k) (X, Γ) is equicontinuous iff E is a topological group whose action on X is jointly continuous. [[Here one needs R. Ellis' joint continuity theorem ([E,1],[A]).]]

(l) (X, Γ) is minimal and equicontinuous (Kronecker) iff E is a topological group whose action on X is jointly continuous and the system (X, Γ) is isomorphic to the homogeneous system $(K/H, \Gamma)$ where K is a compact topological group, H is a closed subgroup of K and Γ is embedded in K as a dense subgroup.

(m) Show that a minimal system (X, Γ) is regular iff it is isomorphic (as a dynamical system) to some (hence every) minimal ideal in its enveloping semigroup. In particular the (non-metrizable) universal minimal system (M, Γ) is regular ([G,1]). ♠

3 The Ellis group of a pointed minimal system

Given a minimal system (X, Γ), we choose a distinguished point x_0 in X such that $ux_0 = x_0$. Our convention is that under a homomorphism a distinguished point goes to a distinguished point. When (X, x_0, Γ) is such a pointed minimal system, its **Ellis group** is $\mathcal{G}(X, x_0) := \{\alpha \in G : \alpha x_0 = x_0\}$. If $(X, x_0, \Gamma) \xrightarrow{\pi} (Y, y_0, \Gamma)$ is a homomorphism of minimal systems then clearly $\mathcal{G}(X, x_0) \subset \mathcal{G}(Y, y_0)$.

The group G is equipped with a compact T_1 topology called the τ-**topology**, with respect to which, all groups of the form $\mathcal{G}(Z, z_0)$ are closed. An elegant way of describing the τ-topology on G, due to J.Auslander, is the following. Consider G as $\text{Aut}(M, \Gamma)$ and define a closure operator for subsets A of G by declaring $\beta \in G$ to be in the τ-closure of A if the set $\text{graph}(\beta)$ is a subset of the closure in $M \times M$ of the set $\bigcup \{\text{graph}(\alpha) : \alpha \in A\}$.

As we have seen a Kronecker system is a homogeneous system K/H where K is a compact group, H a closed subgroup and Γ is embedded in K as a dense subgroup. The main idea of the "algebraic" theory of minimal systems is to view a general minimal system as much as possible as a homogeneous system. For a distal minimal system (X, x_0, Γ) this is literally true since $X = uX = Gx_0$ and we can identify X with the homogeneous space $G/\mathcal{G}(X, x_0)$. For the general pointed minimal system (X, x_0, Γ), uX is a proper subset of X (usually not even a Borel subset), but we can still identify $uX = Gx_0$ with the homogeneous space $G/\mathcal{G}(X, x_0)$. We let the τ-topology on uX be the quotient τ-topology on $G/\mathcal{G}(X, x_0)$ under that identification.

In a way what we do here is to substantiate Mackey's notion of a **virtual group** from ergodic theory. Whereas there the virtual group is just a way of thinking of an ergodic system as a homogeneous space, here we actually realize at least a part of the system, namely uX, as a homogeneous space.

Since for every $v \in J$ we can equally well identify the group vG with $\text{Aut}(M, \Gamma)$ it follows that τ-topologies are defined on the groups vG as well as on the sets vX for all the idempotents in J. In fact it follows that the maps $\alpha \mapsto v\alpha$ and $x \mapsto vx$ from G to vG and from uX to vX respectively, are τ-homeomorphisms.

Exercises:

(1) The operation $A \mapsto \tau\text{-closure} A$ is a closure operator and defines on G a compact T_1 topology weaker than the topology induced on G from M. In fact if α_i is a net in G and $\lim \alpha_i = p$ in M then $\tau\text{-}\lim \alpha_i = up$. The τ-topology on uX does not depend on the point x_0 and is compact and T_1 ([G,1] for exercises (1) and (3)).

(2) The topologies described below on uX, for a minimal (X, Γ), coincide with the τ-topology.

(a) Let Σ denote the collection of continuous pseudometrics on X. For $\sigma \in \Sigma$, $x, y \in X$ and $\epsilon > 0$ set $F_\sigma = \{\inf \sigma(\gamma x, \gamma y) : \gamma \in \Gamma\}$; and $U(x, \epsilon, \sigma) = \{z \in X : F_\sigma(x, z) < \epsilon\}$. Then the family

$$\{uX \cap U(x, \epsilon, \sigma) : x \in uX, \ \epsilon > 0, \ \sigma \in \Sigma\},$$

forms a basis for the τ-topology on uX.

(b) The operation

$$A \mapsto uX \cap (u \circ A) = u(u \circ A),$$

defines the closure operator of the τ-topology.

(c) The collection of sets

$$\left\{ (\bigcup \{\gamma^{-1}\overline{A} : \gamma \in B\}) \cap G : A, B \subset \Gamma, A \text{ infinite}, u \in \overline{B} \right\}$$

(closure in $\beta\Gamma$) form a basis for the τ-topology on G. For a minimal system (X, Γ) a basis for the τ-topology at x is given by the sets

$$(\bigcup \{\gamma^{-1}U : \gamma \in B\}) \cap X,$$

where $B \subset \Gamma$ is such that $u \in \overline{B} \cap M$ and U is a neighborhood of x in X ([EGS,1], section 2).

(3) A subgroup A of G is τ-closed iff there exists a (not necessarily metrizable) minimal pointed system (X, x_0, Γ) with $\mathcal{G}(X, x_0) = A$. [[It is easily seen that $\mathcal{G}(X, x_0)$ is always a τ-closed subgroup. For the other direction consider the natural action of the semigroup $\beta\Gamma$ on the Banach algebra $l^\infty(\Gamma) \cong C(\beta\Gamma)$ as a semigroup of continuous linear operators: $U_p f(q) = f(qp)$, $p, q \in \beta\Gamma$, $f \in C(\beta\Gamma)$. Given a τ-closed subgroup A of G, let $\mathcal{A}(A)$ be the closed Γ-invariant subalgebra

$$\{f \in C(\beta\Gamma) : U_\alpha f = f \ \forall \alpha \in A\}.$$

Now let X be the Gelfand space corresponding to the algebra $\mathcal{A}(A)$ (viewed as the set of continuous multiplicative linear functionals on $\mathcal{A}(A)$) and let x_0 be the multiplicative linear functional defined by evaluation at u. Show that $\mathcal{G}(X, x_0) = A$.]] ♠

For a τ-closed subgroup F of G we let

$$F' := \bigcap \{\tau\text{-closure}V : V \ \tau\text{-open neighborhood of } u \text{ in } F\}.$$

Then F' is a τ-closed normal (in fact characteristic) subgroup of F, characterized as the smallest τ-closed subgroup H of F such that F/H is a compact Hausdorff topological group. Using this characteristic property of F' one proves the following ([G,1], X.4.1).

Lemma 3.1 *Let F be a τ-closed subgroup of G, A, B, τ-closed subgroups of F, with B normal in F, then BA is a τ-closed subgroup and $(BA)'A = B'A$.*

One can iterate this operation to obtain the (possibly transfinite) sequence of "derived" groups $F'' = (F')', \ldots, F^{\iota+1} = (F^\iota)', \ldots$, where for a limit ordinal ν, $F^\nu = \bigcap \{F^\iota : \iota < \nu\}$. For some ordinal this process stabilizes (i.e., there exists a minimal ordinal η such that $F^{\eta+1} = F^\eta$), and we denote $F^\infty = F^\eta$. Notice that iteration in lemma 3.1. yields

$$(BA)^\infty A = B^\infty A.$$

If $F \subset G$ is a τ-closed subgroup, $\mathcal{G}(X, x_0) \subset F$ and $x = px_0$ for $p \in M$ then

$$pF'x_0 = \bigcap\{\tau\text{-closure}(O) : O \text{ is a relative } \tau\text{-neighborhood of } x \text{ in } pFx_0\}.$$

Ellis groups reflect dynamical properties in various ways. Here are some examples.

Exercises:

(4) An extension of minimal systems $(X, \Gamma) \xrightarrow{\pi} (Y, \Gamma)$ is proximal iff $\mathcal{G}(X, x_0) = \mathcal{G}(Y, y_0)$ ([G,1] for exercises (4)–(7)).

(5) It is distal iff for every $y \in Y$, and $x \in \pi^{-1}(y)$, $\pi^{-1}(y) = \mathcal{G}(Y, y)x$ iff for every $y = py_0 \in Y$, p an element of M, $\pi^{-1}(y) = p\pi^{-1}(y_0) = pFx_0$, where $F = \mathcal{G}(Y, y_0)$. In particular (X, Γ) is distal iff $Gx = X$ for some (hence every) $x \in X$.

(6) It is isometric iff it is distal and $\mathcal{G}(Y, y_0)' \subset \mathcal{G}(X, x_0)$. In particular (X, Γ) is equicontinuous iff it is distal and $G' \subset \mathcal{G}(X, x_0)$.

(7) It is RIC iff for every $y = py_0 \in Y$, p an element of M, $\pi^{-1}(y) = p \circ u\pi^{-1}(y_0) = p \circ Fx_0$, where $F = \mathcal{G}(Y, y_0)$. Every distal extension is RIC, hence every distal extension is open. [[Recall that in general $\pi^{-1} : Y \to 2^X$ is an upper-semi-continuous set-valued map and that $\pi : X \to Y$ is open iff $\pi^{-1} : Y \to 2^X$ is continuous.]]

(8) A topological space X is called **maximally compact** if every compact subset of X is closed. Show that G with its τ-topology is not maximally compact. [[Let X be any distal minimal system which is not equicontinuous. Let $A = \mathcal{G}(X, x_0)$ then, since (X, T) is not equicontinuous, G' is not contained in A and $G/A \cong X$ with the quotient τ-topology is not Hausdorff. This in particular, implies the existence of a closed subset $F \subset X$ which is not τ-closed. It follows that $\hat{F} = \{\alpha \in G : \alpha x_0 \in F\}$ is not a τ-closed subset of G. However by exercise (1), $\hat{F}$ is τ-compact. In fact if β_i is a net in $\hat{F}$ then some subnet converges in M, say $\beta_{i'} \to p$. Hence in the τ-topology $\beta_{i'} \to up$ in G and therefore $\beta_{i'}x_0 \to upx_0 = px_0$ in the τ-topology on X. Finally since F is closed, we have $px_0 \in F$ hence $up \in \hat{F}$. Thus $\hat{F}$ is τ-compact but not τ-closed in G.]] ♠

Given a homomorphism $(X, \Gamma) \xrightarrow{\pi} (Y, \Gamma)$ of minimal systems with say $A = \mathcal{G}(X, x_0) \subset F = \mathcal{G}(Y, y_0)$, we would like to apply our "homogeneous space" viewpoint and interpolate in the corresponding extension $G/A \to G/F$ an intermediate homogeneous space G/B (i.e., $A \subset B \subset F$) such that the map $G/B \to G/F$ will be a maximal Hausdorff homogeneous space extension (i.e., the fiber F/B will be Hausdorff in the τ-topology).

Now on the level of groups we know how to do this; just take $B = F'A$. Unfortunately it is not always possible to find an intermediate extension $X \to$

$Z \to Y$ such that $Z \to Y$ is an isometric extension with $\mathcal{G}(Z, z_0) = F'A$. This extension is "virtually" there, as we can see from the group picture, but not necessarily there in reality. In order to unfold this extension on the level of systems we need to modify our systems. We do this via the RIC-shadow diagram.

Let $(X, \Gamma) \xrightarrow{\pi} (Y, \Gamma)$ be a homomorphism of minimal systems, one constructs a commutative diagram of homomorphisms of minimal systems (the **RIC-shadow diagram**),

$$
\begin{array}{ccc}
X & \xleftarrow{\tilde{\theta}} & \tilde{X} \\
{\scriptstyle \pi}\downarrow & & \downarrow{\scriptstyle \tilde{\pi}} \\
Y & \xleftarrow{\theta} & \tilde{Y}
\end{array}
$$

where $\tilde{\pi}$ is RIC and $\theta, \tilde{\theta}$ are proximal (thus we still have $A = \mathcal{G}(X, x_0) = \mathcal{G}(\tilde{X}, \tilde{x}_0)$ and $F = \mathcal{G}(Y, y_0) = G(\tilde{Y}, \tilde{y}_0)$). The concrete description of these objects uses quasifactors and the circle operation:

$$\tilde{Y} = \{p \circ Fx_0 : p \in M\}, \qquad \tilde{X} = \{(x, \tilde{y}) : x \in \tilde{y} \in \tilde{Y}\} \subset X \times \tilde{Y}$$

and

$$\theta(p \circ Fx_0) = py_0, \quad \tilde{\theta}(x, \tilde{y}) = x, \quad \tilde{\pi}(x, \tilde{y}) = \tilde{y}, \quad (p \in M),$$

where $F = \mathcal{G}(Y, y_0)$. In particular θ and $\tilde{\theta}$ are isomorphisms when and only when π is already RIC.

Exercises:
(9) Show that $\tilde{X} = X \vee \tilde{Y}$ (hence $\tilde{X}$ is minimal) and that the extension $\tilde{\pi}$ is RIC ([G,1]). ♠

If we let, for a homomorphism $(X, \Gamma) \xrightarrow{\pi} (Y, T)$,

$$Y^* = \{p \circ \pi^{-1}(y_0) : p \in M\}, \qquad X^* = \{(x, y^*) : x \in y^* \in Y^*\} \subset X \times Y^*$$

and

$$\theta(p \circ \pi^{-1}(y_0)) = py_0, \quad \theta^*(x, y^*) = x, \quad \pi^*(x, y^*) = y^* \quad (p \in M),$$

we get an **O-shadow diagram**

$$
\begin{array}{ccc}
X & \xleftarrow{\theta^*} & X^* \\
{\scriptstyle \pi}\downarrow & & \downarrow{\scriptstyle \pi^*} \\
Y & \xleftarrow{\theta} & Y^*.
\end{array}
$$

(10) Show that π^* is open and θ and θ^* are almost 1-1 extensions. $X^* = X \vee Y^*$ so that X^* is minimal, and θ is an isomorphism iff π is open ([V,2]). ♠

For RIC extensions our group picture of interpolating a minimal group B between the groups $F = \mathcal{G}(Y, y_0)$ and $A = \mathcal{G}(X, x_0)$ such that F/B is a Hausdorff homogeneous space can be realized as the following theorem shows (see e.g., [G,1] and [A]).

Theorem 3.2

Let $(X, \Gamma) \xrightarrow{\pi} (Y, \Gamma)$ be a RIC-extension of minimal systems.

(1) There exists a commutative diagram

$$
\begin{array}{ccc}
 & X & \\
\pi \downarrow & & \searrow \sigma \\
Y & \xleftarrow{\ \ \ \rho\ \ \ } & Z
\end{array}
$$

where ρ is an isometric extension with $B = F'A$, $B = \mathcal{G}(Z, z_0)$, $F = \mathcal{G}(Y, y_0)$, and $A = \mathcal{G}(X, x_0)$.

(2) We have $Q_\pi = R_\sigma$ so that Q_π is an icer and the system $Z = X/Q_\sigma$ is the largest isometric extension of Y under X.

(3) ρ is trivial (i.e., an isomorphism) iff π is a weakly mixing extension iff $Q_\pi = R_\pi$. Moreover in this case, for every $y \in Y$ there exists an idempotent $w \in J$, with $wy = y$, such that for every $x \in \pi^{-1}(y)$ there is a residual subset $L \subset$ closure$(w\pi^{-1}(y))$ with the property that for each $x' \in L$, $\bar{\mathcal{O}}_\Gamma(x, x') = R_\pi$.

(4) For (X, Γ) incontractible—taking (Y, Γ) to be the trivial one point system—Q_X is an icer and the system $Z = X/Q_X$ is the largest equicontinuous factor of X.

(5) For (X, Γ) incontractible, (X, Γ) admits no nontrivial isometric equicontinuous factors iff it is a weakly mixing system iff $Q_X = X \times X$. Moreover in this case there exists an idempotent $w \in J$ such that for every $x \in X$ there is a residual subset $L \subset$ closure wX with $\bar{\mathcal{O}}_\Gamma(x, x') = X \times X$ for each $x' \in L$.

We call an extension $(X, \Gamma) \xrightarrow{\pi} (Y, \Gamma)$ of minimal systems **virtually weakly mixing** if the corresponding RIC-shadow extension $\tilde{X} \xrightarrow{\tilde{\pi}} \tilde{Y}$ is weakly mixing. Thus $(X, \Gamma) \xrightarrow{\pi} (Y, \Gamma)$ is virtually weakly mixing iff $F'A = F$ where

$F = \mathcal{G}(Y, y_0)$ and $A = \mathcal{G}(X, x_0)$ (see exercise 1.(32) for an example of a virtually weakly mixing extension which is not weakly mixing).

Exercises:

(11) Show that for the Morse system (exercise 1.(27),(28)), the extension $\pi = \sigma \circ \rho : X \to Z$, though not weakly mixing, does not admit an intermediate nontrivial isometric extension; i.e., in every diagram $X \overset{\alpha}{\to} U \overset{\beta}{\to} Z$ with $\pi = \beta \circ \alpha$ and β an isometric extension, β is 1-1.

(12) Let $X = \mathbb{T} = \mathbb{R}/\mathbb{Z}$ and $\alpha \in \mathbb{R} \setminus \mathbb{Q}$. Let a and b be the self homeomorphisms of X defined by:

$$a(t) = t + \alpha \ (\mod 1); \qquad b(t) = (t - \frac{[4t]}{4})^2 + \frac{[4t]}{4}, \qquad 0 \le t < 1,$$

and let Γ be the group generated by a and b. Show that the system (X, Γ) is minimal and weakly mixing but the relation Q_X is not an equivalence relation ([M,1]). ♠

4   PI-towers and the structure theorem for minimal systems.

We say that a minimal system (X, Γ) is a **strictly PI system** if there is a (countable) ordinal η and a family of systems $\{(W_\iota, w_\iota)\}_{\iota \le \eta}$ such that (i) W_0 is the trivial system, (ii) for every $\iota < \eta$ there exists a homomorphism $\phi_\iota : W_{\iota+1} \to W_\iota$ which is either proximal or isometric, (iii) for a limit ordinal $\nu \le \eta$ the system W_ν is the inverse limit of the systems $\{W_\iota\}_{\iota < \nu}$, and (iv) $W_\eta = X$. We say that (X, Γ) is a **PI-system** if there exists a strictly PI system $\tilde{X}$ and a proximal homomorphism $\theta : \tilde{X} \to X$.

If in the definition of PI-systems we replace proximal extensions by almost 1-1 extensions we get the notion of **AI-systems**. If we replace the proximal extensions by trivial extensions (i.e., we do not allow proximal extensions at all) we have **I-systems**.

The first structure theorem was proved by Furstenberg for distal systems in 1963 ([F,1]). In our new terminology it says:

Structure theorem for distal systems *A minimal system is distal iff it is an I-system.*

Let us see next how this theorem is used to prove the fixed point theorem of the introduction (theorem 0.2).

Proof of theorem 0.2. By assumption the Γ-action on $X = \text{ext}(Q)$ is distal. If we take X' any minimal subset of X then it is enough to show that $Q' = \overline{\text{co}}(X')$ contains a fixed point. Thus we may assume that X is minimal.

By Furstenberg's theorem X is an I-system. Now the point is that an I-system possesses an invariant probability measure. We see this by transfinite induction. To begin with the system X must have a nontrivial Kronecker factor (otherwise the tower collapses). Such a system is a homogeneous space of a compact group and has its Haar measure. If X_ι is any stage in the I-tower for X with ι a nonlimit ordinal, then $X_\iota \to X_{\iota-1}$ is an isometric extension and we know that there exists a diagram

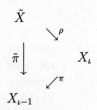

where $\tilde{X} \xrightarrow{\tilde{\pi}} X_{\iota-1}$ is a group extension with some compact group K so that $X_{\iota-1} = \tilde{X}/K$ and $X_\iota = \tilde{X}/H$.

If we assume that $X_{\iota-1}$ admits a Γ-invariant measure then the "Haar lift" of this measure to $\tilde{X}$ is an invariant measure on $\tilde{X}$ and this projects to an invariant measure on X_ι. Clearly the property of possessing an invariant measure lifts to inverse limit of systems. Now transfinite induction establishes the existence of a Γ-invariant probability measure for any I-system, hence for every distal system.

To finish the proof let $B : M(X) \to Q$ be the barycenter map sending a measure on X to its barycenter in Q. This map is affine continuous and Γ-equivariant. Hence the image under B of an invariant measure in $M(X)$ is a fixed point in Q. $\square$

The next structure theorem is due to Veech, 1970, and Ellis, 1973 ([V,1],[E,3]).

Structure theorem for point distal systems *A minimal dynamical system is point distal iff it is an AI-system.*

The fact that every AI-system is point distal is easily proved by transfinite induction on the strict AI-tower (see exercise 1.(23),(25)). In fact this proof yields a dense G_δ subset of distal points.

In his proof of the other direction of the theorem W. Veech assumed that the minimal point distal system has a dense G_δ set of distal points. The later proof of R. Ellis was free of this extra assumption and therefore established the fact that the set of distal points of a point distal system is dense G_δ. This structure theory proof is the only known proof of this simple statement.

Exercises:
(1) A substitution minimal system arising from a primitive substitution of constant length is an AI system. Show that the height of an AI-tower of such a system is finite ([Ma]).
(2) There are examples of such systems which are not strictly AI ([Ma]). ♠

Can one formulate a structure theorem for the general minimal system?

Given a homomorphism $(X, \Gamma) \xrightarrow{\pi} (Y, \Gamma)$ of minimal systems, we would like to find the largest isometric extension of Y under X, say Z_1, then the largest isometric extension of Z_1 under X, say Z_2 etc., until we will no longer be able to form a new isometric extension. Hopefully this process will lead us to a better understanding of the nature of the extension $(X, \Gamma) \xrightarrow{\pi} (Y, \Gamma)$. In particular when Y is the trivial one point system, this may give us the structure of X. Now the first obstruction to this procedure may arise if π is not RIC (X is not incontractible, in the absolute case). Here we will use our RIC-shadow diagram and get a modified homomorphism $X_1 \to Y_1$ which is RIC. Now we can use theorem 3.2 and construct a maximal intermediate isometric extension $X_1 \to Z_1 \to Y_1$. We may then find that the extension $X_1 \to Z_1$ is not RIC and again will be forced to apply our RIC-shadow construction to get a RIC extension $X_2 \to Y_2$, etc. etc.. This procedure can be realized and we then discover that the final obstruction to forming further isometric extensions is relative weak mixing. This (in the absolute case) is the content of the following theorem (Ellis-Glasner-Shapiro, 1975 and McMahon, 1976) ([EGS,1],[M,1]).

Structure theorem for minimal systems *Given a metric minimal system* (X, Γ)*, there exists a countable ordinal* η *and a canonically defined commutative diagram (the canonical PI-Tower)*

where for each $\nu \leq \eta, \pi_\nu$ *is RIC,* ρ_ν *is isometric,* $\theta_\nu, \tilde{\theta}_\nu$ *are proximal and* π_∞ *is RIC and weakly mixing. For a limit ordinal* ν*,* X_ν, Y_ν, π_ν *etc. are the inverse limits (or joins) of* $X_\iota, Y_\iota, \pi_\iota$ *etc. for* $\iota < \nu$*. Thus* X_∞ *is a proximal extension of* X *and a RIC weakly mixing extension of the stictly PI-system* Y_∞*. The homomorphism* π_∞ *is an isomorphism (so that* $X_\infty = Y_\infty$*) iff* X *is a PI-system.*

With a little more effort one can obtain for each positive integer n, an improved PI-tower:

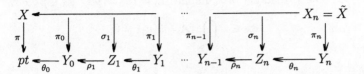

where $\tilde{X} = X_n$ is a proximal extension of X and all the maps π_j are RIC. We call this the *complete canonical tower of order n* associated with (X, Γ) (see [G,5]).

For a distal system (X, Γ) every factor map $X \to Y$ is distal hence RIC and it follows that, in terms of the canonical PI-tower, $X_\infty = Y_\infty = X$. Thus X is an I-system and Furstenberg's theorem follows. Next we would like to deduce the Veech-Ellis theorem from the general structure theorem ([V,2]).

We will need some new definitions and notations. Let $X \xrightarrow{\pi} Y$ be a homomorphism of minimal systems. We say that a point $x \in X$ is **Y-distal** if $P[x] \cap \pi^{-1}(\pi(x)) = \{x\}$. The homomorphism π is called a **point distal homomorphism** if there exists in X a Y-distal point. Whenever $A \xrightarrow{\phi} B$ is a map we let A_ϕ denote the set of ϕ-continuity points in A. Finally recall that a map $\theta : Y \to 2^X$ is called **upper-semi-continuous** if for every convergent sequence $y_i \to y$ we have $\limsup \theta(y_i) \subset \theta(y)$; and that the set of continuity points of such a map is residual.

Our task is to show that if $(X, \Gamma) \xrightarrow{\pi} (Y, \Gamma)$ is a point distal homomorphism then the corresponding RIC-shadow diagram is actually an O-shadow diagram (see exercise 2.(9)). This will follow from the following lemma.

Lemma 4.1 *Let $(X, x_0, \Gamma) \xrightarrow{\pi} (Y, y_0, \Gamma)$ be a point distal homomorphism where x_0 is a Y-distal point. Put $F = \mathcal{G}(Y, y_0)$ and let $\psi : Y \to 2^{2^X}$ be defined by*

$$\psi(y) = \{p \circ Fx_0 : p \in M, \ py_0 = y\}.$$

Then

(1) $\tilde{Y} = \{y \in Y : \bigcap_{A \in \psi(y)} A \neq \emptyset\}$ is residual in Y.

(2) For a residual subset $Z \subset Y$, $z \in Z$ implies $\psi(z) = \{\pi^{-1}(z)\}$.

(3) The quasifactors $\{p \circ Fx_0 : p \in M\}$ and $\{p \circ \pi^{-1}(y_0) : p \in M\}$ coincide.

Proof. (1) Let $\theta : Y \to 2^X \cup \{\emptyset\}$ be defined by

$$\theta(y) = \cap \psi(y) = \bigcap_{A \in \psi(y)} A.$$

We first show that $\tilde{Y} \neq \emptyset$. Let $p \in M$ satisfy $py_0 = y_0$; let $p = v\alpha$ with $v \in J$ and $\alpha \in G$. The Y-distality of x_0 implies $x_0 = vx_0 = p\alpha^{-1}x_0 \in p \circ Fx_0$. Thus

$$x_0 \in \bigcap_{A \in \psi(y_0)} A = \theta(y_0) \neq \emptyset,$$

and $y_0 \in \tilde{Y}$.

Since θ is equivariant $\tilde{Y} = \{y : \theta(y) \neq \emptyset\}$ is invariant and therefore dense in Y. It is easy to check that θ is upper-semi-continuous, hence Y_θ is residual. Since $\{\emptyset\}$ is an isolated point of $2^X \cup \{\emptyset\}$ it follows that $Y_\theta \subset \tilde{Y}$; thus $\tilde{Y}$ is residual.

(2) Let $Z = Y_\theta$ and choose $z \in Z$. By minimality of X, $\cup \Gamma \theta(z)$ is dense in X. Therefore $\limsup_{\gamma z \to z} \theta(\gamma z) = \pi^{-1}(z)$. Since $z \in Y_\theta$ we have $\theta(z) = \bigcap_{A \in \psi(z)} A = \pi^{-1}(z)$. Thus $\psi(z) = \{\pi^{-1}(z)\}$.

(3) Since $\pi^{-1} : Y \to 2^X$ is upper-semi-continuous, $Y_{\pi^{-1}}$ is residual. Hence $Y_\theta \cap Y_{\pi^{-1}}$ is residual and if $z \in Y_\theta \cap Y_{\pi^{-1}}$ with $z = py_0$, $p \in M$ then

$$p \circ Fx_0 = \pi^{-1}(z) = p \circ \pi^{-1}(y_0).$$

Thus the quasifactors coincide. $\square$

Back to the proof of the Veech-Ellis theorem. We now see that (i) the canonical PI-tower for a point distal system (X, Γ) reduces to an AI-tower, and (ii) the canonical proximal homomorphism $X_\infty \to X$ is in fact almost 1-1. The latter fact implies that also X_∞ is a point distal system and therefore our theorem will follow if we show that a point distal system is a PI-system, so that $X_\infty = Y_\infty$ is a strictly AI-system. Now let x_0 be a distal point in X_∞ by theorem 3.2.(3) there exists an idempotent $w \in J$, with $wy_0 = y_0$ where $y_0 = \pi_\infty(x_0)$, and a residual subset $L \subset \text{closure}(w\pi^{-1}(y_0))$ such that for each $x' \in L$, $\bar{O}_\Gamma(x_0, x') = R_{\pi_\infty}$. In particular every such pair (x_0, x') is a proximal pair and as x_0 is a distal point we conclude that $\text{closure}(w\pi^{-1}(y_0)) = \text{closure}(wFx_0)$ is the singleton $\{x_0\}$, where $F = \mathcal{G}(Y_\infty, y_0)$. Thus also $Fx_0 = x_0$, hence $F = A = \mathcal{G}(X_\infty, x_0)$, and since π_∞ is RIC, this means that π_∞ is an isomorphism; i.e., $X_\infty = Y_\infty$. We have shown that a point distal system is a PI-system and our proof of the Veech-Ellis theorem is complete.

Our next application of the general structure theorem is the following characterization of prodal systems due to Bronstein [B,2]. Recall that a system (X, Γ) is called prodal if every topologically transitive subsystem of $X \times X$ with a dense set of almost periodic points is necessarily minimal.

Structure theorem for prodal systems *A minimal dynamical system is prodal iff it is a PI-system.*

Proof. The fact that every PI-system is prodal is easily proved by transfinite induction on the strict PI-tower (exercise 1.(20)). For the other direction we observe that in the canonical PI-tower constructed for the prodal system (X, Γ), the extension $X_\infty \to X$ being proximal, X_∞ is prodal as well (exercise 1.(20)). Now since the extension $X_\infty \xrightarrow{\tilde{\pi}} Y_\infty$ is RIC, the almost periodic points in $R_{\tilde{\pi}}$ are dense. Since $\tilde{\pi}$ is weakly mixing $R_{\tilde{\pi}}$ is topologically transitive. Thus the fact that X_∞ is prodal implies that $R_{\tilde{\pi}}$ is minimal. However the relation $R_{\tilde{\pi}}$ contains the diagonal, whence coincides with the diagonal; i.e., $\tilde{\pi}$ is an isomorphism and $X_\infty = Y_\infty$ is a strictly PI-system. This means that X is a PI-system and the proof is complete. $\square$

Corollary. *A factor of a PI-system is PI.*

Proof. Follows from the fact that the same is true for prodal systems (see exercise 1.(20)). $\square$

Exercises:
(1) Another condition on a minimal system (X, Γ) which is equivalent to prodal is that for some (hence every) point $x \in X$ with $ux = x$, $\mathcal{G}(X, x) \supset G^\infty$. This characterization of PI-systems yields an alternative proof to the preceding corollary ([G,1]). ♠

Finally we have the following considerably simplified structure theorem for normal systems. A minimal system (X, Γ) is called **normal** if its Ellis group is normal: $\mathcal{G}(X, x_0) \lhd G$. It is easy to see that regular systems are normal and, since proximal extensions do not change Ellis group, every proximal extension of a regular system is normal as well.

Structure theorem for normal systems *Let (X, Γ) be a normal system, then there exists a proximal extension $\theta : \tilde{X} \to X$ such that the normal minimal system $\tilde{X}$ has the form: $\tilde{X} \xrightarrow{\phi} Z \xrightarrow{\rho} Y$ where ϕ is virtually weakly mixing, ρ is a group extension and Y is a proximal system. If Γ is abelian then Y is trivial, $X = \tilde{X}$ and Z is the maximal Kronecker factor of X.*

For the proof we refer to [GMS]. Let us remark that the general structure theorem as well as the other structure theorems stated in this section have relativized analogue.

5 The action of $\tau = T \times T^2 \times \cdots T^n$ on points of the diagonal.

In this final section we present a result based on the general structure theorem for minimal $\mathbb{Z}$-systems. The motivation comes from ergodic theory.

H. Furstenberg's multiple recurrence theorem, [F,2] (a direct consequence of which is Szemerédi theorem on arithmetic progressions), is derived from the following statement:

Given an arbitrary ergodic measure preserving system $(X, \mathcal{B}, \mu, T)$, a function $0 \leq f \in L_\infty(\mu)$ not a.e. 0, and k a positive integer,

$$\liminf_{N-M\to\infty} \frac{1}{N-M} \sum_{n=M+1}^{N} \int f(x)f(T^n x)f(T^{2n}x)\cdots f(T^{(k+1)n}x) > 0.$$

Now theorem 10.2 in [F,2] asserts that in this lim inf one can replace the function f by the function

$$\bar{f} = E(f|Y_k),$$

where Y_k is the rank-k distal factor of $(X, \mathcal{B}, \mu, T)$ in its canonical representation as a weakly mixing extension of a distal system, and $E(f|Y_k)$ is the conditional expectation of f with respect to this factor. In fact what Furstenberg shows is that any measure of the form

$$\lim_{N_i-M_i\to\infty} \frac{1}{N_i-M_i} \sum_{n=M_i+1}^{N_i} (T \times T^2 \times \cdots T^{k+1})^n \mu_\Delta^{k+1},$$

where μ_Δ^{k+1} is the diagonal measure on Δ_{k+1}, equals the $k+1$-fold relative product measure over the factor Y_k. In the terminology of [F,4] Y_k, the rank-k distal factor of X, is a *characteristic factor* for τ_{k+1} in the sense that the asymptotic behavior of the transformation $\tau = T \times T^2 \times \cdots T^{k+1}$, acting on points of the diagonal $\Delta_{k+1}(X)$, is determined by the action of τ on $\Delta_{k+1}(Y_k)$. The proof is then concluded by showing that for finite rank distal systems the above lim inf is in fact positive (theorem 11.3 in [F,2]).

Theorem 5.1 below is a topological analogue of this reduction, showing that the maximal rank-k factor in the structure theorem is a "topological characteristic factor" for τ_{k+1}.

The questions whether for each k the weak* limit measure

$$\lim_{N\to\infty} \frac{1}{N} \sum_{n=1}^{N} (T \times T^2 \times \cdots T^k)^n \mu_\Delta^k$$

exists, and if it does is it ergodic, are important open questions in ergodic theory ([F,4]). Proposition 5.4 can be viewed as an answer to a topological analogue of these questions.

Unfortunately the topological multiple recurrence theorem for minimal systems (the topological dynamics equivalent of van der Waerden's theorem) does not seem to follow easily from theorem 5.1. In any case the easy topological dynamics proof of this theorem in [FW] and the elegant algebraic proofs in the literature (see, e.g., [BF] and the even shorter proof in exercise (1) below) make such an effort pointless.

Theorem 5.1 *Let (X,T) be a metric minimal system, $n \geq 1$ a positive integer. Let $\tau = T \times T^2 \times \ldots T^{n+1}$. Let $\tilde{X}$ be the proximal extension of X given by the complete canonical PI-tower of order n, $(*)$, (see the remark following the structure theorem for minimal systems in section 4). Then there exists a dense G_δ subset Ω of $\tilde{X}$, such that for $x \in \Omega$, $L = \bar{\mathcal{O}}_\tau(x, x, \ldots, x)$ is π_n-saturated, where π_n is the projection of $\tilde{X}$ onto Y_n, the rank-n PI-factor of $\tilde{X}$ (L is π_n saturated means that for every $x \in L$ also $\pi_n^{-1}(\pi_n(x)) \subset L$).*

There are two obvious cases when in the complete PI-tower $\tilde{X} = X$. The first is when (X,T) is distal and every factor map is distal hence RIC. The second, is when (X,T) is weakly mixing. In this latter case all π_n's are the trivial map $X \to pt$.

Corollary 5.2 *Let (X,T) be a metric minimal distal system, $n \geq 1$. Let π_n denote the canonical map of X onto its largest class n distal factor. Let $\tau = T \times T^2 \times \cdots T^{n+1}$. Then there exists a dense G_δ subset Ω of X, such that for $x \in \Omega$, $L = \bar{\mathcal{O}}_\tau(x, x, \ldots, x)$ is π_n-saturated.*

Corollary 5.3 *For a metric minimal weakly mixing system (X,T) and an integer $n \geq 1$, almost every point of the diagonal $\Delta_n(X)$ has a dense $\tau = T \times T^2 \times \cdots T^n$ orbit in $X \times X \cdots \times X$.*

The short and elegant proof of the following proposition—which I need for the proof of theorem 5.1—was indicated to me by Robert Ellis. In the exercises at the end of this section I will show how the multiple recurrence theorem and van der Waerden's theorem on arithmetic progressions can be easily derived from proposition 5.4.

Proposition 5.4 *Let (X,T) be a minimal system. For any $n \geq 1$, let $\tau = T \times T^2 \times \cdots T^n$, $\theta = T \times T \times \cdots T$ and denote by $\mathcal{T}$ the group generated by τ and θ. Let Λ be a θ-minimal subsystem of X^n and set $N = \bar{\mathcal{O}}_\tau(\Lambda)$. Then N is $\mathcal{T}$-minimal and the τ-almost periodic points are dense in N.*

Proof. Let $E = E(N, \mathcal{T})$ be the enveloping semigroup of $(N, \mathcal{T})$. Let $\pi_j : N \to X$ be the projection of N on the j-th component, $j = 1, 2, \ldots, n$, and let $\pi_j^* : E(N, \mathcal{T}) \to E(X, \mathcal{T})$ be the corresponding homomorphism of enveloping semigroups. Notice that $E(X, \mathcal{T}) = E(X, T)$. Let now $u \in E(N, \theta)$ be any minimal idempotent in the enveloping semigroup of (N, θ). Choose v a minimal idempotent in the closed ideal $E(N, \mathcal{T})u$; clearly $vu = v$. Set for each j, $u_j = \pi_j^* u$ and $v_j = \pi_j^* v$; then we have $v_j u_j = v_j$. In particular we deduce that v_j is an element of the minimal ideal of $E(X,T)$ which contains u_j. In turn this implies

$$u_j v_j = u_j v_j u_j = u_j,$$

and it follows that also $uv = u$. Thus u is an element of the minimal ideal of $E(N, \mathcal{T})$ which contains v, and therefore u is a minimal idempotent of $E(N, \mathcal{T})$.

To finish the proof, let x be an arbitrary point in Λ and let $u \in E(N, \theta)$ be a minimal idempotent with $ux = x$. By the above argument, u is also a minimal idempotent of $E(N, T)$, whence $N = \bar{\mathcal{O}}_T(x)$ is T-minimal.

The second assertion follows easily. Let $P \subset N$ be any τ-minimal set. Then $\mathcal{O}_\theta(P)$ is a T-invariant subset of N, hence a dense subset of N. Now every point of $\mathcal{O}_\theta(P)$ is τ-almost periodic and we are done. $\square$

Our main tool in proving theorem 5.1 is the following lemma (lemma 3.3 in [G,5]); we refer to [G,5] for the proof.

Lemma 5.5 *Suppose that for each $\nu \in I$, a commutative diagram*

$$
\begin{array}{ccc}
X_\nu & & \\
\pi_\nu \downarrow & \searrow^{\sigma_\nu} & \\
Y_\nu & \xleftarrow{\rho_\nu} & Z_\nu
\end{array}
$$

of minimal systems is given, in which we have for each ν, $B_\nu = F_\nu' A_\nu$, where $A_\nu = \mathcal{G}(X_\nu, x_\nu^0)$, $F_\nu = \mathcal{G}(Y_\nu, y_\nu^0)$, and $B_\nu = \mathcal{G}(Z_\nu, z_\nu^0)$. Let X, Y, Z, be the corresponding product spaces and π, ρ and σ the corresponding product maps. Let N_Y be a closed invariant subset of Y, and put $N_Z = \rho^{-1}[N_Y]$, and $N_X = \pi^{-1}[N_Y]$. Suppose further that each σ_ν is open and that $\overline{JN_X} = N_X$. Let Q be a closed subset of N_X with the property that $\bar{\mathcal{O}}_T(Q) = N_X$ and such that $\cdot$ for every relatively open subset U of Q, $\text{int}_{N_X}(\bar{\mathcal{O}}_T(U)) \neq \emptyset$. Then there exists a dense G_δ subset $\Omega \subset Q$ such that $x \in \Omega$ implies $L = \bar{\mathcal{O}}_T(x)$ is σ-saturated.

Proof of theorem 5.1. We prove theorem 5.1 by induction on $1 \leq k \leq n$. For each k we prove three claims, $1_k, 2_k$ and 3_k. The theorem will follow directly from claim 3_{n-1}. In order to simplify the notation, we denote the $\tilde{X}$ of the complete canonical tower $(*)$ by X.

Claim 1_1. For $\tau = T \times T^2$

$$\bar{\mathcal{O}}_\tau(\Delta_2(X)) = X \times X.$$

Proof. Since

$$(T \times T^2)^n \Delta_2(X) = (I \times T)^n \Delta_2(X) = \{(x, T^n x) : x \in X, m \in \mathbb{Z}\},$$

the claim follows from the minimality of (X, T). $\square$

For a subset $A \subset X$, we let $A^{(k)} = \{(x, x, \ldots, x) \ (k\text{-times}) : x \in A\}$.

Claim 2_1. For $\emptyset \neq U \subset X$ open

$$\text{int}_{X \times X} \bar{\mathcal{O}}_\tau(U^{(2)}) \neq \emptyset.$$

Proof. There exists $J \geq 1$ such that $\Delta_2(X) = \bigcup_{j=1}^{J}(T \times T)^j U^{(2)}$, hence by claim 1_1,

$$X \times X = \bar{\mathcal{O}}_\tau(\Delta_2(X))$$

$$= \text{closure} \bigcup_{m \in \mathbb{Z}} (T \times T^2)^m (\bigcup_{j=1}^{J}(T \times T)^j U^{(2)})$$

$$= \text{closure} \bigcup_{j=1}^{J}(T \times T)^j \mathcal{O}_\tau(U^{(2)}) = \bigcup_{j=1}^{J}(T \times T)^j \bar{\mathcal{O}}_\tau(U^{(2)}).$$

Therefore $\text{int}\bar{\mathcal{O}}_\tau(U^{(2)}) \neq \emptyset$. $\square$

Claim 3_1. *There exists a dense G_δ subset $\Omega_1 \subset X$ such that for $x \in \Omega_1$*

$$L = \bar{\mathcal{O}}_\tau(x,x) \text{ is } \pi_1\text{-saturated.}$$

Proof. We apply lemma 5.5 with $I = \{1,2\}$, to the following diagrams (notations as in $(*)$, section 4):

$$(X, T^\nu)$$

$$(\nu \in I) \qquad \pi_0 \downarrow \qquad \searrow^{\pi_1}$$

$$(Y_0, T^\nu) \xleftarrow{\lambda_1} (Y_1, T^\nu)$$

where Y_0 is the one point trivial system and $\lambda_1 = \rho_1 \circ \theta_1$. Now let $N_Y =$pt and thus $N_Z = Y_1 \times Y_1$ and $N_X = X \times X$. Clearly $JN_X = N_X$. For the set $Q = \Delta_2(X) \subset N_X$, by claims 1_1 and 2_1, the assumptions of lemma 5.5 are satisfied and, identifying X with $\Delta_2(X)$, we get our Ω_1. $\square$

Claim 1_2. *Let $I = \{1,2,3\}$, $\tau = T \times T^2 \times T^3$, and $N_Y = \bar{\mathcal{O}}_\tau(\Delta_3(Y_1))$. Then N_X (which by definition is $\pi_1^{-1}[N_Y]$), coincides with $\bar{\mathcal{O}}_\tau(\Delta_3(X))$.*

Proof. Clearly $\bar{\mathcal{O}}_\tau(\Delta_3(X)) \subset N_X$. For the converse let $(x_1, x_2, x_3) \in N_X$; since

$$(T \times T^2 \times T^3)^n \Delta_3(X) = (I \times T \times T^2)^n \Delta_3(X),$$

it follows from the definition of N_X that sequences $n_i \in \mathbb{Z}$ and $y_i \in Y_1$ exist with $(I \times T \times T^2)^{n_i}(y_i, y_i, y_i) \to (y_1, y_2, y_3)$ where $y_j = \pi_1(x_j), (j = 1,2,3)$. In particular $y_i \to y_1$ and since π_1 is open, we can find a sequence $x_i \in X$ with $(I \times T \times T^2)^{n_i}(x_i, x_i, x_i) \to (x_1, x_2', x_3')$ for some x_2', x_3' such that $\pi_1(x_1, x_2', x_3') = (y_1, y_2, y_3)$. Let now $\Omega_1 \subset X$ be the residual subset obtained in claim 3_1. Thus $z \in \Omega_1$ implies that $\bar{\mathcal{O}}_{T \times T^2}(z, z)$ is π_1-saturated. Since $(\Omega_1)^{(2)}$ is dense in $\Delta_2(X)$, we can choose $x_i' \in \Omega_1$ with

$$(I \times T \times T^2)^{n_i}(x_i', x_i', x_i') \to (x_1, x_2', x_3').$$

Denote $L_i = \bar{O}_{T \times T^2}(x_i', x_i')$ and use the fact that π_1 is open to get:

$$
\begin{aligned}
(x_1, x_2, x_3) &\in \{x_1\} \times \pi_1^{-1}(y_2) \times \pi_1^{-1}(y_3) \\
&= \{x_1\} \times \pi_1^{-1}(\pi_1(x_2')) \times \pi_1^{-1}(\pi_1(x_3')) \\
&= \lim(\{x_i'\} \times \pi_1^{-1}(\pi_1(T^{n_i}x_i')) \times \pi_1^{-1}(\pi_1(T^{2n_i}x_i')) \\
&\subset \lim(\{x_i'\} \times L_i) \\
&= \lim\{x_i'\} \times \bar{O}_{T \times T^2}(x_i', x_i') \\
&= \lim \bar{O}_{I \times T \times T^2}(x_i', x_i', x_i') \\
&\subset \bar{O}_{I \times T \times T^2}(\Delta_3(X)) = \bar{O}_\tau(\Delta_3(X)).
\end{aligned}
$$

$\square$

Claim 2_2. *For $\emptyset \neq U \subset X$ open,*

$$
\mathrm{int}_{N_X} \bar{O}_\tau(U^{(3)}) \neq \emptyset.
$$

Proof. By claim 1_2 and minimality of (X, T) there exists J with

$$
N_X = \bar{O}_\tau(\Delta_3(X)) = \bar{O}_\tau(\bigcup_{j=1}^{J}(T \times T \times T)^j(U^{(3)}))
$$

and one proceeds as in the proof of claim 2_1. $\square$

Claim 3_2. *There exists a dense G_δ subset $\Omega_2 \subset X$ such that for $x \in \Omega_2$*

$$
L = \bar{O}_\tau(x, x, x) \text{ is } \pi_2\text{-saturated.}
$$

Proof. We apply lemma 5.5 with $I = \{1, 2, 3\}$, to the following diagrams (notations as in $(*)$, section 4):

$$
\begin{array}{ccc}
 & (X, T^\nu) & \\
(\nu \in I) & \pi_1 \downarrow \quad \searrow \pi_2 & \\
(Y_1, T^\nu) & \xleftarrow{\quad\lambda_2\quad} & (Y_2, T^\nu)
\end{array}
$$

where $\lambda_2 = \rho_2 \circ \theta_2$. Now let $N_Y = \bar{O}_\tau(\Delta_3(Y_1))$, then $N_X = \pi_1^{-1}[N_Y]$ and $N_Z = \lambda_2^{-1}[N_Y]$. We have to show that $\overline{JN_X} = N_X$. Now by proposition 5.4 the τ almost periodic points are dense in N_Y, i.e., $\overline{JN_Y} = N_Y$. Let $(y_1, y_2, y_3) =$

$v(y_1, y_2, y_3) = (p_1 y_0, p_2 y_0, p_3 y_0) \in JN_Y$, where $v \in J, p_j \in M, p_j = v\alpha_j, \alpha_j \in G, (j = 1, 2, 3)$. Then since π_1 is RIC,

$$
\begin{aligned}
\pi_1^{-1}(y_1, y_2, y_3) &= \pi_1^{-1}(y_1) \times \pi_1^{-1}(y_2) \times \pi_1^{-1}(y_3) \\
&= (v \circ \alpha_1 F_1 x_0) \times (v \circ \alpha_2 F_1 x_0) \times (v \circ \alpha_3 F_1 x_0) \\
&= \lim(T \times T^2 \times T^3)^{n_k}[(\alpha_1 F_1 x_0) \times (\alpha_2 F_1 x_0) \times (\alpha_3 F_1 x_0)],
\end{aligned}
$$

where $n_k \to v$ in $\beta\mathbb{Z}$. Now for every m, $(T \times T^2 \times T^3)^m[(\alpha_1 F_1 x_0) \times (\alpha_2 F_1 x_0) \times (\alpha_3 F_1 x_0)] \subset JN_X$, and we conclude that $\pi^{-1}(y_1, y_2, y_3) \subset \overline{JN_X}$. Since JN_Y is dense in N_Y and since $\pi_1 : N_X \to N_Y$ is open, we conclude that $\pi_1^{-1}[JN_Y]$ is dense in N_X, whence $\overline{JN_X} = N_X$.

For the set $Q = \Delta_3(X) \subset N_X$, by claim 2_2, the assumptions of lemma 3.3 are now satisfied and, identifying X with $\Delta_3(X)$, we get our Ω_2. $\square$

The set of claims $1_2, 2_2, 3_2$ now generalizes, merely by changing the indices and working in the appropriate product spaces, to sets of claims $1_k, 2_k, 3_k, 1 \le k \le n - 1$, and so do their proofs. Thus by induction the following claim is proved.

Claim 3_{n-1}. *Let $\tau = T \times T^2 \times \cdots T^n$. There exists a dense G_δ subset $\Omega \subset X$ such that for $x \in \Omega$*

$$
L = \bar{O}_\tau(x, x, \ldots, x) \text{ is } \pi_{n-1}\text{-saturated.}
$$

Recalling that we used X to denote $\tilde{X}$, we now see that theorem 5.1 is proved. $\square$

Exercises:

(1) Using proposition 5.4 prove the following:

Multiple recurrence theorem *Let (X, T) be a minimal $\mathbb{Z}$-system, U a non-empty open subset of X. Then for every positive integer n there exists a positive integer k with:*

$$
U \cap T^k U \cap T^{2k} U \cap \cdots \cap T^{(n-1)k} U \ne \emptyset.
$$

[[Choose any $x \in U$. By the proof of proposition 5.4, the point

$$
\hat{x} = (x, x, \ldots, x) \in X^n
$$

is an almost periodic point of the system (N, T). Hence the set

$$
N(\hat{x}, \hat{U}) = \{\gamma \in T : \gamma\hat{x} \in \hat{U}\},
$$

where $\hat{U} = U \times U \times U \cdots \times U$, is a syndetic subset of $\mathcal{T}$. In particular there exist $k > 0$ and $l \in \mathbb{Z}$ such that $\tau^{-k}\theta^l \hat{x} \in \hat{U}$. Thus the point $(T^{l-k}x, \ldots, T^{l-k}x)$ is in the set $U \cap T^k U \cap T^{2k}U \cap \cdots \cap T^{(n-1)k}U$.]]

(2) Use the multiple recurrence theorem to prove:

van der Waerden's theorem *If $\mathbb{Z} = \bigcup_{j=1}^m A_j$, then there exists j such that A_j contains arithmetic progressions of every finite length.*

[[Since

$$\sum_{j=1}^m 1_{A_j} \geq 1$$

it follows that there exists j such that the orbit closure $\bar{\mathcal{O}}_T(1_{Aj})$ in $\{0,1\}^\mathbb{Z}$, contains a minimal subset X which is not the trivial set $\{1_\emptyset\}$. The subset $U = \{x \in X : x(0) = 1\}$ is a nonempty open subset of the minimal system (X, T). Applying the multiple recurrence theorem to U we get $x \in X$ with $x(0) = x(k) = x(2k) = \cdots = x(nk) = 1$. Now the fact that $x \in \bar{\mathcal{O}}_T(1_{Aj})$ yields an arithmetic progression of length $n+1$ in A_j. A compactness argument completes the proof.]]

(3) Prove the following extension of van der Waerden's theorem. Call a subset $A \subset \mathbb{Z}$ **big** if there exists $m \in \mathbb{N}$ such that for every $M \in \mathbb{N}$ there exist intervals $(b,c) \subset \mathbb{Z}$ with $c - b \geq M$ and with the property that every subinterval of (b,c) of length m meets A. Clearly when $\mathbb{Z} = \bigcup_{j=1}^m A_j$, at least one of the subsets A_j is big.

Theorem *If $A \subset \mathbb{Z}$ is big, then for every $l \in \mathbb{N}$ there exists $m \in \mathbb{N}$ such that for every $M \in \mathbb{N}$, there exist two intervals $(b_1, c_1), (b_2, c_2)$ of length at least M, so that for any $x_1 \in (b_1, c_1), x_2 \in (b_2, c_2)$ there is an arithmetic progression of length $l, x_1', \ldots, x_l'$ inside A with $|x_1 - x_1'| < m$ and $|x_2 - x_2'| < m$.*

(See [FG]).♠

References

[A] J. Auslander, *Minimal flows and their extensions*, North-Holland, 1988.

[BF] V. Bergelson, H. Furstenberg, N. Hindman, Y. Katznelson, An algebraic proof of van der Waerden's theorem, *L'Enseignement Math.* **35**, 209–215, 1989.

[B,1] I. U. Bronstein, *Extensions of minimal transformation groups*, Sijthoff & Noordhoff, 1979.

[B,2] I. U. Bronstein, A characteristic property of PD-extensions, *Bul. Akad. Stiimce RSS Moldoven*, **3**, 11–15, 1977 (Russian).

[E,1] R. Ellis, Locally compact transformation groups, *Duke Math. Journal*, **24**, 119–125, 1957.

[E,2] R. Ellis, *Lectures in topological dynamics*, Benjamin, New York, 1969.

[E,3] R. Ellis, The Veech structure theorem, *Trans. Amer. Math. Soc.*, **186**, 203–218, 1973

[EGS,1] R. Ellis, S. Glasner, L. Shapiro, Proximal-Isometric flows, *Advances in Math.* **17**, 213–260, 1975.

[EGS,2] R. Ellis, S. Glasner, L. Shapiro, Algebraic equivalents of flow disjointness, *Illinois J. Math.*, **20**, 354–360, 1976.

[F,1] H. Furstenberg, The structure of distal flows, *Amer. J. Math.*, **85**, 477–515, 1963.

[F,2] H. Furstenberg, Ergodic behavior of diagonal measures and a theorem of Szemerédi, *J. d'Analyse Math.*, **31**, 204–256, 1977.

[F,3] H. Furstenberg, *Recurrence in ergodic theory and combinatorial number theory*, Princeton Univ. Press, Princeton, N.J., 1981.

[F,4] H. Furstenberg, Nonconventional ergodic averages, *Proc. Symp. Pure Math.*, **50**, 43–56, 1990.

[FG] H. Furstenberg and E. Glasner, Subset dynamics and van der Waerden's theorem, *Contemporary Math.*, **215**, 197–203, 1998.

[FW] H. Furstenberg and B. Weiss, Topological dynamics and combinatorial number theory, *J. d'Analyse Math.*, **34**, 61–85, 1978.

[G,1] S. Glasner, *Proximal flows*, Springer Verlag, Lecture Notes in Math. 517, 1976.

[G,2] E. Glasner, Book review, *Bull. Amer. Math. Soc.*, **21**, 316–319, 1989.

[G,3] E. Glasner, Regular PI metric flows are equicontinuous, *Proc. Amer. Math. Soc.*, **114**, 269–277, 1992.

[G,4] E. Glasner, Minimal nil-transformations of class two, *Israel J. of Math.*, **81**, 31–51, 1993.

[G,5] E. Glasner, Topological ergodic decompositions and applications to products of powers of a minimal transformation, *J. Anal. Math.*, **64**, 241–262. 1994.

[GMS] E. Glasner, M. K. Mentzen, A. Siemaszko, A natural family of factors for minimal flows, *Contemporary Math.*, **215**, 19–42, 1998.

[GW,1] S. Glasner and B. Weiss, On the construction of minimal skew products, *Israel J. Math.*, **34**, 321–336, 1979.

[GW,2] S. Glasner and B. Weiss, Interpolation sets for subalgebras of $l^\infty(\mathbb{Z})$, *Israel J. Math.*, **44**, 345–360, 1983.

[H] G. A. Hedlund, Fuchsian groups and transitive horocycle, *Duke Jour. of Math.*, **2**, 530–542, 1936.

[J] A. del Junco, A simple measure preserving transformation with trivial centralizer, *Pacific Jour. Math.*, **79**, 357–362, 1978.

[K] M. Keane, Generalized Morse sequences, *Zeit. Wahrsch.*, **10**, 335–353, 1968.

[KR] H. Keynes and J. B. Robertson, Eigenvalues theorems in topological transformation groups, *Trans. Amer. Math. Soc.*, **139**, 359–369, 1968.

[M,1] D. C. McMahon, Weak mixing and a note on the structure theorem for minimal transformation groups, *Illinois J. of Math.*, **20**, 186–197, 1976.

[M,2] D. C. McMahon, Relativized weak mixing of uncountable order, *Can.J. of Math.*, **32**, 559–566, 1980.

[Ma] J. C. Martin, Substitution minimal flows, *Amer. J. of Math.*, **93**, 503–526, 1971.

[MP] N. G. Markley and M. E. Paul, Almost automorphic minimal sets without unique ergodicity, *Israel J. Math.*, **34**, 259–272, 1979.

[P] K. Petersen, *Ergodic theory*, Cambridge Univ. Press, Cambridge, 1983.

[Q] M. Queffélec, *Substitution dynamical systems—spectral analysis*, Springer Verlag, Lecture Notes in Math. 1294, 1987.

[V,1] W. A. Veech, Point-distal flows, *Amer. J. Math.*, **92**, 205–242, 1970.

[V,2] W. A. Veech, Topological dynamics, *Bull. Amer. Math. Soc.*, **83**, 175–242, 1977.

[Vr] J. de Vries, *Elements of topological dynamics*, Kluwer Academic Publishers, 1993.

[W] S. Williams, Toeplitz minimal flows which are not uniquely ergodic, *Zeit. Wahrsch.*, **67**, 95–107, 1984.

Orbit Properties of Pseudo-homeomorphism Groups of a Perfect Polish Space and their Cocycles

Valentin Ya. Golodets, Vyacheslav M. Kulagin,
and Sergey D. Sinel'shchikov

B. Verkin Institute for Low Temperature Physics and Engineering
National Academy of Sciences of Ukraine
Kharkov, Ukraine

Abstract. Let Γ be a countable group of pseudo-homeomorphisms of a perfect Polish space X. In a well known work by Sullivan, Weiss, and Wright, the orbit properties of such groups are described. We consider ergodic cocycles for such group actions (in other terms, cocycles with dense ranges) with values in Polish groups. The uniqueness theorem for such cocycles is proved, specifically, that modulo a meagre subset of X, any two ergodic cocycles α and β are weakly equivalent (α is cohomologous to β up to automorphisms of a full group normalizer $N[\Gamma]$). This result is applied to establish the outer conjugacy of countable groups of pseudo-homeomorphisms from the normalizer $N[\Gamma]$ of the full group $[\Gamma]$. Another application here is to the subrelations of ergodic countable equivalence relations on a perfect Polish space. In particular, a complete classification of normal ergodic subrelations is obtained, which presents some new information on the orbit structure for groups of pseudo-homeomorphisms.

1 Introduction

The problem of studying cocycles of automorphism groups on a measure space permanently attracted the attention of specialists because such cocycles carry an important information about a dynamical system. Primarily, this field was discovered by G. Mackey [11], who introduced the basic notions and constructions, which allow one to use cocycles for producing profound invariants of group actions. Among those, one should mention the skew product and the so called Mackey action associated to a cocycle. That was an essential impetus for extensive research on cocycles; by now it advanced appreciably in measure theoretic ergodic theory. In particular, R. Zimmer in his important works used cocycles to introduce the notion of amenability [13] and the property T [14] for group actions. The classification results for cocycles in terms of Mackey actions was obtained by a number of authors (see [7], [8], [9]), starting from the so called uniqueness theorem for cocycles with dense ranges in amenable groups.

A very natural idea is to transfer this class of results onto cocycles of dynamical systems of a different nature, specifically for topological dynamics. A good subject in this context is a group of pseudo-homeomorphisms of a Polish space, which has already been investigated in the work of D. Sullivan, B. Weiss, and J. Wright [12]. This paper contains a description of orbit properties of such groups, which to some extent look like those in measurable dynamics of amenable transformation groups. It certainly hints that a worthwhile theory of cocycles of such dynamical systems could be developed. This is just the problem we deal with in the present work, which is approached here by considering the ergodic cocycles (i.e., cocycles with dense range). It is still open to work out a good approach for studying more general cocycles in this setting.

Our main result is the uniqueness theorem for ergodic cocycles of a countable group of pseudo-homeomorphisms with values in a Polish group (theorem 4.3). As an immediate consequence, we obtain the outer conjugacy for countable subgroups of a full group normalizer (theorem 5.2). That sort of problem was originally considered by A. Connes and W. Krieger [3] for individual automorphisms normalizing a type II full group on a measure space, who found a complete system of invariants for outer conjugacy in this case; a similar problem for type III full groups was solved in [1], [2]. In fact, the outer conjugacy problem is a part of a more general problem in ergodic theory, specifically, the problem of classification of measure space transformations. It should be noted, that, unlike the measure-theoretic case, no reference to the amenability properties is involved. Another of our subjects here is the study of subrelations of generic equivalence relations on Polish spaces. Similar problems in measurable dynamics were considered by J. Feldman, C. Sutherland, and R. Zimmer [6], and after that also by some other authors [4], [5]. In this context, we get a complete classification of normal ergodic subrelations, thus producing analogues of the results of [6]. For that, we adopt the methods used before while proving the uniqueness theorem for cocycles in measurable dynamics, and those applied in [12], to our case.

It is interesting to observe that D. Sullivan, B. Weiss, and J. Wright [12] introduced a new class of generic dynamical systems. The results on their orbit properties were applied to considering the monotone complete C*-algebras. In particular, they solved a well known isomorphism problem for such algebras. We don't intend to go into that class of problems here. However, our results could be also helpful in considering the following questions concerning monotone complete C*-algebras: outer conjugacy for automorphisms of such algebras (see also remark 5.2), and studying their subalgebras, in particular the finite index subalgebras. It is still open to apply the groupoid techniques in this context.

2 Preliminaries

Let X be a perfect Polish space, that is a complete separable metric space without isolated points. Throughout this paper we consider only this kind of space. Note that a dense G_δ subset of such a space is a perfect Polish space again. A Borel bijection Θ of X is said to be a pseudo-homeomorphism of X, provided that A is meagre if and only if $\Theta(A)$ is meagre. Note that each homeomorphism is a pseudo-homeomorphism.

Let $\mathcal{R}$ be an equivalence relation on X. For any $A \subset X$ the saturation of A by $\mathcal{R}$ is the set $\mathcal{R}[A] = \{y \in X : y \overset{\mathcal{R}}{\sim} x \text{ for some } x \in A\}$. As in [12], we say that $\mathcal{R}$ is a *countable generic equivalence relation* if it satisfies the following conditions: $\mathcal{R}$ is a Borel subset of $X \times X$; each equivalence class of $\mathcal{R}$ is countable; the saturation $\mathcal{R}[E]$ of any meagre set $E \subset X$ is meagre.

Given a countable group Γ of pseudo-homeomorphisms of X, one may consider the equivalence relation $\mathcal{R}_\Gamma$ generated by Γ: $\mathcal{R}_\Gamma = \{(x, \gamma x) : x \in X, \gamma \in \Gamma\}$. It is easy to see that in this case $\mathcal{R}_\Gamma$ is a countable generic equivalence relation. As it was shown in [12], by neglecting meagre sets the study of countable equivalence relations reduces to considering countable groups of homeomorphisms. These are just the subjects of sections 2,3,4,5. It will be implicit in the subsequent considerations that everything is done modulo meagre sets.

Let Γ be a countable group of pseudo-homeomorphisms on X. We say that this action is *ergodic* if for some $x \in X$, the orbit Γx is dense in X. Equivalently, each Γ-invariant Borel subset of X is either meagre or the complement of a meagre set. One can also find some other equivalent definitions in [12]. The ergodicity of a generic countable equivalence relation $\mathcal{R}$ on X is defined as the ergodicity of a countable transformation group that generates $\mathcal{R}$.

PROPOSITION 2.1 *Let A_n ($n \in \mathbb{N}$) be a countable family of Borel subsets of X. Then there exists a dense G_δ-subset Y of X, such that $A_n|_Y$ is open in Y for all $n \in \mathbb{N}$.*

Proof: Each A_n has the Baire property, so we can write $A_n = (G_n \backslash P_n) \cup R_n$, where G_n is open set and P_n, R_n are meagre sets (see [10]). Let $P = (P_1 \cup R_1) \cup (P_2 \cup R_2) \cup \ldots$, and $\widetilde{Y} = X \backslash P$. Then, for each n, $A_n \cap \widetilde{Y} = G_n \cap \widetilde{Y}$. $\widetilde{Y}$ also has the Baire property, so $\widetilde{Y} = Y \cup S$, where S is a meagre set and Y is a G_δ-subset of X (see [10]). It is clear that Y is dense in X and $A_n|_Y$ is open in Y for all $n \in \mathbb{N}$. ∎

It follows from Proposition 2.1 that we can always suppose, without loss of generality, that X is totally disconnected.

A Borel function $f : X \to \mathbb{R}$ is called Γ-*almost invariant* if, given any $\gamma \in \Gamma$, $f(\gamma y) = f(y)$ for all y in some dense G_δ-subset $Y \subset X$.

The following proposition is standard:

PROPOSITION 2.2 *Let Γ be a countable group of homeomorphisms of X. Γ is ergodic if and only if every Γ-almost invariant Borel function $f : X \to \mathbb{R}$ is constant modulo a meagre subset of X.*

3 Cocycles: a background

Now we turn to studying cocycles of generic countable equivalence relations. For that, some terminology related to group actions is required. In particular, these are the notions of full group and normalizer of a full group. Some auxiliary results are also presented. Our attention here is basically concentrated on ergodic cocycles.

Let Γ be a countable group of homeomorphisms on X.

Definition 3.1 A pseudo-homeomorphism h of X belongs to the full group $[\Gamma]$, if there exist a sequence of pairwise disjoint clopen subsets $\{K_j\}$ ($j = 1, 2, \dots$) of X and a sequence $\{\gamma_j\}$ ($j = 1, 2, \dots$) in Γ such that $\bigcup K_j$ is dense in X and $hx = \gamma_j x$ for each $x \in K_j$. If $\bigcup K_j = X$, then h is said to be strongly Γ-decomposable over X.

An application of Proposition 2.1 and the properties of pseudo-homeomorphisms readily yields the following:

PROPOSITION 3.1
i) $h \in [\Gamma]$ if and only if there exists a Γ-invariant, dense G_δ-set $Y \subset X$, such that $h|_Y$ is a homeomorphism of Y onto Y and $h|_Y$ is strongly Γ-decomposable over Y.

ii) The set $[\Gamma]$ is a group of pseudo-homeomorphisms.

Definition 3.2 The set of such pseudo-homeomorphism Θ that $\Theta[\Gamma]\Theta^{-1} = [\Gamma]$ is called the normalizer $N[\Gamma]$ of the full group $[\Gamma]$.

Let $\mathcal{R}$ be a countable generic equivalence relation. Let G be a Polish group.

Definition 3.3 A Borel map $\phi : \mathcal{R} \to G$ is called a cocycle of $\mathcal{R}$ with values in the Polish group G if for some $\mathcal{R}$-invariant dense G_δ-subset Y of X $\phi(x, y)\phi(y, z) = \phi(x, z)$ for all $(x, y), (y, z) \in \mathcal{R}|_{Y \times Y}$.

The set of all cocycles of $\mathcal{R}$ is denoted by $Z^1(\mathcal{R}, G)$. We identify two cocycles if they differ only on a meagre set. In the context of this work, only orbit cocycles are considered, that is, cocycles of equivalence relations as in the above definition. If a countable group Γ acts as homeomorphisms on X, then any cocycle of $\mathcal{R}_\Gamma = \{(x, \gamma x) : x \in X, \gamma \in \Gamma\} \subset X \times X$ can be also treated as a cocycle of the Γ-action.

Two cocycles $\alpha, \beta \in Z^1(\mathcal{R}, G)$ are said to be *cohomologous*, if there exist an $\mathcal{R}$-invariant dense G_δ-subset Y of X and a Borel function $f : Y \to G$ such that $\alpha(x, y) = f(x)\beta(x, y)f(y)^{-1}$ for all $(x, y), (y, z) \in \mathcal{R}|_{Y \times Y}$. The set of cohomology classes of cocycles is denoted by $H^1(\mathcal{R}, G)$. We say that a cocycle $\sigma \in Z^1(\mathcal{R}, G)$ *cobounds* if it cohomologous to the identity cocycle.

Remark 3.4 If $\phi \in Z^1(\mathcal{R}_\Gamma, G)$ then there exists a Γ-invariant dense G_δ-set $Y \subset X$ such that for each $\gamma \in \Gamma$ and for each open set $F \subset G$ the set $\{y \in Y : \phi(y, \gamma y) \in F\}$ is clopen in Y.

Let $\alpha \in Z^1(\mathcal{R}, G)$. Consider the map $\gamma : X \times G \to X \times G$ given by $\gamma(x, g) = (\gamma x, \alpha(x, \gamma x)g)$. It is not hard to verify that this map is a pseudo-homeomorphism of $X \times G$ and thus we obtain an action of Γ on $X \times G$ by pseudo-homeomorphisms. This action is called a skew product action and is denoted by $\Gamma(\alpha)$. By virtue of Remark 3.4, one may view the skew product action as an action by homeomorphisms on $X \times G$.

Definition 3.5 A cocycle $\alpha \in Z^1(\mathcal{R}, G)$ is called ergodic, if the skew product action $\Gamma(\alpha)$ on $X \times G$ is ergodic.

Remarks 3.6
1. The ergodicity of α is a property of the class α in $H^1(\mathcal{R}, G)$.
2. If $\alpha \in Z^1(\mathcal{R}_\Gamma, G)$ is ergodic then the action of Γ on X is necessarily ergodic.

The proof of the following lemma is straightforward.

LEMMA 3.2 *Let $\alpha \in Z^1(\mathcal{R}_\Gamma, G)$ be an ergodic cocycle.*

(i) There exists a Γ-invariant dense G_δ-set $Y \subset X$ such that for each $y \in Y$, for each open $U \in G$ there exists $g_0 \in U$ with the orbit $\Gamma(\alpha)(y, g_0)$ is dense in $X \times G$.

(ii) Let $\Delta \subset X$ be the Γ-orbit of $t_0 \in Y$, where Y as above. For every open $A \in X$ and every open $U \in G$ there exists $t_k \in \Delta$ with $t_k \in A$ and $\alpha(t_0, t_k) \in U$.

The theorem below answers the question about existence of ergodic cocycles with values in a given countable group G. Since any two ergodic actions of countable groups are orbit equivalent [12], it suffices to produce a cocycle for some action of a countable group.

THEOREM 3.3 *For any countable group G there exists an ergodic cocycle $\phi \in Z^1(\mathcal{R}, G)$, with $\mathcal{R}$ being an ergodic generic countable equivalence relation.*

Proof: The group G acts on the space $\{0, 1\}^G$ in a natural way. Let $X = Y^{\mathbb{Z}}$. Then we can consider two actions on X:
(1) The action of $\mathbb{Z}$ by powers of the homeomorphism T: $(Tx)_i = x_{i+1}$; this action is ergodic.
(2) The action of G: $(gx)_i = gx_i$.
Evidently, these actions commute. Let Γ be the group of homeomorphisms, generated by the actions of G and $\mathbb{Z}$ on X. Let us define the cocycle $\phi : \mathcal{R}_\Gamma \to G$,

$$\phi(gT^n x, x) = g, \qquad x \in X, \ g \in G.$$

It is easy to verify that the cocycle ϕ is really ergodic. ∎

COROLLARY 3.4 *Given any Polish group G, then there exists an ergodic cocycle $\phi \in Z^1(\mathcal{R}, G)$, where $\mathcal{R}$ is an ergodic generic countable equivalence relation.*

Proof: Let H be some dense countable subgroup of G. It is easy to verify that every ergodic $\phi \in Z^1(\mathcal{R}, H)$ with H being treated as a discrete group, is also ergodic as a cocycle with values in G.   ∎

4   Weak equivalence

Let Γ be a countable group of homeomorphisms on X.

Definition 4.1 Two cocycles $\phi, \psi \in Z^1(\mathcal{R}_\Gamma, G)$ are called weakly equivalent if there exists $\Theta \in N[\Gamma]$ such that the cocycles ϕ and $\psi \circ (\Theta \times \Theta)$ are cohomologous.

Before formulating our main theorem about weak equivalence of ergodic cocycles we provide several auxiliary results.

LEMMA 4.1 *Let Γ be an ergodic countable group of homeomorphisms on X, G a Polish group, $\phi \in Z^1(\mathcal{R}_\Gamma, G)$ an ergodic cocycle. Suppose we are given $g \in G$, a neighborhood V of the identity in G, and a dense orbit $\Delta = \Gamma x_0$ in X, such that for some $g_0 \in G$ the orbit $\Gamma(\phi)(x_0, g_0)$ is dense in $X \times G$ (cf. lemma 3.2). Let A, B be nonempty disjoint clopen sets in X. Then there exist a Γ-invariant, dense G_δ-set $Y \subset X$, and a homeomorphism h of Y with the following properties: $Y \supset \Delta$, $h = h^{-1}$, h interchanges $A \cap Y$ and $B \cap Y$, h is the identity on $Y \backslash (A \cup B)$ and is strongly Γ-decomposable over Y, and $\phi(x, hx) \in Vg$ for all $x \in A \cap Y$.*

Proof: Let us enumerate the countable sets $A \cap \Delta$ and $B \cap \Delta$. Let x_1 be the first term in the enumeration of $A \cap \Delta$. Since ϕ is ergodic, there exists $\gamma_1 \in \Gamma$ with $\gamma_1 x_1 \in B$ and $\phi(x_1, \gamma_1 x_1) \in Vg_1$. Let A_1 be a clopen neighborhood of x_1 such that A_1 is a proper subset of A, $\gamma_1 A_1$ is a proper subset of B, and $\phi(x, \gamma_1 x) \in Vg_1$ for all $x \in A_1$. Let $B_1 = \gamma_1 A_1$.

Let x_2 be the first term in the enumeration of $B \cap \Delta$, which is not in B_1. Using the same argument as above, we find $\gamma_2 \in B_2$ and a clopen neighborhood B_2 of x_2 such that B_2 is a proper subset of $B \backslash B_1$, $A_2 = \gamma_2 B_2$ is a proper subset of $A \backslash A_1$, and $\phi(x, \gamma_2^{-1} x) \in Vg_1$ for all $x \in A_2$.

Proceed in this way to produce two sequences $\{A_k\}$ and $\{B_k\}$ ($k = 1, 2, \ldots$), each of which consists of pairwise disjoint sets, and the sequence $\{\gamma_k\} \subset \Gamma$. We have $\Delta \cap A \subset \bigcup_{k=1}^{\infty} A_k$ and $\Delta \cap B \subset \bigcup_{k=1}^{\infty} B_k$. Let $S =$

$R_\Gamma((A \cup B) \backslash (\bigcup_{k=1}^{\infty} A_n \cup B_n))$, and let $Y = X \backslash S$. Clearly, Y is a Γ-invariant dense G_δ set. We define h by:

$$hx = \gamma_{2k-1} x, \ x \in A_{2k-1} \cap Y$$

$$hx = \gamma_{2k-1}^{-1} x, \ x \in B_{2k-1} \cap Y$$

$$hx = \gamma_{2k}^{-1} x, \ x \in A_{2k} \cap Y$$

$$hx = \gamma_{2k} x, \ x \in B_{2k} \cap Y$$

$$hx = x, \ x \in Y \backslash (A \cap B) \qquad \blacksquare$$

LEMMA 4.2 *Let Γ be a countable group of homeomorphisms acting ergodically on X. Let $\phi \in Z^1(\mathcal{R}_\Gamma, G)$, where G is a Polish group. Then ϕ is cohomologous to some cocycle $\psi \in Z^1(\mathcal{R}_\Gamma, G)$ with values in a given countable dense subgroup H of G.*

Proof: Choose a fundamental system $\{W_i\}$ of neighborhoods of the identity of G with the following properties: $W_i = W_i^{-1}$ and $W_{i+1} \cdot W_{i+1} \cdot W_{i+1} \subset W_i$, $i \in \mathbb{N}$.

We denote by $\underline{\alpha}$ the element $(\alpha_1, \alpha_2, \ldots, \alpha_n)$ of $\{0, 1\}^n$.

We use the technique developed in [12] to construct an action of the group $\bigoplus_{\mathbb{N}} \mathbb{Z}_2$ that generates the same equivalence relation as Γ. According to [12, lemma 1.7, th. 1.8], one may assume without loss of generality that for every $n \in \mathbb{N}$ there exist a family $\{K^n(\underline{\alpha}) : \underline{\alpha} \in \{0, 1\}^n\}$ of pairwise disjoint clopen sets and a family $\{h_n\}$ of mutually commuting homeomorphisms of X, with the following properties:

(i) $h_n = h_n^{-1}$; each h_n is strongly Γ-decomposable over X and each $\gamma \in \Gamma$ is strongly Ω-decomposable over X, with Ω being the group generated by $\{h_n\}_{n=1}^{\infty}$.

(ii) $h_1^{\alpha_1} h_2^{\alpha_2} \ldots h_n^{\alpha_n} K^n(\underline{0}) = K^n(\underline{\alpha})$ for every $\underline{\alpha} \in \{0, 1\}^n$, and $K^n(\underline{0}) = K^{n+1}(0, 0, \ldots, 0, 1) \cup K^{n+1}(0, 0, \ldots, 0)$.

(iii) $\bigcup_{\underline{\alpha} \in \{0,1\}^n} K^n(\underline{\alpha}) = X$.

(vi) $\{t_0, t_1, \ldots\}$ is such a Γ-orbit that $t_0 \in K^n(\underline{0})$ for all $n \in \mathbb{N}$.

For every $n \in \mathbb{N}$ form the map $\delta_n : K^{n-1}(\underline{0}) \to K^n(\underline{0})$ $(K^0(\underline{0}) = X)$ given by

$$\delta_n x = h_n^{-1} x, \qquad x \in K^n(0, \ldots, 0, 1)$$

$$\delta_n x = x, \qquad x \in K^n(0, \ldots, 0)$$

Let $\lambda_n = \delta_n \cdot \ldots \cdot \delta_1$.

Let Ω_k be the finite group $\{h_1^{\alpha_1} \ldots h_k^{\alpha_k} : \underline{\alpha} \in \{0, 1\}^k\}$.

It suffices to define the cocycle ψ on points of the form $(\delta_n x, x)$, $x \in K^{n-1}(\underline{0})$, $n \in \mathbb{N}$. Proceed by induction in n and describe the nth step. By virtue of 2.1 we may assume that the map $\phi(t, \omega t) : X \to G$ is continuous in t for each fixed $\omega \in \Omega$.

Let $g(\underline{\alpha}) = \phi(t_0, h_1^{\alpha_1} \dots h_n^{\alpha_n} t_0)$. By continuity of $\phi(t, h_1^{\alpha_1} \dots h_n^{\alpha_n} t)$ in t one can find an open neighborhood O_n of t_0 such that $\phi(t, h_1^{\alpha_1} \dots h_n^{\alpha_n} t) \in g(\underline{\alpha})W_{n+1}$ for all $\underline{\alpha} \in \{0, 1\}^n$, $t \in O_n$. Let $A_n = \bigcup_{\omega \in \Omega_n} \omega O_n$. Let V_{n+1} be a neighborhood of the identity in G with $g(\underline{\alpha})^{-1} V_{n+1} g(\underline{\alpha}) \subset W_n$ for all $\underline{\alpha} \in \{0, 1\}^n$. Approximate the map $\phi(\delta_n x, x)$, $x \in K^{n-1}(\underline{0})$, by the map $\psi(\delta_n x, x)$ with values in H so that $\phi(\delta_n x, x)^{-1} \psi(\delta_n x, x) \in V_n$.

Let $p_n(x) = \phi(\lambda_n x, x)^{-1} \psi(\lambda_n x, x)$. Observe that $\lambda_n \omega x = \lambda_n x$ for $\omega \in \Omega_k$, $n \geq k$. Therefore

$$p_n(\omega x)^{-1} \phi(\omega x, x) p_n(x) =$$
$$= \psi(\lambda_n \omega x, \omega x)^{-1} \phi(\lambda_n \omega x, \omega x) \phi(\omega x, x) \phi(\lambda_n x, x)^{-1} \psi(\lambda_n x, x) =$$
$$= \psi(\lambda_n \omega x, x)^{-1} \psi(\lambda_n x, x) = \psi(\omega x, x) \qquad (1)$$

Let $x \in \bigcap_{i=n+1}^{n+k} A_i$. Then

$$p_{n+k}(x) = \phi(\lambda_{n+k} x, x)^{-1} \psi(\lambda_{n+k} x, x) =$$
$$= \phi(\lambda_{n+k-1} x, x)^{-1} \phi(\delta_{n+k} \lambda_{n+k-1} x, \lambda_{n+k-1} x)^{-1} \cdot$$
$$\cdot \psi(\delta_{n+k} \lambda_{n+k-1} x, \lambda_{n+k-1} x) \psi(\lambda_{n+k-1} x, x) \subset$$
$$\subset \phi(\lambda_{n+k-1} x, x)^{-1} V_{n+k} \psi(\lambda_{n+k-1} x, x) =$$
$$= \phi(\lambda_{n+k-1} x, x)^{-1} V_{n+k} \phi(\lambda_{n+k-1} x, x) \phi(\lambda_{n+k-1} x, x)^{-1} \psi(\lambda_{n+k-1} x, x) \subset$$
$$\subset W_{n+k} W_{n+k} W_{n+k} p_{n+k-1}(x) \subset W_{n+k-1} p_{n+k-1}(x) \subset \dots \subset W_n p_n(x).$$

Note that $\bigcap_{i=n+1}^{n+k} A_i$ is an open set which contains $\{\omega t_0 : \omega \in \Omega_{n+1}\}$, so we conclude that the sequence $\{p_n\}$ converges on a dense G_δ-set $Y \subset X$, to some Borel map p. In view of (1) we obtain

$$p(x)^{-1} \phi(x, y) p(y) = \psi(x, y) \qquad \text{for all} \quad (x, y) \in R_\Gamma|_{Y \times Y} \qquad \blacksquare$$

THEOREM 4.3 *Let Γ be a countable ergodic group of homeomorphisms of X, and G a Polish group. Suppose that the cocycles $\phi, \psi \in Z^1(\mathcal{R}_\Gamma, G)$ are ergodic. Then there exist a Γ-invariant dense G_δ-set $Y \subset X$, $\Theta \in N[\Gamma]$ and a Borel function $f : Y \to G$ such that $\Theta|_Y$ is a homeomorphism of Y, and*

$$\phi(\gamma y, y) = f(\gamma y)^{-1} \psi(\Theta \gamma y, \Theta y) f(y) \qquad \text{for all } y \in Y, \gamma \in \Gamma$$

Proof: By virtue of lemma 4.2 one can suppose that the cocycles ϕ and ψ take values in a dense countable subgroup $H \subset G$. Moreover, we can assume by proposition 2.1 that for each $h \in H$ and for each $\gamma \in \Gamma$ the sets $\{x \in X : \phi(\gamma x, x) = h\}$, $\{x \in X : \psi(\gamma x, x) = h\}$, and $\{x \in X : \gamma x = x\}$ are clopen.

LEMMA 4.4 _Under the conditions of theorem 4.3, let_ $\Delta = \{t_0, t_1, \dots\}$ _and_ $\widetilde{\Delta} = \{\widetilde{t}_0, \widetilde{t}_1, \dots\}$ _be dense orbits of the_ Γ-_action, such that for some_ $g_0, \widetilde{g}_0 \in G$ _each of the orbits_ $\Gamma(\phi)(t_0, g_0)$, $\Gamma(\psi)(\widetilde{t}_0, \widetilde{g}_0)$ _is dense in_ $X \times G$.

Let $\{S_n\}$ _and_ $\{\widetilde{S}_n\}$ $(n \in \mathbb{N})$ _be monotone decreasing sequences of clopen neighborhoods of_ t_0 _and_ $\widetilde{t}_0$ _respectively, such that_ $S_n \subset B(t_0, 1/n)$, $\widetilde{S}_n \subset B(\widetilde{t}_0, 1/n)$, _where_ $B(x, r)$ _denotes the ball of radius_ r _centered at_ x, _and_ $t_n \notin S_{2n-1}$, $\widetilde{t}_n \notin \widetilde{S}_{2n}$.

Then the following is true:

1. _There exist two_ Γ-_invariant_ G_δ-_sets_ T _and_ $\widetilde{T}$ _with_ $\Delta \subset T$, $\widetilde{\Delta} \subset \widetilde{T}$, _and two sequences_ $\{h_n\}$, $\{\widetilde{h}_n\}$, _where each_ h_n _and each_ $\widetilde{h}_n$ _is a homeomorphism of_ T _and_ $\widetilde{T}$, _respectively,_ $h_n = h_n^{-1}$, $\widetilde{h}_n = \widetilde{h}_n^{-1}$, _each_ h_n, $\widetilde{h}_n$ _is strongly_ Γ-_decomposable over_ T _and_ $\widetilde{T}$, _respectively._

2. _For every_ n _there exist two families_ $\{K^n(\underline{\alpha})\}$ _and_ $\{\widetilde{K}^n(\underline{\alpha})\}$, $\underline{\alpha} \in \{0, 1\}^n$, _each of which consists of pairwise disjoint clopen subsets in_ T _and_ $\widetilde{T}$, _respectively, so that the union of the subsets is_ T _and_ $\widetilde{T}$, _respectively._

3. $K^n(\alpha_1, \dots, \alpha_n) = K^{n+1}(\alpha_1, \dots, \alpha_n, 0) \cup K^{n+1}(\alpha_1, \dots, \alpha_n, 1)$

   $\widetilde{K}^n(\alpha_1, \dots, \alpha_n) = \widetilde{K}^{n+1}(\alpha_1, \dots, \alpha_n, 0) \cup \widetilde{K}^{n+1}(\alpha_1, \dots, \alpha_n, 1)$

4. $K^n(\underline{0}) \subset S_n \cap T$, $t_0 \in K^n(\underline{0})$, _for all_ n,

   $\widetilde{K}^n(\underline{0}) \subset \widetilde{S}_n \cap \widetilde{T}$, $\widetilde{t}_0 \in \widetilde{K}^n(\underline{0})$ _for all_ n.

5. $h_1^{\alpha_1} \dots h_n^{\alpha_n} K^n(\underline{0}) = K^n(\underline{\alpha})$

   $\widetilde{h}_1^{\alpha_1} \dots \widetilde{h}_n^{\alpha_n} \widetilde{K}^n(\underline{0}) = \widetilde{K}^n(\underline{\alpha})$,

   for all $\alpha \in \{0, 1\}^n$.

6. $\{t_0, \dots, t_n\} \subset \{h_1^{\alpha_1} \dots h_{2n}^{\alpha_{2n}}(t_0) : \underline{\alpha} \in \{0, 1\}^{2n}\}$

   $\{\widetilde{t}_0, \dots, \widetilde{t}_n\} \subset \{\widetilde{h}_1^{\alpha_1} \dots \widetilde{h}_{2n}^{\alpha_{2n}}(\widetilde{t}_0) : \underline{\alpha} \in \{0, 1\}^{2n}\}$

7. _There exists a finite set_ $Q_n \subset H$ _and a clopen set_ $A_n \supset \{h_1^{\alpha_1} \dots h_n^{\alpha_n}(t_0) : \underline{\alpha} \in \{0, 1\}^n\}$, $A_n \subset T$, _with_ $\phi(x, h_1^{\alpha_1} \dots h_n^{\alpha_n} x) \in Q_n$ _for all_ $\underline{\alpha} \in \{0, 1\}^n$, $x \in A_n$.

8. $\phi(\delta_n x, x)^{-1} \psi(\widetilde{\delta}_n y, y) \in V_n$ _for all_ $x \in K^{n-1}(\underline{0})$, _and_ $y \in \widetilde{K}^{n-1}(\underline{0})$, _where_ δ_n _is the same map as in the proof of lemma 4.2,_ V_n _is a neighborhood of the identity of_ G _with_ $g V_n g^{-1} \subset W_n$ _for all_ $g \in Q_1^{-1} \dots Q_{n-1}^{-1}$, $(n \geq 2)$, $V_1 = W_1$.

Proof:

We proceed by induction. At each step a meagre set is to be discarded. To avoid some technical complications, we suppose that everything happens on a space T, which is a Γ-invariant dense G_δ-subset of X, and all 'bad' sets are already removed. The same assumption is also made about $\widetilde{T}$.

Let $D = S_1$, $E = X \backslash S_1$. Then $t_0 \in D$, $t_1 \in E$. Let $t_1 = \gamma_1 t_0$, $\phi(t_0, t_1) = g_1$ and let O_1 be a clopen neighborhood of t_0 with the following properties: O_1 is a proper subset of D, $\gamma_1 O_1$ is a proper subset of E, $\phi(x, \gamma_1 x) = g_1$ for all $x \in O_1$. Lemma 4.1 implies that there exists a homeomorphism h_1 of T such that h_1 is strongly Γ-decomposable over T, $h_1 = h_1^{-1}$, h_1 interchanges D and E, and

$h_1 x = \gamma_1 x$, $x \in O_1$,

$h_1 x = \gamma_1^{-1} x$, $x \in \gamma_1 O_1$,

$\phi(x, h_1 x) \in U_1 g_1$ for all $x \in D$, where U_1 is such a neighborhood of the identity of G that $g_1^{-1} U_1^{-1} U_1 g_1 \subset V_1$.

Let $A_1 = O_1 \cup \gamma_1 O_1$, $Q_1 = \{g_1, g_1^{-1}, e\}$, $K^1(0) = D$, $K^1(1) = E$.

Let $\widetilde{D} = \widetilde{S}_1$, $\widetilde{E} = X \backslash \widetilde{S}_1$. Arguing as above we deduce that there exists a homeomorphism $\widetilde{h}_1$ of $\widetilde{T}$ such that $\widetilde{h}_1$ is strongly Γ-decomposable over $\widetilde{T}$, $\widetilde{h}_1 = \widetilde{h}_1^{-1}$, $\widetilde{h}_1$ interchanges $\widetilde{D}$ and $\widetilde{E}$, and $\psi(y, \widetilde{h}_1 y) \in U_1 g_1$ for all $y \in \widetilde{D}$. We set up $\widetilde{K}^1(0) = \widetilde{D}$, $\widetilde{K}^1(1) = \widetilde{E}$. Clearly, $\phi(\delta_1 x, x)^{-1} \psi(\widetilde{\delta}_1 y, y) \in g_1^{-1} U_1^{-1} U_1 g_1 \subset V_1$ for all $x \in T$, $y \in \widetilde{T}$.

Let us now suppose that we have constructed $\{h_k\}_{k=1}^n$, $\{\widetilde{h}_k\}_{k=1}^n$, and the families $\{K^k(\underline{\alpha})\}$, $\{\widetilde{K}^k(\underline{\alpha})\}$, $\underline{\alpha} \in \{0,1\}^k$, $k = \overline{1,n}$. We describe the $(n+1)$th step. Let us assume that n is odd. Let $\widetilde{t}_m$ be the first term in the enumeration of $\widetilde{\Delta}$ with $\widetilde{t}_m \notin \{\widetilde{h}_1^{\alpha_1} \ldots \widetilde{h}_n^{\alpha_n} t_0 : \alpha \in \{0,1\}^n\}$. $\widetilde{t}_m \in \widetilde{K}^n(\beta)$ for some $\beta \in \{0,1\}^n$. Let $\widetilde{c} = \widetilde{h}_1^{\beta_1} \ldots \widetilde{h}_n^{\beta_n} \widetilde{t}_m \in \widetilde{K}^n(\underline{0})$, $\widetilde{g}_m = \widetilde{\psi}(\widetilde{t}_0, \widetilde{c})$. Apply the same argument as in [12,lemma 1.7] and at the first step to construct $\widetilde{h}_{n+1}$, $\{\widetilde{K}^{n+1}(\underline{\alpha}) : \underline{\alpha} \in \{0,1\}^{n+1}\}$, so that $\psi(x, \widetilde{h}_{n+1} x) \in U_{n+1} \widetilde{g}_m$ for all $x \in \widetilde{K}^{n+1}(\underline{0})$, where U_{n+1} is a neighborhood of the identity of G with $\widetilde{g}_m^{-1} U_{n+1}^{-1} U_{n+1} \widetilde{g}_m \in V_{n+1}$. By lemma 3.2 there exists $t_j \in (K^n(\underline{0}) \backslash S_{n+1}) \cap \Delta$ such that $\phi(t_0, t_j) = g_j \in U_{n+1} \widetilde{g}_m$. Let $t_j = \gamma t_0$. There exists a clopen neighborhood O_{n+1} of t_0 such that $\phi(t, \gamma_j t) = g_j$ for all $t \in O_{n+1}$, and the set $Q_n = \{\phi(t, h_1^{\alpha_1} \ldots h_n^{\alpha_n} \gamma_j t) : \underline{\alpha} \in \{0,1\}^n, t \in O_{n+1}\}$ is finite.

Just as it was done at the first step, we establish that there exists a strongly Γ-decomposable over T homeomorphism h_{n+1}, interchanging $K^{n+1}(\underline{0})$ and $K^{n+1}(0, \ldots, 0, 1)$, $h_{n+1} = h_{n+1}^{-1}$,

$h_{n+1} x = \gamma_j x$, $x \in O_{n+1}$

$h_{n+1} x = \gamma_j^{-1} x$, $x \in \gamma_j O_{n+1}$

$\phi(x, h_{n+1} x) \in U_{n+1} \widetilde{g}_m$ for all $x \in K^n(\underline{0})$.

Let $A_{n+1} = \bigcup_{\underline{\alpha} \in \{0,1\}^{n+1}} h_1^{\alpha_1} \ldots h_{n+1}^{\alpha_{n+1}} O_{n+1}$.

One can readily verify that all the objects we have just produced satisfy the conditions of the lemma.

If n is even, the construction is made in a similar way, but starting from a selection of $t_m \in \Delta$, where m is the first term in the enumeration of Δ such that $t_m \notin \{h_1^{\alpha_1} \dots h_n^{\alpha_n} t_0 : \alpha \in \{0,1\}^n\}$. ∎

Proof of the theorem:

Consider $\bigoplus_{\mathbb{N}} \mathbb{Z}_2$ as subspace of the Cantor space $\prod_{\mathbb{N}} \mathbb{Z}_2$ with the induced topology. Let Θ_1 be a homeomorphism from Δ onto $\bigoplus_{\mathbb{N}} \mathbb{Z}_2$, given by:

$$\Theta_1(h_1^{\alpha_1} \dots h_m^{\alpha_m} t_0) = (\alpha_1, \dots, \alpha_m, 0, 0, \dots), \qquad \underline{\alpha} \in \{0,1\}^m$$

(see th 1.8 of [12]); let Θ_2 be the homeomorphism from $\bigoplus_{\mathbb{N}} \mathbb{Z}_2$ onto $\widetilde{\Delta}$ given by:

$$\Theta_2(\alpha_1, \dots, \alpha_m, 0, 0, \dots) = \widetilde{h}_1^{\alpha_1} \dots \widetilde{h}_m^{\alpha_m} \widetilde{t}_0, \qquad \underline{\alpha} \in \{0,1\}^m.$$

Thus we have a homeomorphism $\Theta = \Theta_1 \Theta_2$ from Δ onto $\widetilde{\Delta}$.

By virtue of Lavrentiev's Theorem [10] there exist (dense) G_δ-subsets $Y_1 \supset \Delta$ and $Y_2 \supset \widetilde{\Delta}$ of T and $\widetilde{T}$ respectively, and an extension of Θ to a homeomorphism of Y_1 onto Y_2 (which we also denote by Θ). Now notice that one can assume in lemma 4.4 that $\Delta = \widetilde{\Delta}$, $t_0 = \widetilde{t}_0$. Then, with $Y_3 = Y_1 \cap Y_2$, we obtain the homeomorphism $\Theta : Y_3 \to Y_3$. Evidently, we may assume that Y_3 is Γ-invariant, h_k, $\widetilde{h}_k$-invariant for all $k \in \mathbb{N}$. Now the homeomorphism $\Theta h_1^{\alpha_1} \dots h_m^{\alpha_m} \Theta^{-1}$ coincides with $\widetilde{h}_1^{\alpha_1} \dots \widetilde{h}_m^{\alpha_m}$ on a dense (countable) subset of Y_3, hence $\Theta h_1^{\alpha_1} \dots h_m^{\alpha_m} \Theta^{-1} y = \widetilde{h}_1^{\alpha_1} \dots \widetilde{h}_m^{\alpha_m} y$ for all $y \in Y_3$. Just as it was done in [12, th. 1.8] we establish that the action of $\{\Theta \gamma \Theta^{-1}\}_{\gamma \in \Gamma}$ generates the same equivalence relation as Γ, hence $\Theta r \Theta^{-1} \in [\Gamma]$ for each $r \in [\Gamma]$, that is $\Theta \in N[\Gamma]$. Let us consider the cocycle ϕ_1 of $\mathcal{R}|_{Y_3}$ given by $\phi_1(x, y) = \psi(\Theta x, \Theta y)$, $(x, y) \in \mathcal{R}|_{Y_3}$. Note that Θ maps $K^n(\underline{\alpha})$ onto $\widetilde{K}^n(\underline{\alpha})$, $\alpha \in \{0,1\}^n$, $n \in \mathbb{N}$. Now it follows from lemma 4.4 (condition 8) that $\phi(\delta_n x, x)^{-1} \phi_1(\delta_n x, x) \in V_n$ for all $x \in K^{n-1}(\underline{0})$. By our construction, $\phi(\delta_n x, x) \in Q_n$ for all $x \in A_n \cap Y_3$. Applying the same argument as in the proof of lemma 4.2, we conclude that the cocycles ϕ and ϕ_1 are cohomologous. ∎

Remark 4.2: It should be noted that any ergodic cocycle admits a kind of reduction to an ergodic cocycle with values in a discrete countable group. Indeed, every $\phi \in Z^1(\mathcal{R}, G)$ is cohomologous to a cocycle taking values in a dense countable subgroup (lemma 4.2). On the other hand, let $\psi \in Z^1(\mathcal{R}, G)$ be ergodic with H being treated as a discrete group, then ψ is also ergodic as a cocycle with values in G. Now, of one assumes ϕ to be ergodic as well as a G-valued cocycle, it follows from theorem 4.3 that ϕ and ψ are weakly equivalent. This sort of reduction is not accessible, in general, in the measure theoretic approach.

Remark 4.3: In the case when G is a countable discrete group, the relation between the cocycles ϕ and ψ in the statement of theorem 4.3 can be written in a simpler form: $\phi(\gamma y, y) = \psi(\Theta_1 \gamma y, \Theta_1 y)$, with $\Theta_1 \in N[\Gamma]$. To show this, we make the following observation. If $\sigma \in Z^1(\mathcal{R}_\Gamma, G)$ is an ergodic cocycle, with G as above, then any Borel function $f : Y \to G$ admits a representation in the form: $f(y) = \sigma(\tau y, y)$ for some $\tau \in [\Gamma]$. This fact is readily applicable to the function f in the statement of the theorem, and so the initial relation transforms to $\phi(\gamma y, y) = \psi(\Theta_1 \gamma y, \Theta_1 y)$, where $\Theta_1 = \Theta \tau^{-1}$.

5   Outer conjugacy

This section is devoted to studying the outer conjugacy of outer actions of countable groups from the full group normalizers. Just as in the work of A. Connes and W. Krieger, we present a complete system of invariants for this kind of conjugacy.

Let Γ be a countable group of homeomorphisms of X. The following lemma asserts that in the case when Γ-action is ergodic, any element of $N[\Gamma]$ must be either purely outer or purely inner.

LEMMA 5.1 *If the Γ-action is ergodic, then for any $\tau \in N[\Gamma]$, the set $A = \bigcup_{\gamma \in \Gamma} \{x \in X : \tau x = \gamma x\}$ is either meagre or the complement of a meagre set.*

Definition 5.1 Let π_1, π_2 be actions of a countable group G as homeomorphisms of X, such that $\pi_1(g), \pi_2(g) \in N[\Gamma]$ for all $g \in G$. The actions π_1 and π_2 are called outer conjugate if there exist a Γ-invariant dense G_δ-subset Y of X and a homeomorphism Θ of Y with

$$\pi_1(g)y = \Theta^{-1} \pi_2(g) \tau \Theta y,$$

where $\Theta \in N[\Gamma]$, $\tau = \tau(g) \in [\Gamma]$ for all $g \in G$, $y \in Y$.

THEOREM 5.2 *Let Γ be an ergodic group of homeomorphisms of X. The actions π_1, π_2 with $\pi_1(g), \pi_2(g) \in N[\Gamma]$ for all $g \in G$, are outer conjugate if and only if*

$$\{g \in G : \pi_1(g) \in [\Gamma]\} = \{g \in G : \pi_2(g) \in [\Gamma]\}.$$

Proof: The 'only if' part is obvious, so we prove the 'if' part. Let H_i be a group of homeomorphisms, generated by $\pi_i(G)$ and Γ ($i = 1, 2$). Let $\mathcal{R}_1$ be an equivalence relation, generated by H_1. Since the groups H_1, H_2 are ergodic, by theorem 1.8 [12] there exist a Γ-invariant dense G_δ-set $X_1 \subset X$

and a homeomorphism Q of X_1 with $[H_1] = Q[H_2]Q^{-1}$. Let us consider $G_0 = \{g \in G : \pi_1(g) \in [\Gamma]\} = \{g \in G : \pi_2(g) \in [\Gamma]\}$. Then G_0 is a normal subgroup of G. Let $p : G \to G/G_0$ be a natural projection. Let us define cocycles ϕ and ψ of $R_1|_{X_1 \times X_1}$ with values in G/G_0 by:

$$\phi(\pi_1(g)\gamma x, x) = p(g)$$

$$\psi(Q\pi_2(g)\gamma Q^{-1}x, x) = p(g) \qquad g \in G, \; \gamma \in \Gamma \; x \in X_1$$

The definitions are correct in view of 5.1.

The proof of the following lemma is straightforward from the definitions and lemma 5.1.

LEMMA 5.3 *Let* $h \in [H_1]$ *and* $\psi(hx, x) = p(g)$ *for all* $x \in X_1$. *Then there exists* $\omega \in [\Gamma]$ *such that* $hx = Q\pi_2(g)\omega Q^{-1}x$ *for all* x *from some dense* G_δ-*subset* X_2 *of* X_1.

It is not difficult to see that ϕ, ψ are ergodic. Theorem 4.3 implies that there exist a Γ-invariant dense G_δ-subset Y of X_1 and a homeomorphism $F \in N[H_1]$ of Y with $\phi(x, y) = \psi(Fx, Fy)$ for all $(x, y) \in R_1|_{Y \times Y}$ (see also Remark 4.6). Then for each $\tilde{\gamma} \in [\Gamma]$, $y \in Y$ one has $\psi(F\tilde{\gamma}F^{-1}y, y) = p(e)$, where e is the identity of G. By lemma 5.3, $Q^{-1}F\tilde{\gamma}F^{-1}Qy = \omega y$ for some $\omega \in [\Gamma]$. A similar argument as in lemma 5.3 applied to the cocycle ϕ allows one to establish that $F^{-1}Q\tilde{\gamma}Q^{-1}F \in [\Gamma]$ for any $\tilde{\gamma} \in \Gamma$, so $Q^{-1}F \in N[\Gamma]$. Besides that, for all $y \in Y$, $\psi(F\pi_1(g)F^{-1}y, y) = p(g)$, hence $F\pi_1(g)F^{-1}y = Q\pi_2(g)\tau Q^{-1}y$ for $\tau \in [\Gamma]$. By letting $\Theta = Q^{-1}F$ we obtain: $\pi_1(g)y = \Theta^{-1}\pi_2(g)\tau\Theta y$, $\tau = \tau(g) \in [\Gamma]$, as required. ∎

Remark 5.2. It should be noted that the result of D. Sullivan. B. Weiss, and J. Wright [12] implies the isomorphism of all the AW*-cross-product algebras arising from a countable group of pseudo-homeomorphisms on perfect Polish space. On the other hand, an automorphism from the full group normalizer could be used to generate a larger cross-product. If two such automorphisms are outer conjugate in the sense of this work, it is natural to expect that the associated cross-products are isomorphic.

6   Subrelations of generic equivalence relations

D. Sullivan, B. Weiss, and J. Wright [12] produced a complete description of ergodic countable equivalence relations up to orbit equivalence. In this chapter we turn to a more detailed studying of orbit properties for such relations. We follow the ideas of [6] to develop the theory of subrelations of ergodic generic countable equivalence relations. Some proofs which are very close to those in [6] are omitted.

From now on $\mathcal{R}$, $\mathcal{S}$ stand for countable generic equivalence relations on X. To begin with, we recall some definitions, which are convenient while working with equivalence relations. For a countable generic equivalence relation $\mathcal{F}$, generated by a countable group Γ of homeomorphisms of X, we denote $\text{Int}\mathcal{F} = [\Gamma]$, $\text{Aut}\mathcal{F} = N[\Gamma]$. If $\mathcal{F}_1 \subseteq \mathcal{F}_2$ we denote $\text{Aut}_{\mathcal{F}_2}\mathcal{F}_1 = \text{Aut}\mathcal{F}_1 \cap \text{Int}\mathcal{F}_2$. Let $[\sigma]$ stand for the cohomology class of $\sigma \in Z^1(\mathcal{S}, G)$ and $\Sigma(N)$ for the full permutation group on $\{0, 1, \dots, N-1\}$ with $N < \infty$, or on $\{0, 1, \dots\}$ with $N = \infty$.

Throughout this section $\mathcal{S}$ will be ergodic. Suppose that $\mathcal{R} \subseteq \mathcal{S}$. $\mathcal{R}$ induces an equivalence relation on each $\mathcal{S}$-class $\mathcal{S}[x]$ $(x \in X)$; let $J(x)$ be the quotient $\mathcal{S}[x]/\mathcal{R}$. It follows from proposition 2.2 and the ergodicity of $\mathcal{S}$ (see also [6]) that $|J(x)|$ is constant on X modulo a meagre set. The cardinal $N = |J(x)|$ is called the *index* of $\mathcal{R} \subseteq \mathcal{S}$. From now on, by J we denote the set $\{0, \dots, N-1\}$ for $N < \infty$ or $\{0, 1, \dots\}$ for $N = \infty$. The same ideas as in [6] allow one to deduce the existence of choice functions, that is a collection of Borel functions $\{\phi_j\}_{j=0}^{N-1}$, $\phi_j : X \to X$, such that $\{\mathcal{R}[\phi_j(x)], 0 \le j \le N-1\}$ is a partition of $\mathcal{S}[x]$ for all $x \in X$ outside of a meagre set.

The index cocycle $\sigma \in Z^1(\mathcal{S}, \Sigma(N))$ of the pair $\mathcal{R} \subseteq \mathcal{S}$ is defined by $\sigma(x,y)(i) = (j)$ if $\mathcal{R}[\phi_i(y)] = \mathcal{R}[\phi_j(x)]$. The cohomology class of σ does not depend on a selection of choice functions. Let us check the converse. Suppose $\tilde{\sigma}(x,y) = \nu(x)^{-1}\sigma(x,y)\nu(y)$. We wish to show that $\tilde{\sigma}$ is also an index cocycle of $\mathcal{R} \subseteq \mathcal{S}$ associated to some family of choice functions. Indeed, let $\tilde{\phi}_j(x) = \phi_{\nu(x)(j)}(x)$. Evidently $\{\mathcal{R}[\tilde{\phi}_j(x)]\}_{j=0}^{N-1}$ is a partition of $\mathcal{S}[x]$. To see that $\tilde{\phi}_j$ is Borel, note that $\tilde{\phi}_j^{-1}(A) = \bigcup_n (\phi_n^{-1}(A) \cap \{x \in X : \nu(x)(j) = (n)\})$, for any $A \subset X$. Clearly, $\tilde{\sigma}$ is the index cocycle defined by the choice functions $\{\tilde{\phi}_j\}_{j=0}^{N-1}$.

LEMMA 6.1 *Let $\{\phi_j\}_{j=0}^{N-1}$ be choice functions of the pair $\mathcal{R} \subseteq \mathcal{S}$, and suppose $\mathcal{R}$ is ergodic. Then there exist a dense $\mathcal{S}$-invariant G_δ-set $Y \subset X$ and choice functions $\{\psi_j\}_{j=0}^{N-1}$ for $\mathcal{R} \subseteq \mathcal{S}$ such that for all $j \in \{0, 1, \dots, N-1\}$*

(i) ψ_j is a homeomorphism of Y.

(ii) $(\phi_j(x), \psi_j(x)) \in \mathcal{R}$ for all $x \in Y$.

Proof: Let $\Gamma = \{\gamma_k\}_{k=1}^{\infty}$ be a countable group of homeomorphisms that generates $\mathcal{S}$. For each fixed j, let
$$\tilde{E}_1 = \{x \in X : \phi_j x = \gamma_1 x\}, \; E_1 = \tilde{E}_1,$$

$$\dots$$

$$\tilde{E}_k = \{x \in X : \phi_j x = \gamma_k x\}, \; E_k = \tilde{E}_k \backslash (E_1 \cup \dots \cup E_{k-1}).$$

Observe that $\{E_k\}_{k=1}^{\infty}$ is a countable Borel partition of X. We may assume that all the sets E_k, $A_k = \phi_j(E_k)$ are clopen. Let $\{A_k\}_{k=1}^{\infty}$ be a partition of X

consisting of clopen sets. Using [12, lemma 1.6] one can show that there exist a S-invariant dense G_δ-set $Y \subset X$ and a family $\{h_k\}_{k=1}^\infty$ with the following properties: $h_k \in \text{Int}\mathcal{R}$, h_k interchanges $\widetilde{A}_k \cap Y$ and $A_k \cap Y$, $h_k = h_k^{-1}$, $h_k = id$ on $Y \backslash (\widetilde{A}_k \cup A_k)$ for all k. Define $\psi_j x = h_k \phi_j x$ for $x \in Y \cup E_k$. On readily verifies that $\{\psi_j\}$, $j = 0, \ldots, N-1$, form a desired family of choice functions. ∎

Let $\mathcal{R} \subseteq S$ be given on X with the index cocycle σ. Let us define an equivalence relation $\widehat{\mathcal{R}} = S \times_\sigma J$ on $X \times J$ by:
$$((x,i),(y,j)) \in \widehat{\mathcal{R}} \iff (x,y) \in S, \text{ and } \sigma(x,y)(j) = (i).$$
When σ_1, σ_2 are two index cocycles for $\mathcal{R} \subseteq S$ it is clear that $\widehat{\mathcal{R}}_1$ is isomorphic to $\widehat{\mathcal{R}}_2$.

Let Γ be a countable group of homeomorphisms which generates S. Then $\widehat{\mathcal{R}}$ is generated by the Γ-action on $X \times J$ given by $\gamma(x,i) = (\gamma x, \sigma(\gamma x, x)(i))$.

Note, that our construction of the equivalence relation $S \times_\sigma J$ is also valid for any cocycle $\sigma \in Z^1(S, \Sigma(N))$.

PROPOSITION 6.2 $\mathcal{R}$ *is ergodic if and only if* $\widehat{\mathcal{R}}$ *is ergodic.*

Proof: Let $\widehat{\mathcal{R}}$ be ergodic. Suppose the contrary: there exists an open nontrivial (non-comeagre) set $A \subset X$ with $\mathcal{R}[A] = A$. Then the set $B = \bigcup_{j \in J} (\phi_j^{-1}(A) \times \{j\}) \subset X \times J$ is nontrivial and Borel. We show that $\widehat{\mathcal{R}}[B] = B$.

Indeed, let $(\phi_j^{-1}(x), j) \overset{\widehat{\mathcal{R}}}{\sim} (y, k)$, for $x \in A$. Then $\mathcal{R}[x] = \mathcal{R}[\phi_k(y)]$, hence $\phi_k(y) \in A$, that is $(y, k) \in B$ and we obtain the contradiction.

Conversely, suppose now that $\mathcal{R}$ is ergodic. According to lemma 6.1 we may assume that each choice function is a homeomorphism of X. We claim that the orbit of $(x, 0)$ is dense in $X \times J$. In fact, let $(y, j) \in X \times J$ and U be a neighborhood of y in X. The set $\phi_j^{-1}(\mathcal{R}[\phi_0(x)])$ is dense in X, because $\mathcal{R}[\phi_0(x)]$ is. Hence there exists $z \in U \cap \phi_j^{-1}(\mathcal{R}[\phi_0(x)])$, then $(z, j) \overset{\widehat{\mathcal{R}}}{\sim} (x, i)$ as required. ∎

The following theorem is proved in exactly the same way as [6, theorem 1.6].

THEOREM 6.3 *Let S on X be given. Then*

(i) if $\sigma \in Z^1(S, \Sigma(J))$ is such that $S \times_\sigma J$ is ergodic, σ is the index cocycle for some ergodic subrelation $\mathcal{R}$ of S.

(ii) if $\mathcal{R}, \widetilde{\mathcal{R}}$ are ergodic subrelations of S having index cocycles $\sigma \in Z^1(S, \Sigma(J))$, $\widetilde{\sigma} \in Z^1(S, \Sigma(\widetilde{J}))$, then $\mathcal{R}$ and $\widetilde{\mathcal{R}}$ are conjugate under $\text{Aut}(S)$ (respectively $\text{Int}(S)$) if and only if $J = \widetilde{J}$ and $[\sigma] = [\widetilde{\sigma}]$ up to $\text{Aut}(S)$ (respectively $[\sigma] = [\widetilde{\sigma}]$).

PROPOSITION 6.4 *Suppose $N < \infty$ is the index of $\mathcal{R} \subseteq \mathcal{S}$. Then there exists a partition $\{X_k\}_{k=0}^M$ of X modulo a meagre set, $M \leq N - 1$, such that each X_k is open, $\mathcal{R}[X_k] = X_k$, and $\mathcal{R}$ is ergodic on X_k for each $k \in \overline{0, M}$.*

Proof: Let $x \in X$ with $\mathcal{S}[x]$ is dense in X. Let us consider the following sets: $A_j = \mathcal{R}[\phi_j(x)]$. We claim that $\bigcup_0^{N-1} \overline{A_j} = X$. For every $y \in X$ there exists a sequence $x_n \to y$, $\{x_n\}_{n=1}^\infty \subset \mathcal{S}[x]$. Since $N < \infty$, there exist $j \in J$ and a subsequence $x_{n_i} \to y$ with $\{x_{n_i}\}_{i=1}^\infty \subset A_j$, that is $y \in \overline{A_j}$. Set up $X_k = \overline{A_k}$ for each non-meagre $\overline{A_k}$. In view of 2.1 we may assume that each X_k is open. ∎

PROPOSITION 6.5 *Let σ be an index cocycle for $\mathcal{R} \subseteq \mathcal{S}$. σ is cohomologous to the identity cocycle if and only if modulo a meagre set there exists a partition $\{X_j\}_{j=0}^{N-1}$ of X such that X_j is clopen, $\mathcal{R}[X_j] = X_j$, and $\mathcal{R}|_{X_j} = \mathcal{S}|_{X_j}$ for all $j \in \{0, 1, \dots, N-1\}$.*

For a proof of this proposition, the reader can consult [6] for a very similar argument.

7 Normal subrelations

The notion of normality for subrelations of ergodic measurable equivalence relations was introduced in [6]. In our context it looks very much like that: A subrelation $\mathcal{R} \subseteq \mathcal{S}$ is called *normal* in $\mathcal{S}$ if the restriction $\sigma|_{\mathcal{R}}$ of an index cocycle σ to $\mathcal{R}$ cobounds. Also one has a similar criterion for normality:

PROPOSITION 7.1 *For $\mathcal{R} \subseteq \mathcal{S}$ the following statements are equivalent:*
(i) $\mathcal{R}$ is normal in $\mathcal{S}$.
(ii) There exist choice functions $\{\phi_j\}$ with $\phi_j \in \mathrm{Ends}\mathcal{R}$ for all j.

Now for an arbitrary ergodic normal subrelation $\mathcal{R} \subseteq \mathcal{S}$ we construct what is called the 'quotient'. For this, as in [6], we produce a group Q, together with the quotient homomorphism (more precisely, the quotient cocycle) $\Theta : \mathcal{S} \to Q$, so that $\mathrm{Ker}\,\Theta = \mathcal{R}$ and Θ is class-surjective. The class-surjectivity of Θ means that, modulo meagre sets, for every $q \in Q$ and every $x \in X$ there exists $y \in X$ with $\Theta(x, y) = q$. These objects have the 'universal property', that is for any countable group $\widetilde{Q}$ and cocycle $\widetilde{\Theta} : \mathcal{S} \to \widetilde{Q}$ with $\mathrm{Ker}\,\widetilde{\Theta} \supseteq \mathcal{R}$ there exists a homomorphism $\pi : Q \to \widetilde{Q}$ with $\pi \circ \Theta = \widetilde{\Theta}$.

Let $\{\phi_j\}$ be a family of choice functions with $\phi_j \in \mathrm{Auts}\mathcal{R}$ for all j. For every $j, k \in J$ define the Borel map $j * k : X \to J$ by $(j * k)(x) = l$ if $\mathcal{R}[\phi_j(\phi_k(x))] = \mathcal{R}[\phi_l(x)]$. This definition is correct because of our selection of choice functions. Moreover, it follows from the ergodicity of $\mathcal{R}$ that the map

$(j * k)(x)$ is constant modulo a meagre set. Thus one obtains the operation '∗' on J. Use the same arguments as in [6] to establish that the set J is equipped with a group structure via this operation and that the quotient group $Q = S/\mathcal{R}$, together with Θ, have the universal property.

Note also, that if for ergodic $\mathcal{R} \subseteq S$ there exist some universal Θ and Q as above, then $\mathcal{R}$ is normal in S.

In the case of a finite index normal subrelation $\mathcal{R}$ of S (one may even require that $\mathcal{R}$ has only finitely many ergodic components), the quotient $S/\mathcal{R}$ can be produced for any such $\mathcal{R}$ in a manner of [6]. This quotient is a groupoid with finite unit space.

THEOREM 7.2 *Let S be given on X as above. Then for every countable group Q, there exists a normal ergodic subrelation $\mathcal{R} \subseteq S$ with $S/\mathcal{R}$ being isomorphic to Q. Moreover, $\mathcal{R}$ is unique up to automorphisms of S.*

Proof: Let Q be given. By virtue of 3.3 there exists an ergodic cocycle $\sigma \in Z^1(S, Q)$. Let us define $x \overset{\mathcal{R}}{\sim} y$, if $x \overset{S}{\sim} y$ and $\sigma(x, y) = e$, where e is the identity of Q. Clearly, $\mathcal{R}$ is an ergodic subrelation of S. Then it is easy to see that Q is the quotient group with the quotient homomorphism σ.

Suppose now that $\widetilde{\mathcal{R}}$ is another normal ergodic subrelation of S with $S/\widetilde{\mathcal{R}} = Q$. Let $\widetilde{\sigma}$ be the quotient cocycle, which is also the index cocycle for $\widetilde{\mathcal{R}} \subseteq S$. It is not difficult to verify that $\widetilde{\sigma}$ is ergodic. Hence by theorem 4.3 $[\sigma] = [\widetilde{\sigma}]$ up to Aut$S$ and theorem 6.3 implies that $\mathcal{R}$ and $\widetilde{\mathcal{R}}$ are conjugate under Aut(S). ∎

References

[1] S. I. Bezuglyi and V. Ya. Golodets, Groups of measure space transformations and invariants of outer conjugation for automorphisms from normalizers of type III full groups, *J. Funct. Anal.*, **60** (1985), 341-369.

[2] S. I. Bezuglyi and V. Ya. Golodets, Type III_0 transformations of measure space and outer conjugacy of countable amenable groups of automorphisms, *J. Operator Theory*, **21** (1989), 3-40.

[3] A. Connes and W. Krieger, Measure space automorphisms, the normalizers of their full groups, and approximate finiteness, *J. Funct. Anal.*, **24** (1977), 336-352.

[4] A. Danilenko, Quasinormal subrelations of ergodic equivalence relations, *Proc. Amer. Math. Soc.*,, **126** (1998), 3361-3370.

[5] A. L. Fedorov and B. Z. Rubshtein, Admissible subgroups of full ergodic groups, *Ergod. Th. & Dynam. Sys.*, **16** (1996), 1221-1239.

[6] J. Feldman, C. E. Sutherland, and R. J. Zimmer, Subrelations of ergodic equivalence relations, *Ergod. Th. & Dynam. Sys.*, **9** (1989), 239-269.

[7] V. Ya. Golodets and S. D. Sinel'shchikov, Classification and structure of cocycles of amenable ergodic equivalence relations, *Journal of Funct. Anal.*, **121** (1994), 455-485.

[8] V. Ya. Golodets and Sinel'shchikov, Outer conjugacy for actions amenable groups, *Publ. Res. Inst. Math. Sci.*, **23** (1987), 737-769.

[9] V. Ya. Golodets and Sinel'shchikov, Amenable ergodic actions of groups, and images of cocycles, *Soviet Math. Dokl.*, **41** (1990), 523-526.

[10] K. Kuratowski, Topology, Academic Press, New York, 1966.

[11] G. Mackey, Ergodic theory and virtual groups, *Math. Ann.*, **166** (1966), 187-207.

[12] D. Sullivan, B. Weiss, and J. D. M. Wright, Generic dynamics and monotone complete C^*-algebras, *Trans. Amer. Math. Soc.*, **295** (1986), 795-809.

[13] R. Zimmer, Amenable ergodic group actions and an application to Poisson boundaries of random walks, *J. Funct. Anal.*, **27** (1978), 350-372.

[14] R. Zimmer, Orbit equivalence and rigidity for ergodic actions of Lie groups, *Ergod. Theory and Dyn. Syst.*, **1** (1981), 237-253.

Descriptive Dynamics

Alexander S. Kechris
Department of Mathematics
Caltech
Pasadena, CA 91125

The purpose of the following informal lectures is to give a brief introduction to *descriptive dynamics*, which I understand here to be the *descriptive theory of Polish group actions*. I will concentrate on the foundations, and hopefully at a level accessible to anyone with a basic knowledge of descriptive set theory. I will illustrate some of the main methods used in this area, including Baire category arguments and various implementations of the "changing the topology" technique. A general reference for the results discussed in this paper is Becker-Kechris [1996].

Acknowledgments. This paper is based on the text of a series of lectures that I was scheduled to give at the International Workshop on Descriptive Set Theory and Dynamical Systems, CIRM, Luminy, July 1-5, 1996. Unfortunately, at the last moment, a family emergency prevented me from attending the meeting, but A. Louveau has kindly stepped in and delivered these lectures. (Except for Lecture IV for which there was not sufficient time.) I would like to thank him for taking over this, despite his heavy load of duties during this meeting. Research and preparation of this paper were partially supported by NSF Grant DMS 96-19880.

Lecture I

A Polish Groups

Classically in various branches of dynamics one studies actions of the group of integers ($\mathbb{Z}$), reals ($\mathbb{R}$), Lie groups, or even more generally (second countable) locally compact groups. We expand here this scope by considering the more comprehensive class of *Polish groups*, which seems to be the widest class of well-behaved (for our purposes) groups and which includes practically every type of topological group we are interested in.

Definition. A *Polish group* is a topological group whose topology is Polish (i.e., separable, completely metrizable).

Here are some examples of Polish groups (for which more details are given in Becker-Kechris [1996] and Kechris [1995]).

Examples.

1. (Second countable) locally compact groups;

2. Separable Banach spaces (under +);

3. Various groups of symmetries of mathematical objects. Here are some concrete examples:

(a) S_∞, the infinite symmetric group of all permutations of $\mathbb{N}$, with the topology of pointwise convergence.

(b) Consider an arbitrary countable structure

$$\mathcal{A} = \langle A, f, g, \dots, R, S, \dots \rangle$$

consisting of a countable set A equipped with certain operations $f, g, \dots$ and certain relations $R, S, \dots$ (each of varying numbers of arguments). Typical examples include: groups $\langle G, \cdot \rangle$, fields $\langle F, +, \cdot \rangle$, ordered groups $\langle G, \cdot, \le \rangle$, graphs $\langle V, E \rangle$, etc. Let $\mathrm{Aut}(\mathcal{A})$ be the automorphism group of $\mathcal{A}$. This is a Polish group, again with the topology of pointwise convergence. When A is infinite we can take, without loss of generality, $A = \mathbb{N}$ and in this case $\mathrm{Aut}(\mathcal{A})$ is a closed subgroup of S_∞. Conversely, every closed subgroup of S_∞ is of that form for an appropriate $\mathcal{A}$ (with $A = \mathbb{N}$).

(c) $U(H)$, the unitary group of a separable Hilbert space H, with the weak (or equivalently strong) topology.

(d) $H(X)$, the homeomorphism group of a compact metrizable space X, with the uniform topology.

(e) $\mathrm{Iso}(X, d)$, the isometry group of a complete separable metric space (X, d), with the pointwise convergence topology.

(f) $\mathrm{Aut}(X, \mu)$ (resp. $\mathrm{Aut}^*(X, \mu)$), the group of measure preserving (resp. nonsingular, i.e., null set preserving) transformations of a standard probability measure space (X, μ) (X is a *standard Borel space*, i.e., a Polish space equipped with its σ-algebra of Borel sets and μ is a Borel probability measure on X). These can be viewed, by the usual association of a unitary operator to each transformation, as closed subgroups of $U(L^2(X, \mu))$, up to (topological group) isomorphism.

We often consider subclasses of Polish groups with various additional nice properties. These can have algebraic flavor, as, for example, the classes of abelian, nilpotent, and solvable groups, or topological flavor, as, for example, the classes of locally compact, admitting invariant metric (which must necessarily be complete), or admitting *complete* left-invariant metric Polish groups. See Becker [1998] for a more complete exposition. Here are the inclusions among these classes: abelian $\subsetneq$ nilpotent $\subsetneq$ solvable, abelian $\subsetneq$ inv. metric, l.c. $\cup$ solvable $\cup$ inv. metric $\subsetneq$ l.inv.complete metric $\subsetneq$ Polish.

Another interesting class of Polish groups consists of those which have a countable local basis at 1 consisting of open subgroups. These are exactly the closed subgroups of S_∞ or equivalently the automorphism groups of countable structures (up to isomorphism). See Becker-Kechris [1996].

Finally, there is a universal Polish group:

Theorem (Uspenskii [1986]). *The group* $H([0,1])^{\mathbf{N}}$ *of homeomorphisms of the Hilbert cube is a universal Polish group, i.e., every Polish group is isomorphic to a closed subgroup of it.*

B   Polish and Borel G-spaces

Definition. Let G be a Polish group. A *Polish G-space* is a Polish space X together with a continuous action of G on X. A *Borel G-space* is a standard Borel space X together with a Borel action of G on X.

There are two basic facts concerning Polish and Borel G-spaces.

Theorem (Effros [1965]; see also Becker-Kechris [1996]). *Let G be a Polish group and X a Polish G-space. For each $x \in X$, the following are equivalent, where $G_x = \{g : g \cdot x = x\}$ is the stabilizer of x;*

(i) $gG_x \mapsto g \cdot x$ *is a homeomorphism of the Polish space G/G_x onto $G \cdot x$ (or equivalently the map $g \mapsto g \cdot x$ is open from G onto $G \cdot x$);*

(ii) $G \cdot x$ *is not meager in its relative topology;*

(iii) $G \cdot x$ *is G_δ in X.*

Corollary. *If $G \cdot x$ is not meager (in X), then it is G_δ in X.*

Theorem (D. Miller [1977]; see also Kechris [1995]). *Let G be a Polish group and X a Borel G-space. Then G_x is closed and $G \cdot x$ is Borel in X.*

However, in general, the *orbit equivalence relation*

$$xE_G^X y \Leftrightarrow xE_G y \Leftrightarrow \exists g(g \cdot x = y)$$

is (analytic but) not Borel.

Given a closed subgroup $G \subseteq H$ of a Polish group H, there is a canonical "minimal" way to extend a given G-action to an H-action, called the *induced action*. This construction is quite useful in showing that various properties of Polish groups with respect to their actions are hereditary, i.e., are inherited by their closed subgroups.

Theorem (Mackey [1966]; see also Becker-Kechris [1996]). *Let G, H be Polish groups with G a closed subgroup of H. Let X be a Borel G-space. There is a unique, up to Borel H-isomorphism, Borel H-space $\tilde{X}$ and Borel G-injection*

$i : X \to \tilde{X}$ such that for every Borel H-space Y and Borel G-map $f : X \to Y$, there is a unique Borel H-map $\tilde{f} : \tilde{X} \to Y$ so that $\tilde{f} \circ i = f$. (If X is a G-space and Y is an H-space, a map f is a G-map if $f(g \cdot x) = g \cdot f(x)$, for $g \in G, x \in X$.)

The space $\tilde{X}$ is denoted by $H \times_G X$ and can be realized as follows:

$\tilde{X} = (H \times X)/G :=$ the orbit space of the action of G on $H \times X$ given by $g \cdot (h, x) = (g \cdot h, g \cdot x)$, with the quotient Borel structure; the action of H on $\tilde{X}$ is given by

$$h \cdot [h', x] = [h'h^{-1}, x]$$

and

$$i(x) = [1, x].$$

Moreover, identifying X with $i(X)$, note that every H-orbit of Y contains a unique G-orbit of X, a fact which is often quite useful.

One also has the analog of the preceding theorem in the topological context.

Theorem (Hjorth; see Becker-Kechris [1996]). *Let G, H be Polish groups with G a closed subgroup of H. Let X be a Polish G-space. There is a unique, up to H-homeomorphism, Polish H-space $\tilde{X}$ and G-homeomorphism $i : X \to \tilde{X}$ such that for every Polish H-space Y and continuous G-map $f : X \to Y$, there is a unique continuous H-map $\tilde{f} : \tilde{X} \to Y$ so that $\tilde{f} \circ i = f$.*

The space $\tilde{X} = H \times_G X$ is defined exactly as before with the quotient topology now. It also turns out that $i(X)$ is closed in $\tilde{X}$.

C Universal G-spaces

Definition. Let G be a Polish group. A Borel G-space $\mathcal{U}$ is *universal* if every Borel G-space can be Borel G-embedded into $\mathcal{U}$.

It is easy to see that such a space is unique up to Borel G-isomorphism, if it exists. We will denote it by $\mathcal{U}_G$.

Theorem (Mackey [1962], Varadarajan [1963]). *If G is Polish locally compact, then $\mathcal{U}_G$ exists and can be realized as a compact Polish G-space (i.e., the space acted upon is compact metrizable).*

This was extended recently to arbitrary Polish groups.

Theorem (Becker-Kechris [1996]). *If G is Polish, then $\mathcal{U}_G$ exists and can be realized as a compact G-space.*

There are several realizations of the universal G-space $\mathcal{U}_G$ which are useful in different circumstances.

Realization 1. Let G be a Polish group and $\mathcal{F}(G)$ the standard Borel space of closed subsets of G with the Effros Borel structure. Let G act on $\mathcal{F}(G)$ by left-translation:

$$g \cdot F = gF.$$

Then the infinite produce G-space

$$\mathcal{U}_G^1 = \mathcal{F}(G)^{\mathbf{N}}$$

is universal; see Becker-Kechris [1996].

Proof. In fact we will show that any Borel action of G on a separable metrizable space Borel G-embeds into $\mathcal{U}_G^1$. Fix a countable open basis $\{U_n\}$ for X. Clearly it separates points. For $A \subseteq G$, let

$$
\begin{aligned}
E(A) \ = \ \{g \in G : &\text{For every open nbhd} \\
&V \text{of } g, \ V \cap A \text{ is not meager}\}.
\end{aligned}
$$

Then $E(A)$ is closed and, if A has the Baire property, $A \Delta E(A)$ is meager. Put

$$
\begin{aligned}
\pi = (\pi_n) &: X \to \mathcal{F}(G)^{\mathbf{N}}, \\
\pi_n(x) &= E(\{g : g \cdot x \in U_n\})^{-1}.
\end{aligned}
$$

$\dashv$

Notice that if G is locally compact Polish and we give $\mathcal{F}(G)$ the Fell topology, then $\mathcal{U}_G^1$ becomes a compact Polish G-space.

Realization 2. Let G be a Polish group and d a left-invariant compatible metric, $d < 1$. Let

$$\mathcal{L}_d(G) \equiv \mathcal{L}(G) = \{f : G \to [0,1] : |f(g) - f(h)| \le d(g,h)\}$$

with the topology it inherits as a subset of $[0,1]^G$ (with the product topology), so it is compact metrizable. Let G act on $\mathcal{L}(G)$ by left-shift

$$g \cdot f(h) = f(g^{-1}h).$$

Then $\mathcal{L}(G)$ is a compact Polish G-space and

$$\mathcal{U}_G^2 = \mathcal{L}(G)^{\mathbf{N}}$$

is universal.

Proof. Embed $\mathcal{F}(G)$ into $\mathcal{L}(G)$ by

$$F \mapsto (g \mapsto d(g, F)).$$

⊣

Realization 3 (Gao [1996]). Let G be a Polish group and d a left-invariant compatible metric, $d < 1$. Fix a dense set $\{g_n\}$ in G. It is well-known that the map $\pi : G \to [0,1]^{\mathbf{N}}$, given by $\pi(g) = (d(g, g_n))_{n \in \mathbf{N}}$, is a homeomorphism. Denote by $\bar{G}$ the closure of $\pi(G)$ in $[0,1]^{\mathbf{N}}$. Then one can extend to $\bar{G}$ the left-translation action of G on itself, by defining

$$g \cdot x = \lim_n \pi(gh_n)$$

for $x \in \bar{G}, h_n \in G, \pi(h_n) \to x$. It turns out that $\bar{G}$ with this action is a compact Polish G-space.

Consider $\mathcal{K}(\bar{G})$, the hyperspace of all compact subsets of $\bar{G}$ with the Vietoris topology, which is a compact metrizable space. G acts continuously on it by $g \cdot K = \{g \cdot x : x \in K\}$ and so $\mathcal{K}(\bar{G})$ is a compact Polish G-space. Let

$$\mathcal{U}_G^3 = \mathcal{K}(\bar{G})^{\mathbf{N}}.$$

Since the map $F \in \mathcal{F}(G) \mapsto \overline{\pi(F)} \in \mathcal{K}(\bar{G})$ is a Borel G-embedding, it follows that $\mathcal{U}_G^3$ is universal. ⊣

Realization 4, for Polish locally compact G (Mackey [1962], Varadarajan [1963]). Let G be Polish locally compact, μ_G its Haar measure and put

$$\mathcal{U}_G^4 = B_1(L^{\infty}(G, \mu_G), w^*),$$

the unit ball of $L^{\infty}(G, \mu_G)$ with the weak*-topology, and G acting on $\mathcal{U}_G$ by left-shift again. Then $\mathcal{U}_G^4$ is a compact Polish G-space which is universal.

Proof. Fix a Borel G-space $X \subseteq [0,1]$. Embed it into $\mathcal{U}_G^4$ by

$$x \mapsto f_x(g) = g \cdot x.$$

⊣

D Applications

We will now present some applications of the universal space.

1. Actions of S_∞ and model theory

Consider structures of the form $\mathcal{A} = \langle \mathbb{N}, R_1, R_2, \cdots \rangle$, with $R_n \subseteq \mathbb{N}^n$ ("hypergraphs"). Each such structure can be identified with an element of

$$\mathcal{U}_{S_\infty} = \prod_n 2^{\mathbb{N}^n}.$$

S_∞ acts on this space by $g \cdot \langle \mathbb{N}, R_1, R_2, \cdots \rangle = \langle \mathbb{N}, R_1', R_2', \ldots \rangle$, where

$$R_n(a_1, \cdots, a_n) \Leftrightarrow R_n'(g(a_1), \cdots, g(a_n)).$$

This is called the *logic action*. Clearly $\mathcal{A}, \mathcal{A}' \in \mathcal{U}_{S_\infty}$ belong to the same orbit iff $\mathcal{A} \cong \mathcal{A}'$, i.e., the orbits are simply the isomorphism classes. One can use Realization 1 to show that this is a universal S_∞-space (see Becker-Kechris [1996]). So if X is an arbitrary Borel S_∞-space, there is a Borel S_∞-embedding $f : X \to \mathcal{U}_{S_\infty}$, and so in particular $f(X)$ is an invariant under $\cong$ Borel subset of $\mathcal{U}_{S_\infty}$. By a theorem of Lopez-Escobar (see, e.g., Kechris [1995]), membership in any Borel invariant subset of $\mathcal{U}_{S_\infty}$ can be defined by a countable set of axioms, in an appropriate logical language, called $L_{\omega_1 \omega}$, which allows, beyond the usual logical operations, $\wedge, \vee, \neg, \exists, \forall$, countable infinitary conjunctions and disjunctions. (A typical example of such axioms are those for the torsion-free abelian groups.) Therefore, any Borel S_∞-action is Borel isomorphic to the logic action on the models of a countable theory (in $L_{\omega_1 \omega}$) and this establishes a basic connection between S_∞-actions and model theory, which motivates a lot of work in this area. For more on this subject, see Becker-Kechris [1996].

2. Tarski's Theorem

Suppose G is an arbitrary (abstract) group and X an arbitrary G-space. Given $A, B \subseteq X$ we say that A, B are *equivalent by finite decomposition*, in symbols $A \sim B$, if there are partitions $A = \bigcup_{i=1}^n A_i$, $B = \bigcup_{i=1}^n B_i$ and $g_i \in G$, with $g_i \cdot A_i = B_i$. We say that X is *paradoxical* if $X \sim A \sim B$, with $A \cap B = \emptyset$. A well-known result in the theory of paradoxical decompositions is the following:

Theorem (Tarski; see Wagon [1993]). *For any group G and any G-space X, the following are equivalent:*

(i) *There is a finitely additive probability measure on X (defined for all subsets of X) which is G-invariant;*

(ii) *X is not paradoxical.*

The problem has been raised (see, e.g., Wagon [1993]) whether there is an analog of Tarski's theorem for ordinary, countably additive measures. The context is as follows.

Assume $(X, \mathcal{S})$ is a measurable space and G acts on X preserving $\mathcal{S}$, i.e., $g \in G, A \in \mathcal{S} \Rightarrow g \cdot A \in \mathcal{S}$. Put for $A, B \in \mathcal{S}$:

$$A \sim_\infty B \quad \Leftrightarrow \quad \text{there are partitions } A = \bigcup_{i=1}^{\infty} A_i, B = \bigcup_{i=1}^{\infty} B_i \text{ with}$$
$$A_i, B_i \in \mathcal{S}, \text{ and } g_i \in G \text{ such that } g_i \cdot A_i = B_i.$$

We say that X is *countably paradoxical* if $X \sim_\infty A \sim_\infty B$ with $A, B \in \mathcal{S}$, $A \cap B = \emptyset$. Is it true that X is not countably paradoxical iff there is a (countably additive) probability measure on $(X, \mathcal{S})$ which is G-invariant?

In this generality it turns out that the answer is negative (see Wagon [1993]), but one can use the existence of universal actions and a theorem of Nadkarni [1990] to show that one gets a positive answer in regular situations.

Theorem (Becker-Kechris [1996]). *Let G be a Polish group and X a Borel G-space. Then the following are equivalent:*

(i) *There is a Borel probability measure on X which is G-invariant;*

(ii) *X is not countably paradoxical (for the class of Borel sets).*

Nadkarni [1990] essentially proves this result in the case G is countable. For the general case, the existence of a Polish universal space shows that every Borel G-space is Borel isomorphic to a continuous action of G on a Borel set in a Polish space. One can then apply Nadkarni's theorem to a countable subgroup of G and a straightforward continuity argument to complete the proof.

3. Embedding Polish G-spaces

Now let X be a Polish G-space (G a Polish group). Is it possible to G-embed *topologically* X into a compact Polish G-space Y? The answer is positive if G is locally compact (see deVries [1978] and Megrelishvili [1989]). Solving a related old problem in the theory of transformation groups, Megrelishvili [1988] showed that the answer is in general negative. Very recently, Scarr [1998] has in fact shown that for a Polish group G, G is locally compact iff every Polish G-space can be G-embedded topologically into a compact Polish G-space. Considering the embedding in the universal space given by

Realization 2, Hjorth and Kechris (see Hjorth [1999a]) proved the following facts:

Theorem (Hjorth-Kechris; see Hjorth [1999a]). *Let G be a Polish group and let $\mathcal{U}_G^2$ be the universal G-space defined in C, Realization 2. Then $\mathcal{U}_G^2$ is a compact Polish G-space and for any Polish G-space X there is a G-embedding $\rho : X \to \mathcal{U}_G^2$ such that*

(i) $\rho(X)$ *is G_δ in $\mathcal{U}_G^2$ (so Polish).*

(ii) $\rho : X \to \rho(X)$ *is open.*

(iii) $\rho : X \to \rho(X)$ *is Baire class 1.*

(So π just misses being a homeomorphism, as expected by Megrelishvili's result.)

Proof. Let ρ be the composition of the embeddings of Realizations 1,2, i.e., $\rho = (\rho_n)$, where

$$\rho_n(x) = (g \mapsto d(g, \{h : h \cdot x \in U_n\}^{-1})),$$

with $\{U_n)$ an open basis for X (where $d(g, \emptyset) = 1$).

We first check that $\rho : X \to \rho(X)$ is open. Fix open $U \subseteq X$ in order to show that $\rho(U)$ is open in $\rho(X)$. Let $x \in U$. Then $1 \cdot x \in U$, so find $1 > \varepsilon > 0$ and n such that $x \in U_n$ and for any h with $d(1, h) < \varepsilon$ we have $h^{-1} \cdot U_n \subseteq U$. Consider then the open hbhd $\{\rho(y) : |\rho_n(y)(1) - \rho_n(x)(1)| < \varepsilon\}$ of $\rho(x)$ in $\rho(X)$. It is enough to show it is contained in $\rho(U)$. Fix y in that nbhd. Then $\rho_n(y)(1) = d(1, \{h : h \cdot y \in U_n\}^{-1}) < \varepsilon$ as $\rho_n(x)(1) = 0$ (since $1 \cdot x \in U_n$). So let h be such that $h \cdot y \in U_n$ and $d(1, h^{-1}) < \varepsilon$, so $d(1, h) < \varepsilon$. Then $h^{-1} \cdot U_n \subseteq U$, and so $y \in U$.

For the proof that ρ is Baire class 1 it is enough to check that for each $n, g \in G$, $a \in \mathbb{R}$, the set $\{x : \rho_n(x)(g) < a\}$ is open and the set $\{x : \rho_n(x)(g) > a\}$ is F_σ. The first is straightforward from the definition. For the second, notice that

$$\rho_n(x)(g) > a \quad \Leftrightarrow \quad d(g, \{h : h \cdot x \in U_n\}^{-1}) > a$$
$$\Leftrightarrow \quad \exists \text{ rational } b > a (B_g^d(b)^{-1} \cdot x \subseteq U_n^c),$$

where $B_g^d(b) = \{h : d(h, g) < b\}$ and $U_n^c = X \setminus U_n$.

Finally it remains to shows that $\rho(X)$ is G_δ, i.e., Polish in its relative topology. Equivalently if τ^* is the topology on X obtained by transferring the topology of $\rho(X)$ to X via ρ^{-1} we have to show that τ^* is Polish. Since ρ is open, if τ is the topology on X, clearly $\tau \subseteq \tau^*$.

Now a basis for τ^* consists of the sets of the form

$$U_n \cap M_{m_1,g_1,a_1} \cap \ldots \cap M_{m_k,g_k,a_k},$$

where

$$M_{m,g,a} = \{x : \rho_m(x)(g) > a\}.$$

(since the sets of the form $\{x : \rho_m(x)(g) < a\}$ are in τ). We can denote this set by $< n; m_1, g_1, a_1; \ldots ; m_k, g_k, a_k >$. To show that (X, τ^*) is Polish, by the Choquet Criterion it is enough to show that II wins the strong Choquet game for this space (see Kechris [1995]).

We describe his strategy below:

If I plays $x_1, < n^1; m_1^1, g_1^1, a_1^1; \ldots >$, then $\rho_{m_j^1}(x)(g_j^1) > a_j^1$. Fix $\bar{a}_j^1 > a_j^1$ with $\rho_{m_j^1}(x)(g_j^1) > \bar{a}_j^1$ and $\bar{n}^1$ with $x_1 \in U_{\bar{n}^1} \subseteq \overline{U_{\bar{n}^1}} \subseteq U_{n^1}$ and $\mathrm{diam}(U_{\bar{n}^1}) < \frac{1}{2}$ (in some complete metric for (X, τ)). II responds by playing $< \bar{n}^1; m_1^1, g_1^1, \bar{a}_1^1; \ldots >$. Next I plays $x_2, < n^2; m_1^2, g_1^2, a_1^2; \ldots >$. Then define $< \bar{n}^2; m_1^2, g_1^2, \bar{a}_1^2; \ldots >$ as before, except that $\mathrm{diam}(U_{\bar{n}^2}) < \frac{1}{3}$, and let II play $< \bar{n}^2; m_1^2, g_1^2, \bar{a}_1^2; \ldots >$, etc. Clearly $x_i \to x \in \bigcap_k U_{n^k}$ (in τ). Also $x_i \in \{x : \varphi_{m_1^1}(x)(g_1^1) > \bar{a}_1^1\}$ for all i, thus $B_{g_1^1}^d(\bar{a}_1^1)^{-1} \cdot x_i \subseteq U_{m_1^1}^c\}$, so $B_{g_1^1}^d(\bar{a}_1^1)^{-1} \cdot x \subseteq U_{m_1^1}^c$ and thus $\rho_{m_1^1}(x)(g_1^1) \geq \bar{a}_1^1 > a_1^1$, i.e., $x \in M_{m_1^1,g_1^1,a_1^1}$. Similarly we see that $x \in < n^1; m_1^1, g_1^1, a_1^1; \ldots >$ and also $x \in < n^k; m_1^k, g_1^k, a_1^k; \ldots >$ for each k, i.e., x belongs in the intersection of all open sets played by I (and II), so this intersection is non-$\emptyset$ and II won. $\dashv$

The following was stated as an open problem at the time of the workshop:

Problem. *Is there a universal Polish G-space, i.e., a Polish G-space in which every Polish G-space can be G-embedded topologically?*

The answer was known to be positive for locally compact G (see deVries [1975]). Very recently Hjorth [1999b] solved this, affirmatively again, for *any* Polish group G.

Lecture II

A   An Equivariant Version of Kuratowski's Theorem

We will discuss here the "changing the topology" idea, which is quite useful in numerous contexts. We first recall a classical result of Kuratowski, which has many applications in descriptive set theory (see, for example, Kechris [1995]). It shows that, in some sense, Borel sets can be thought as clopen and similarly Borel functions as continuous.

Theorem (Kuratowski).

(i) *Let X be a standard Borel space, $A \subseteq X$ a Borel set. Then there is a Polish topology τ_A, generating the Borel structure of X, with A clopen in (X, τ_A). If (X, τ) is given as a Polish space, we can also make sure that $\tau_A \supseteq \tau$.*

(ii) *If X is a standard Borel space, Y a Polish space and $f : X \to Y$ is Borel, then there is a Polish topology τ_f generating the Borel structure of X, so that $f : (X, \tau_f) \to Y$ is continuous. Again if (X, τ) is given as a Polish space, we can take $\tau_f \supseteq \tau$.*

This result easily extends to deal simultaneously with countably many A's or f's.

We now have the following G-version of this result.

Theorem (Becker-Kechris). *Let G be a Polish group. Let X be a Borel G-space, Y a Polish G-space, and $f : X \to Y$ a Borel G-map. Then there is a Polish topology τ_f on X giving its Borel structure, so that (X, τ_f) is a Polish G-space and $f : (x, \tau_f) \to Y$ is continuous. If, moreover, (X, τ) is a Polish G-space, then we can take $\tau_f \supseteq \tau$.*

Proof. The main tool is the concept of Vaught transform, which plays an important role in descriptive dynamics, and is an application of Baire Category techniques.

Definition. Let G be a Polish group and X a G-space. For $P \subseteq X$, $U \subseteq G$ open non-$\emptyset$, we define the *Vaught transforms*

$$P^{\Delta U} = \{x \in X : \exists^* g \in U(g \cdot x \in P)\},$$
$$P^{*U} = \{x \in X : \forall^* g \in U(g \cdot x \in P)\},$$

where "$\exists^* g \in U$" means "there exist nonmeager many $g \in U$" and "$\forall^* g \in U$" means "there exist comeager many $g \in U$".

An important point is that for X a Borel G-space and P Borel, $P^{\Delta U}$ and P^{*U} are Borel as well.

Lemma 1. *Let G be a Polish group and Y a Polish G-space. If $\mathcal{V}$ is a basis for Y and $\mathcal{U}_1$ a nbhd basis of $1 \in G$ the set*

$$\mathcal{V}^{\Delta \mathcal{U}_1} = \{V^{\Delta N} : V \in \mathcal{V}, N \in \mathcal{U}_1\}$$

is a basis for Y.

Proof. It is easy to see that every set in $\mathcal{V}^{\Delta\mathcal{U}_1}$ is open. Let now $W \subseteq Y$ be open and $x \in W$. Let $x \in V \in \mathcal{V}$, $N \in \mathcal{U}_1$ be such that $N^{-1} \cdot V \subseteq W$. It is enough to show that $x \in V^N \subseteq W$. To use this notice that $\{g : g \cdot x \in V\}$ is an open nbhd of $1 \in G$, so it intersects N, thus $x \in V^{\Delta N}$. Now let $y \in V^{\Delta N}$, so that $\exists^* g \in N(g \cdot y \in V)$. Then for some $g \in N$, $g \cdot y \in V$, so $y \in g^{-1} \cdot V \subseteq N^{-1} \cdot V \subseteq W$.

$\dashv$

Lemma 2 (Becker-Kechris [1996]). *Let G be a Polish group and X a Borel G-space. Let $\mathcal{B}$ be a countable collection of Borel sets in X and $\mathcal{U}$ a countable basis for G. Then there is a Polish topology $\tau_{\mathcal{B},\mathcal{U}}$ on X generating its Borel structure, such that the G-action on X is continuous and $\mathcal{B}^{\Delta\mathcal{U}} = \{B^{\Delta U} : B \in \mathcal{B}, U \in \mathcal{U}\} \subseteq \tau_{\mathcal{B},\mathcal{U}}$. If moreover, (X, τ) is given as a Polish G-space, then one can take $\tau_{\mathcal{B},\mathcal{U}} \supseteq \tau$.*

Let now $\mathcal{B} = f^{-1}(\mathcal{V}) = \{f^{-1}(V) : V \in \mathcal{V}\}$ and put $\tau_f = \tau_{\mathcal{B},\mathcal{U}}$. Since $f^{-1}(V^{\Delta N}) = [f^{-1}(V)]^{\Delta N}$, it follows that $f^{-1}(\mathcal{V}^{\Delta\mathcal{U}_1}) \subseteq \mathcal{B}^{\Delta\mathcal{U}} \subseteq \tau_{\mathcal{B},\mathcal{U}}$, where $\mathcal{U}_1 = \{N \in \mathcal{U} : 1 \in N\}$, and so by Lemma 1, $f : (X, \tau_f) \to Y$ is continuous and we are done.

$\dashv$

We will now discuss two particular cases of this result.

Corollary (Becker-Kechris [1996]). *Let G be a Polish group. If X is a Borel G-space, there is a Polish topology τ on X, giving its Borel structure, so that (X, τ) is a Polish G-space. Equivalently, any Borel G-space is Borel isomorphic to a Polish G-space.*

The proof is obtained by applying the theorem to the trivial G-space $Y = \{y_0\}$, $g \cdot y_0 = y_0$, $f(x) = y_0 \in Y$. This corollary follows immediately from Kuratowski's Theorem for countable G, and was proved by Wagh [1988] for $G = \mathbb{R}$. It was raised as an open problem by D. Miller [1977] and, for locally compact G, by Ramsay [1985].

Corollary (Becker-Kechris [1996]). *Let G be a Polish group and X a Borel G-space. If $A \subseteq X$ is an invariant Borel set, there is a Polish topology τ_A, giving its Borel structure, so that (X, τ_A) is a Polish G-space and A is clopen in τ_A. If (X, τ) is given as a Polish G-space, then one can take $\tau_A \supseteq \tau$.*

Proof. Apply the theorem for $X, Y = \{0, 1\}$, the trivial action of G on Y, $g \cdot y = y$, and $f : X \to Y$ defined by $f(x) = 1$ iff $x \in A$.

$\dashv$

This is again due to Kuratowski for G countable and to Sami [1994] for $G = S_\infty$. Some of Sami's ideas here have found their way in the proof of these more general results.

B   An Application

Let us first mention one more corollary of the preceding theorem. (For the Fell topology on the space of closed subsets of a locally compact Polish space, see Kechris [1995].)

Corollary. *Let G be Polish locally compact, X a Borel G-space. Put on $\mathcal{F}(G)$ the Fell topology, so it is a compact Polish G-space under the conjugation action*

$$g \cdot F = gFg^{-1}.$$

Then there is a Polish topology τ on X, generating its Borel structure, so that (X, τ) is a Polish G-space and the map $x \mapsto G_x$ is continuous from (X, τ) into $\mathcal{F}(G)$.

Proof. Simply notice that $G_{g \cdot x} = gG_x g^{-1}$ and apply the theorem.   ⊣

We will use this last corollary to provide a somewhat simplified and streamlined proof of the following descriptive strengthening of a measure theoretic result of Feldman-Hahn-Moore [1979]. This version of the proof was motivated by a communication of Ramsay.

Theorem. (Kechris [1992]). *Let G be a Polish locally compact group and X a Borel G-space. Then there exists a Borel lacunary complete section for X, i.e., a Borel set $S \subseteq X$ meeting every orbit of X, and such that for some open nbhd U of $1 \in G$ and all $s \in S$, $U \cdot s \cap S = \{s\}$. In particular, the intersection of S with every orbit is countable.*

Proof. By the preceding corollary, we can assume that X is a Polish G-space and $x \mapsto G_x$ is continuous. Fix a compatible metric d on X, a compact nbhd Λ of $1 \in G$ and a compact symmetric nbhd Δ of $1 \in G$ with $\Delta^2 \subseteq \Lambda$. Denote by Δ^0 the interior of Δ. Put

$$R_\Delta(x, y) \Leftrightarrow \exists g \in \Delta(g \cdot x = y).$$

Following Forrest [1974], let for $\varepsilon > 0$,

$$A_\varepsilon = \{x \in X : \forall g \in \Lambda(d(g \cdot x, x) \leq \varepsilon \Rightarrow g \in \Delta^0 G_x)\}.$$

Using the continuity of $x \mapsto G_x$ it is not hard to see that A_ε is open and then an argument by contradiction shows that $X = \bigcup_{n>0} A_{1/n}$. It is also easy to check that $R_\Delta | B$ is an equivalence relation, for any set $B \subseteq A_\varepsilon$ with $d(B) \leq \varepsilon$. So we can find a countable sequence of open subsets of X, $\{U_n\}$, such that R_Δ is a closed equivalence relation on each U_n. (It is closed, since Δ is compact.) So $R_n = R_\Delta | U_n$ is smooth, i.e., there is a Borel function $f_n : U_n \to [0, 1]$ with $xR_ny \Leftrightarrow f_n(x) = f_n(y)$ (see Kechris [1995]).

We next want to actually show that R_n has a Borel transversal, say T_n. If this is found, then let

$$S = \bigcup_n (T_n \setminus \bigcup_{m<n} [U_m]),$$

where $[U_m] = $ saturation of U_m, which easily works.

To prove the existence of T_n we use some descriptive set theory. Although in general it is not true that a closed equivalence relation has a Borel transversal, there are special circumstances under which this happens and these are satisfied here.

Definition. Let E be a Borel equivalence relation on the standard Borel space X. We say that E is *idealistic* if one can assign in a Borel way to each E-equivalence class C a σ-ideal I_C of subsets of C with $C \notin I_C$. (In a "Borel way" means that if $A \subseteq E$ is Borel, so is $\{x : A_x \in I_{[x]_E}\}$.)

Theorem (Kechris [1995]). *Let E be a smooth Borel equivalence relation on a standard Borel space X. If E is idealistic, then E has a Borel transversal.*

Finally, notice that each R_n is idealistic, since for each R_n-equivalence class C we can choose $x \in C$ and define:

$$B \in I_C \Leftrightarrow \{g : g \cdot x \in B\} \text{ is meager.}$$

Notice that this is independent of x, by the translation invariance of meagerness. Also $C \notin I_C$, since no open set in G is meager. ⊣

Lecture III

We will study from now on the orbit equivalence relation E_G of a group action and the orbit space $X/G = X/E_G$.

A   Complexity of the Orbit Equivalence Relation

For G a Polish group and X a Borel G-space, let

$$x E_G^X y \Leftrightarrow x E_G y \Leftrightarrow \exists g(g \cdot x = y)$$

be the orbit equivalence relation. Recall that every orbit, i.e., every E_G-equivalence class, is Borel. However we have:

Theorem (Folklore). *The equivalence relation E_G is analytic but not in general Borel.*

Proof. Consider the logic action on structures of the form $\langle \mathbb{N}, R \rangle$, i.e., on the space $2^{\mathbb{N}^2}$. It can be shown that the corresponding E_G is not Borel (see, e.g., Kechris [1995]).   ⊣

Under what circumstances is E_G actually Borel? Here are some well-known special cases (see Kechris [1995]).

Proposition. *E_G is Borel if the action is free or if G is locally compact.*

The following result characterizes when E_G is Borel.

Theorem (Becker-Kechris [1996]). *Let G be a Polish group and X a Borel G-space. Then the following are equivalent:*

(i) *E_G is Borel;*

(ii) *$x \mapsto G_x$ (from X into $\mathcal{F}(G)$) is Borel.*

Idea of the Proof. (ii) $\Rightarrow$ (i) is classical. We sketch the proof that (i) $\Rightarrow$ (ii), which is motivated by model theoretic ideas. For each $x \in X$, we fix a countable Boolean algebra $\mathcal{B}_x$ of Borel sets, which depends "uniformly in a Borel way" (this is where the Borelness of E_G is used) on x, such that $G \cdot x \in \mathcal{B}_x, \mathcal{B}_x^{\Delta \mathcal{U}} \subseteq \mathcal{B}_x$ (with $\mathcal{U}$ a countable basis for G), and the topology generated by $\mathcal{B}_x$ is Polish. By Becker-Kechris [1996], the topology τ_x generated by $\mathcal{B}_x^{\Delta \mathcal{U}}$ is Polish and the action is continuous for (X, τ_x). Since $G \cdot x = (G \cdot x)^{\Delta G} \in \tau_x$, by Effros' Theorem the map $g \mapsto g \cdot x$ from G onto $G \cdot X$ is open (for τ_x restricted to $G \cdot x$). From this one can check the following formula: For $W, V \subseteq G$ open,

$$WG_x \cap V \neq \emptyset \Leftrightarrow \exists U \in \mathcal{U}, U \subseteq W \forall B \in \mathcal{B}_x (x \in B^{\Delta U} \Rightarrow x \in B^{\Delta V}),$$

which easily implies the Borelness of $x \mapsto G_x$.   ⊣

Although in general E_G is not Borel, one has the following "approximation".

Theorem (Becker-Kechris [1996]). *Let G be a Polish group and X a Borel G-space. There is a sequence $\{X_\alpha\}_{\alpha < \omega_1}$ of Borel sets such that $X = \bigcup_{\alpha < \omega_1} X_\alpha$ is a partition of X, each X_α is invariant, $E_G | X_\alpha$ is Borel (in fact in a "uniform in α" way) and moreover we have the following cofinality property: If $A \subseteq X$ is invariant Borel and $E_G | A$ is Borel, then for some $\alpha < \omega_1$, $A \subseteq \bigcup_{\beta \leq \alpha} X_\beta$.*

The proof of this result uses the descriptive set theory of co-analytic sets.

B   The Topological Vaught Conjecture

This is a basic question concerning the "effective" or "definable" cardinality of the orbit space X/G.

The Topological Vaught Conjecture (D. Miller, 1977). *Let G be a Polish group and X a Polish G-space. Either E_G has countably many classes or else there is a Cantor set $C \subseteq X$ such that $x, y \in C$, $x \neq y \Rightarrow \neg x E_G y$ (in which case we say that E_G has perfectly many classes).*

By TVC(G) we abbreviate the statement that the above holds for the Polish group G, and we let

$$\text{TVC} \Leftrightarrow \forall G \ \text{TVC}(G).$$

By the results in Lecture II, TVC(G) is equivalent to its formulation for Borel G-spaces.

The TVC generalizes the famous *Vaught Conjecture* (VC) in model theory, which is the assertion that a first-order theory has either countably many or continuum many countable models (up to isomorphism). By Lecture I this is a special case of TVC(S_∞).

Both TVC and VC are open. We discuss below some progress that has been achieved to date.

First we remark that the analog of the TVC fails for analytic equivalence relations. However it holds for co-analytic equivalence relations. In particular it holds for Borel ones.

Theorem (Silver [1980]). *Let X be a Polish space and E a co-analytic equivalence relation on X. Then either E has countably many classes or else perfectly many classes.*

Idea of the Proof (due to Harrington). By a standard result of Mycielski, Kuratowski (see Kechris [1995]) an equivalence relation on a "reasonable" topological space S which is meager (in S^2) has perfectly many classes.

One now defines a new second countable "reasonable" topology τ on X, extending its given Polish topology, which is generated by a suitably chosen countable family of analytic sets. This is a version of the so-called Gandy-Harrington topology, which is defined using concepts of effective descriptive set theory. Then let $W = \bigcup_{C \in X/E} \text{Int}^\tau(C)$. If $X = W$, there are clearly only countably many classes of E. Otherwise, let $U = X \setminus W \neq \emptyset$. Magically it turns out that U is open (it is clearly closed) in τ and then one can check that $E|U$ is meager in (U^2, τ^2), so it has perfectly many classes. $\dashv$

Corollary. *If E_G is Borel, for example if the action is free or G is locally compact, then E_G has countably many of perfectly many classes.*

Corollary. *TVC(G) holds for Polish locally compact G.*

In another direction, but still making use of Silver's Theorem, Sami [1994] proved TVC(G) for every abelian Polish group G. This was extended by

Hjorth-Solecki [1999], who proved that TVC(G) holds for all nilpotent Polish groups and all Polish groups admitting an invariant metric. Finally, very recently, Becker [1998] proved TVC(G) for all Polish G which admit a complete left-invariant metric, and Hjorth [1997c] proved TVC(G) for all Polish G with no closed subgroup which has S_∞ as a quotient, which is the widest class of groups known to satisfy the TVC to date.

C   Glimm-Effros Dichotomies

A basic problem concerning a given G-space X is the "classification" of members of X up to orbit equivalence by "invariants". This is a special case of the more general problem of classifying elements of a given standard Borel space X up to some equivalence relation E defined on that space.

Definition. Let E, E' be two equivalence relations on standard Borel spaces X, X'. We say that E is *Borel reducible* to E', in symbols.

$$E \leq_B E'$$

if there is Borel $f : X \to X'$ such that

$$xEy \Leftrightarrow f(x)E'f(y).$$

Letting $\tilde{f}([x]_E) = [f(x)]_{E'}$ it is clear that $\tilde{f} : X/E \to X'/E'$ is an "embedding" of X/E into X'/E'.

Intuitively, $E \leq_B E'$ can be interpreted as meaning any one of the following:

(i) E has a simpler classification problem than E': any invariants for E' work for E as well (after composing with f).

(ii) One can classify E-equivalence classes by invariants which take the form of E'-equivalence classes.

(iii) The quotient space X/E "Borel embeds" into the quotient space X'/E', so X/E has "definable cardinality" less than or equal to that of X'/E'.

Notation. If the function f above is actually 1-1 we put $E \sqsubseteq_B E'$. If it is moreover continuous, we let $E \sqsubseteq_c E'$.

In this notation, we can restate Silver's Theorem as follows: Let for each set A, $\Delta(A)$ be the equality relation on A. Then for every co-analytic equivalence relation on a Polish space X, we have that exactly one of the following holds:

(I) $E \leq_B \Delta(\mathbb{N})$;

(II) $\Delta(2^{\mathbf{N}}) \sqsubseteq_c E$.

Definition. Let E be a Borel equivalence relation on a standard Borel space X. We call E *concretely classifiable* or *smooth* if $E \leq_B \Delta(Y)$ for some standard Borel space Y (or equivalently if E has a countable Borel separating family $\{A_n\}$, i.e., $xEy \Leftrightarrow \forall n(x \in A_n \Leftrightarrow y \in A_n)$).

So if E is smooth, there is a Borel function $f : X \to Y$ (Y a standard Borel space) such that

$$xEy \Leftrightarrow f(x) = f(y).$$

Thus we can classify elements of X, up to E-equivalence, by invariants, computed in a Borel way, which are members of some standard Borel space.

Note the following equivalent formulation in the case of actions.

Theorem (Burgess [1979]; see also Kechris [1995]). *Let G be a Polish group and X a Borel G-space. Then if E_G is Borel, E_G is smooth iff E_G has a Borel transversal.*

This is a special case of the last theorem discussed in Lecture II.

Definition. E_0 is the following equivalence relation on $2^{\mathbf{N}}$:

$$xE_0y \Leftrightarrow \exists n \forall m \geq n(x(m) = y(m)).$$

This is (essentially) the equivalence relation induced by the odometer map and can be thought of as the combinatorial version of the classical Vitali equivalence relation on $[0,1] : xE_Vy \Leftrightarrow \exists q \in \mathbb{Q}(q + x = y)$.

We now have

The Glimm-Effros Dichotomy (Effros [1965], [1981]). *Let G be a Polish group and X a Polish G-space for which E_G is F_σ. Then exactly one of the following holds:*

(I) *E_G is smooth.*

(II) *$E_0 \sqsubseteq_c E_G$.*

Alternatives (I) and (II) also have the following equivalents:

(I) (a) all the orbits are G_δ;

    (b) all the orbits are locally closed (i.e., the difference of two closed sets);

(II) There is an E_G-ergodic non-atomic probability measure on X.

One can derive the preceding result from the following theorem, which can be proved by a combinatorial construction.

Theorem (Becker-Kechris [1996]). *Let G be any group acting by homeomorphisms on a Polish space X. Assume there is a dense orbit and E_G is meager. Then $E_0 \sqsubseteq_c E_G$.*

To see how to prove the Glimm-Effros Dichotomy from this, consider the *generic ergodic decomposition* of X, i.e., define the following equivalence relation:

$$x \tilde{E}_G y \Leftrightarrow \overline{[x]_{E_G}} = \overline{[y]_{E_G}}.$$

It is easy to see that $\tilde{E}_G \ (\supseteq E_G)$ is a G_δ equivalence relation, whose equivalence classes are (of course) G_δ sets on which the G-action is minimal (i.e., all orbits are dense). If $E_G = \tilde{E}_G$, then every orbit is G_δ, so the map $x \mapsto \overline{[x]_{E_G}}$ (from X into the standard Borel space $\mathcal{F}(X)$ of all closed subsets of X with the Effros Borel structure) shows that E_G is smooth. Otherwise, one $\tilde{E}_G$-equivalence class, say C, contains at least two orbits and since every orbit if F_σ and dense in C, it follows, from the Baire Category Theorem, that every orbit in C is meager, so by the Kuratowski-Ulam Theorem, $E_G|C$ is meager (in C^2), so $E_0 \sqsubseteq_c E_G|C$ and thus $E_0 \sqsubseteq_c E_G$.

In 1990 the Glimm-Effros Dichotomy has been extended to the general context of Borel equivalence relations.

Theorem (Harrington-Kechris-Louveau [1990]). *Let E be a Borel equivalence relation on a Polish space X. Then exactly one of the following holds:*

(I) *E is smooth;*

(II) *$E_0 \sqsubseteq_c E$.*

Moreover, (I) is equivalent to the existence of a Polish topology σ on X, extending its given topology, so that E is closed in σ, and (II) is equivalent to the existence of an E-ergodic, non-atomic probability Borel measure on X.

The proof uses the "change of topology" idea. One defines, as in the proof of Silver's Theorem, a new topology τ, using effective descriptive set theory, and then considers E in (X^2, τ^2): If E is closed in (X^2, τ^2), then it turns out that E is smooth, while otherwise one can show that $E_0 \sqsubseteq_c E$.

It follows that one has a full Glimm-Effros Dichotomy for E_G, provided that it is Borel. However, any reasonable form of Glimm-Effros Dichotomy for general Borel or Polish G-spaces fails, if in the first alternative we require the classifying invariants to be members of a standard Borel space (even for $G = S_\infty$). However, we have the following version of this dichotomy for arbitrary Borel G-spaces, in which the classifying invariants are now replaced by countable transfinite sequences of 0's and 1's. This is motivated by a classical result of Ulm on classifying countable abelian p-groups (see Fuchs [1970]).

Theorem (Becker, Hjorth-Kechris [1995]). *Let G be a Polish group and X a Polish G-space. Then exactly one of the following holds:*

(I) E_G *can be classified (definably) by invariants which are countable transfinite sequences of 0's and 1's (Ulm-type classification);*

(II) $E_0 \sqsubseteq_c E_G$.

In another direction, one can recover the original Glimm-Effros Dichotomy, for arbitrary Polish G-spaces, by considering restricted classes of groups G, which are somehow nicer.

To motivate the next definition, notice that if X is a Polish G-space with E_G in F_σ and the action is minimal, then if one orbit is non-meager, the action is actually transitive. This, by using the generic ergodic decomposition, implies the Glimm-Effros Dichotomy in the following strong form: either every orbit is G_δ or $E_0 \sqsubseteq_c E$.

Definition. We say that a Polish group G is a *GE-group* if every minimal Polish G-space with a non-meager orbit is transitive.

Definition. We say that the Polish group G satisfies the *strong Glimm-Effros Dichotomy* if for every Polish G-space X one of the following holds: every orbit is G_δ or $E_0 \sqsubseteq_c E_G$.

Thus every GE-group satisfies the strong Glimm-Effros Dichotomy. The following are known to be GE-groups:

(i) locally compact groups (since then E_G is F_σ);

(ii) (Hjorth-Solecki [1999]) nilpotent or having an invariant metric groups;

(iii) (Hjorth [1996]) countable products of locally compact Polish groups.

Solecki has shown that every GE-group admits a complete left-invariant metric. However Hjorth-Solecki [1999] have found examples of solvable Polish groups (of rank 2), which fail to satisfy the strong Glimm-Effros Dichotomy. On the other hand, Hjorth-Solecki [1999] and Kechris showed that it holds even for Polish groups admitting a complete left-invariant metric for all *free* actions.

Finally, very recently, Becker showed that although Polish groups which admit complete left-invariant metrics fail to satisfy the strong Glimm-Effros

Dichotomy, they still satisfy the Glimm-Effros Dichotomy, in fact, in the following form.

Theorem (Becker [1998]). *Let G be a Polish group admitting a complete left-invariant metric. Let X be a Polish G-space. Then either every orbit is $\mathbf{\Pi}^0_\omega$ or else $E_0 \sqsubseteq_c E$. Since in the first alternative E_G is actually Borel, we have that either E_G is smooth or else $E_0 \sqsubseteq_c E_G$, so the Glimm-Effros Dichotomy holds for such G.*

Lecture IV

Turbulence

We will describe here very recent work of G. Hjorth [1997a] (see also Kechris [1997] for an exposition of Hjorth's results).

Definition. Let E be an equivalence relation on a standard Borel space X. We say that E *admits classification by countable structures* if there is a Borel map assigning to each $x \in X$ a countable structure $\mathcal{A}_x$ (with domain $\mathbb{N}$ in some countable language) such that

$$x E y \Leftrightarrow \mathcal{A}_x \cong \mathcal{A}_y,$$

i.e., invariants are countable structures up to isomorphism.

It is easy to see that if E is smooth, then it admits classification by countable structures (but the converse easily fails). Also by Lecture I, E admits classification by countable structures iff $E \leq_B E^Y_{S_\infty}$ for some Borel S_∞-space Y.

Examples.

1. Let D be the set of all ergodic $T \in \text{Aut}(X, \mu)$ which have discrete spectrum. Let E_C be conjugacy in $\text{Aut}(X, \mu)$ restricted to D. (So E_C is induced by the conjugacy action of $\text{Aut}(X, \mu)$ on D.) By the Halmos-von Neumann Theorem, for $S, T \in D$:

$$S E_C T \Leftrightarrow \{\lambda \in \mathbb{T} : \lambda \text{ is an eigenvalue of } S\}$$
$$= \{\lambda \in \mathbb{T} : \lambda \text{ is an eigenvalue of } T\}.$$

Restricting ourselves for convenience to the set of $T \in D$ with infinite spectrum (otherwise we have to modify appropriately what follows), we can find a Borel function $f : D \to \mathbb{T}^{\mathbb{N}}$ such that $f(T) = \{x_n\}$, where $\{x_n\}$ is a 1-1 enumeration of the set of eigenvalues of T. Let S_∞ act on $\mathbb{T}^{\mathbb{N}}$ by $g \cdot \{x_n\} = \{x_{g(n)}\}$. Then

$$SE_C T \Leftrightarrow f(x) E_{S_\infty} f(y),$$

i.e., $E_C \leq_B E_{S_\infty}$ (for this action), so E_C admits classification by countable structures.

2. (This comes from recent work of Giordano-Putnam-Skau [1995]). Consider minimal $f \in H(2^{\mathbf{N}})$. Let

$$f E g \Leftrightarrow \exists h \in H(2^{\mathbf{N}})(h \text{ maps the}$$
$$\text{orbits of } f \text{ onto the orbits of } g).$$

Then Giordano-Putnam-Skau show (among other things) that one can assign to each $f \in H(2^{\mathbf{N}})$ a countable partially ordered group with distinguished order unit, $\mathcal{A}_f$, such that

$$f E g \Leftrightarrow \mathcal{A}_f \cong \mathcal{A}_g;$$

so E admits classification by countable structures.

We now consider the following general question: Given a Polish G-space X, when does E_G admit classification by countable structures?

Definition. Let G be a Polish group and X a Polish G-space. Fix an open nonempty set $U \subseteq X$ and a symmetric open nbhd V of $1 \in G$. The (U, V)-*local graph* is the following symmetric, reflexive relation on U:

$$x R_{U,V} y \Leftrightarrow x, y \in U \text{ and } \exists g \in V(g \cdot x = y).$$

The (U, V)-*local orbit* of $x \in U$, $\mathcal{O}(x, U, V)$, is the connected component of x in this graph.

Definition. The Polish G-space X is *turbulent* if every orbit is dense and meager, and every local orbit is somewhere dense (i.e., its closure has nonempty interior).

Examples. Let $\mathbb{R}^{<\mathbf{N}} \subseteq G \subsetneq \mathbb{R}^{\mathbf{N}}$ be a Polishable subgroup, i.e., a Borel subgroup of $\mathbb{R}^{\mathbf{N}}$ which is Borel isomorphic to a Polish group. An example of such a G is ℓ^p, for $1 \leq p < \infty$. Then the translation action of G on $\mathbb{R}^{\mathbf{N}}$ is turbulent (when G is viewed as a Polish group). Similarly for many Polishable subgroups of $\mathbb{Z}_2^{\mathbf{N}}$. On the other hand Hjorth and Kechris have shown that any closed subgroup of a countable product of locally compact groups and closed subgroups of S_∞ *never* has turbulent actions.

We now have:

Theorem (Hjorth [1997a]; see also Kechris [1996]). *Let G be a Polish group and X a turbulent Polish G-space. Then E_G^X does not admit classification by countable structures.*

In fact, if we call a Polish G-space *generically turbulent*, if its restriction to an invariant dense G_δ is turbulent, we have the following characterization.

Theorem (Hjorth [1997a]). *Let G be a Polish group and X be a Polish G-space, such that X has a dense orbit and every orbit is meager. Then the following are equivalent:*

(i) *X is generically turbulent;*

(ii) *If Y is a Borel S_∞-space and $f : X \to Y$ is Baire measurable, such that $xE_Gy \Rightarrow f(x)E_{S_\infty}F(y)$, then F maps a comeager set into a single E_{S_∞}-class, i.e., E_G is generically E_{S_∞}-ergodic.*

One now has the following dichotomy, at least for GE-groups.

Theorem (Hjorth [1997a]). *Let G be a GE-group. Then exactly one of the following happens for each Polish G-space X:*

(I) *E_G^X admits classification by countable structures;*

(II) *There is a turbulent Polish G-space Y with $E_G^Y \leq_B E_G^X$.*

This is proved by an appropriate "changing the topology" technique. Using model theoretic ideas – an analog of the Scott analysis of countable structures – one assigns in a Borel way to each $x \in X$ a countable structure $\mathcal{A}_x$ (a partial ordering with some additional relations) and an $L_{\omega_1\omega}$ sentence σ_x which is a weak version of a Scott sentence of $\mathcal{A}_x$, such that $xE_G^Xy \Rightarrow \mathcal{A}_x \cong \mathcal{A}_y \Rightarrow \sigma_x = \sigma_y$. If $xE_G^Xy \Leftrightarrow \sigma_x = \sigma_y$, then clearly E_G^X admits classification by countable structures, so we have alternative (I). Otherwise, one can define a new Polish topology on an invariant Borel set $Y \subseteq X$, containing x, y with $\sigma_x = \sigma_y$ but $x \not\!\!E_G^Xy$, so that the action of G on Y is continuous and turbulent, thus we have alternative (II).

The following application of these results has been observed by Kechris: Measure equivalence and conjugacy on $U(H)$ do not admit classification by countable structures (in contrast with the examples above concerning automorphism with countable discrete spectrum). It was conjectured that conjugacy on $\mathrm{Aut}(X, \mu)$ does not admit classification by countable structures. This has now been proved by Hjorth [1997b]. At the time of the workshop it was also open whether conjugacy on $U(H)$, $\mathrm{Aut}(X, \mu)$ is generically turbulent. It has been recently proved by Kechris-Sofronidis [1997] that indeed conjugacy on $U(H)$ is generically turbulent but this is still open for $\mathrm{Aut}(X, \mu)$.

A somewhat weaker version of the preceding dichotomy has been also proved by Hjorth for arbitrary Polish groups G. Write, for equivalence relations E, F on X, Y resp.,

$$E \leq_{p\Delta_2^1} F$$

iff there is a provably Δ_2^1 function $f : X \to Y$ with $xEy \Leftrightarrow f(x)Ff(y)$.

Theorem (Hjorth [1997a]). *Let G be a Polish group. Then exactly one of the following holds for each Polish G-space X:*

(I) $E_G^X \leq_{p\Delta_2^1} E_{S_\infty}^Z$, *for some Borel S_∞-space Z;*

(II) *There is a turbulent Polish G-space Y with $E_G^Y \leq_B E_G^X$.*

(I) in the preceding theorem essentially says that E_G^X admits classification by countable structures, albeit in a somewhat weaker form, since the classifying map is $p\Delta_2^1$ but not necessarily Borel. Thus, intuitively speaking, this shows that even for arbitrary Polish groups G, the precise obstruction for classifying orbit equivalence relations E_G by countable structures is turbulence. It would be nice to replace "$p\Delta_2^1$" by "Borel", so that one has the full Hjorth dichotomy valid for arbitrary (not just GE groups), but this is still open.

References

H. Becker [1998], Polish group actions, dichotomies and generalized elementary embeddings, *J. Amer. Math. Soc.*, **11**, 397-449.

H. Becker and A.S. Kechris [1996], *The Descriptive Set Theory of Polish Group Actions*, London Math. Soc. Lecture Notes, **232**, Cambridge Univ. Press, Cambridge.

J.P. Burgess [1979], A selection theorem for group actions, *Pac. J. Math.*, **80**, 333-336.

J. deVries [1975], Universal topological transformation groups, *General Top. and its Appl.*, **5**, 107-122.

J. deVries [1978], On the existence of G-compactifications, *Bull. Acad. Polonaise des Sc., Série des sc. math., ast. et phys.*, **XXVI (3)**, 275-280.

E.F. Effros [1965], Transformation groups and C^*-algebras, *Ann. of Math.*, **81**, 38-55.

E.G. Effros [1981], Polish transformation groups and classification problems, in: Rao and McAuley, eds., *General Topology and Modern Analysis*, Academic Press, New York, 217-227.

J. Feldman, P. Hahn and C.C. Moore [1979], Orbit structure and countable sections for actions of continuous groups, *Adv. in Math.*, **26**, 186-230.

P.H. Forrest [1974], Virtual subgroups of $\mathbb{R}^n$ and $\mathbb{Z}^n$, *Adv. in Math.*, **3** 187-207.

L. Fuchs [1970], *Infinite Abelian Groups*, Academic Press, New York.

S. Gao [1996], A universal Polish G-space with respect to Baire class 1 G-embeddings, preprint.

T. Giordano, I.F. Putnam and C.F. Skau [1995], Topological orbit equivalence and C^*-crossed products, *J. Reine Angew. Math.*, **469**, 51-111.

L. Harrington, A.S. Kechris and A. Louveau [1990], A Glimm-Effros dichotomy for Borel equivalence relations, *J. Amer. Math. Soc.*, **3**, 903-928.

G. Hjorth [1996], Products of locally compact groups, preprint.

G. Hjorth [1997a], Classifications and orbit equivalence relations, preprint.

G. Hjorth [1997b], On the isomorphism problem for measure preserving transformations, preprint.

G. Hjorth [1997c], Vaught's Conjecture on analytic sets, preprint.

G. Hjorth [1999a], Sharper changes in topologies, *Proc. Amer. Math. Soc.*, **127 (1)**, 271-278.

G. Hjorth [1999b] A universal Polish *G*-space, *Topology Appl.*, **91 (2)**, 141-150.

G. Hjorth and A.S. Kechris [1995], Analytic equivalence relations and Ulm-type classifications, *J. Symb. Logic*, **60**, 1273–1300.

G. Hjorth and S. Solecki [1999], Vaught's Conjecture and the Glimm-Effros property for Polish transformation groups, *Trans. Amer. Math. Soc.*, to appear.

A.S. Kechris [1992], Countable sections for locally compact group actions, *Erg. Th. and Dyn. Syst.*, **12**, 283–295.

A.S. Kechris [1995], *Classical Descriptive Set Theory*, Graduate Texts in Mathematics, **156**, Springer-Verlag, New York.

A.S. Kechris [1997], Actions of Polish groups and classification problems, preprint.

A.S. Kechris and N.E. Sofronidis [1997], A strong generic ergodicity property of unitary conjugacy, preprint.

G.W. Mackey [1962], Point realizations of transformation groups, *Ill. J. Math.*, **6**, 327–335.

G.W. Mackey [1966], Ergodic theory and virtual groups, *Math. Annalen*, **166**, 187–207.

M. Megrelishvili [1988], A Tikhonov *G*-space that does not have compact *G*-extensions and *G*-linearizations, *Uspekhi Mat. Nauk*, **43**, 145–146.

M. Megrelishvili [1989], Compactification and factorization in the category of *G*-spaces, *Categorical topology and its relation to analysis, algebra and combinatorics (Prague 1988)*, World Sci. Publishing, Teaneck, NJ, 220-237.

D.E. Miller [1977], On the measurability of orbits in Borel actions, *Proc. Amer. Math. Soc.*, **63**, 165–170.

M.G. Nadkarni [1990], On the existence of a finite invariant measure, *Proc. Indian Acad. Sci., Math. Sci.* **100**, 203–220.

A. Ramsay [1985], Measurable group actions are essentially Borel actions, *Israel J. Math.*, **51**, 339–346.

R.L. Sami [1994], Polish group actions and the Vaught conjecture, *Trans. Amer. Math. Soc.*, **34**, 335–353.

T. Scarr [1998], Topological and Borel classifications of Polish *G*-spaces, preprint.

J.H. Silver [1980], Counting the number of equivalence classes of Borel and coanalytic equivalence relations, *Ann. Math. Logic*, **18**, 1-28.

V.V. Uspenskii [1986], A universal topological group with a countable base, *Funct. Anal. and Its Appl.*, **20**, 160–61.

V.S. Varadarajan [1963], Groups of automorphisms of Borel spaces, *Trans. Amer. Math. Soc.*, **109**, 191–220.

R.L. Vaught [1974], Invariant sets in topology and logic, *Fund. Math.*, **82**, 269–283.

V.M. Wagh [1988], A descriptive version of Ambrose's representation theorem for flows, *Proc. Indian Acad., Sci. Math. Sci.*, **98**, 101-108.

S. Wagon [1993], *The Banach-Tarski Paradox*, Cambridge Univ. Press, Cambridge.

Polish Groupoids

Arlan B. Ramsay
Department of Mathematics
University of Colorado
Boulder, CO 80309

0   Introduction

This chapter has two themes. The primary one is closely related to the topic
of the workshop, being the idea of groupoids as generalizing transformation
groups and as a tool for thinking about transformation groups. Questions
about orbit spaces or ergodicity, for example, are natural for groupoids. The
secondary theme is minor from the viewpoint of the workshop, but it seems
worthwhile to mention some other situations where groupoids can be used.

The first section gives some basic information about Polish spaces. The
second one provides a definition of groupoid and a few examples. Then we
continue with a third section about groupoids related to foliations of manifolds and a fourth section about groupoids in which the unit space has a
cover by open invariant sets on which the groupoid structure is given by a
transformation group. For the final section we concentrate on dichotomies.

1   Polish Spaces

Although many readers may know good sources for information on Polish
spaces, it won't hurt to mention books by Bourbaki [Bou], Kechris [Ke],
Kuratowski [Ku], and Srivistava [Sr], for those not familiar with them.

Many spaces that arise naturally in mathematics are Polish, and this is
certainly true for groupoids in the analytic setting. Besides analysis itself, we
include in this class anything involving spaces of functions, most problems
having a physical motivation, and most questions deriving primarily from
geometry.

One reason for the abundance of Polish spaces is that if X is a second
countable locally compact Hausdorff space, then X is Polish. To verify this
fact, it is helpful to construct a continuous function that 'goes to ∞ at ∞'
on X. Combining such a function with a suitable sequence of compactly
supported continuous functions will give a homeomorphism of X onto a closed
set in $\mathbb{R}^\infty$. Manifolds constitute an important special class of second countable
locally compact Hausdorff spaces.

It is also true that every G_δ set in a Polish space is Polish, which helps
to make Polish spaces common. The fact that a countable product of Polish spaces is Polish makes them still more flexible to work with than locally
compact second countable spaces. Of course locally compact spaces are the
easiest places to do measure theory in some ways, so the flexibility of Polish
spaces has a price. One example of this is the fact that locally compact second
countable groups always have a faithful unitary representation, the regular
representation on the Hilbert space of functions square integrable relative to
Haar measure, and hence a faithful family of irreducible unitary representa-

tions [Ma1]. It would be useful to have better techniques for doing analysis on Polish spaces, including, for example, some replacement for unitary representations for Polish groups. An alternative approach could be to look for a restricted class of Polish spaces in which the groups do have an adequate supply of unitary representations. To be useful, such a class should include many examples that arise naturally.

Polish spaces have yet another feature that is useful when the basic interest lies in their Borel sets, a common occurrence. That feature is that we are free to enlarge the topology for convenience, according to the following well-known lemma [Ke, p. 83; Sr, p. 93]:

Lemma A. If $(X, \mathcal{T})$ is a Polish space and $\mathcal{C}$ is a countable collection of Borel sets in X, then there is a Polish topology $\mathcal{T}'$ that contains $\mathcal{T} \cup \mathcal{C}$ and generates the same σ-algebra of Borel sets as $\mathcal{T}$.

A set together with a σ-algebra of subsets is called a Borel space, and one that is isomorphic to the Borel space associated to a Polish space is called *standard*.

Suppose that $(X, \mathcal{B})$ is a standard Borel space and Γ is a countable group of Borel automorphisms of $(X, \mathcal{B})$. Using Lemma A, it is possible to show that there is a Polish topology on X that generates the given σ-algebra of Borel sets and for which all the elements of Γ are homeomorphisms [Ke, p. 84]. This result is a very special case of a result of Becker and Kechris [B-K, page 58].

2   Groupoids

There are several motivating examples for the definition of the algebraic notion of *groupoid*. We begin with one close to the origins of group theory.

Suppose that Y is a set that has a function p mapping it onto another set X. Let G be the set of bijections having as domain one of the sets $p^{-1}(x)$ for $x \in X$ and as range one of the same. It is clear that G is closed under forming the composition of two of its elements if one of them maps $p^{-1}(x_1)$ to $p^{-1}(x_2)$ and the other maps $p^{-1}(x_2)$ to $p^{-1}(x_3)$. Composition of functions is always associative, and it is clear that each element of G has left and right identities and an inverse, all contained in G. These are the basic properties of groupoids. We may write $\mathrm{Perm}(Y, p, X)$ or $\mathrm{Perm}(Y, p)$ for G. In the terminology introduced below, $\mathrm{Perm}(Y, p)$ is a groupoid on X. In the case that X has only one element, $\mathrm{Perm}(Y, p)$ is simply the group of all permutations of Y.

In this kind of example, we can add structures to all the sets and requirements on all the mappings. These conditions can involve Borel, measure theoretic, topological, metric, or differentiable structures, allowing us to obtain examples of many types of groupoids. To be more specific, Y could be a vector bundle over X, or a covering space. We could also have a measure on X and a measure on each set $p^{-1}(x)$. In those more specific circumstances we would limit our attention to mappings from one $p^{-1}(x)$ to another that preserve the relevant structure: vector space, inner product, measure, measure class, etc., getting different groupoids.

Algebraically speaking, we can give several abbreviated definitions of the term *groupoid*. One is that a groupoid is a small abstract category with all maps invertible. Expanding this somewhat, we can say that a groupoid is a set in which some pairs of elements have a product defined, satisfying these three conditions: (a) whenever a triple product is defined the associative law holds; (b) every element has a left identity and a right identity; (c) every element has an inverse. Since we want to think of elements of groupoids as being generalizations of mappings, we usually picture them as arrows. Following that intuition, we give one formal definition, but the reader is encouraged to formulate equivalent definitions. The letters r and s used in the definition are intended to suggest 'range' and 'source', so we refer to them as the *range* and *source* mappings.

Definition. If G is the set of edges and X is the set of vertices of a directed graph, and γ is an edge from x to y, write $s(\gamma)$ for x and write $r(\gamma)$ for y. Also, write $\gamma : x \to y$. Then write $G^{(2)}$ for the set $\{(\gamma, \gamma') \in G^2 : s(\gamma) = r(\gamma')\}$. We say that G *is a groupoid on* X to mean that there is a function defined from $G^{(2)}$ to G, with the value at (γ, γ') denoted by $\gamma\gamma'$, so that the following properties are true:

(a) If $s(\gamma) = r(\gamma')$ and $s(\gamma') = r(\gamma'')$, then $(\gamma\gamma')\gamma'' = \gamma(\gamma'\gamma'')$.

(b) For each $x \in X$, there is an element i_x of G such that $s(\gamma) = x$ implies $\gamma i_x = \gamma$ and $r(\gamma) = x$ implies $i_x\gamma = \gamma$.

(c) For each γ in G, there is a γ' in G such that $\gamma\gamma' = i_{r(\gamma)}$ and $\gamma'\gamma = i_{s(\gamma)}$.

We remark that the γ' of (c) is unique, and denote it by γ^{-1}. The *inversion* map, taking γ to γ^{-1}, is clearly its own inverse.

For a second example of a groupoid, let E be an equivalence relation on a set X. For $(x, y) \in E$, let the source be y and let the range be x (so that 'composition' reads from right to left). Define $(x, y)(y, z) = (x, z)$, $i_x = (x, x)$ and $(x, y)^{-1} = (y, x)$. The product is defined because E is transitive, and the necessary properties are easy to verify.

Notice that for every groupoid, G, the pair (r, s) maps G onto an equivalence relation $E \subseteq X \times X$ and is a groupoid homomorphism. We call E the *equivalence relation associated with* G. The groupoid G is called *principal* iff

(r, s) is one-to-one, although there are topological conditions for a topological groupoid to be principal.

Our third example of a groupoid is simply a disjoint union of groups, one for each point of a set X, so that $r(\gamma) = s(\gamma)$ for every γ.

There is a sense in which the pure algebra of groupoids is somewhat trivial, because every one is built in a standard way from its associated equivalence relation and a disjoint union of groups, known as the *stabilizer subgroupoid* of G, which is just defined as $\{\gamma : r(\gamma) = s(\gamma)\}$. Once topology, Borel structure or measures come into the picture, groupoids are much more complex.

For our fourth example, we start with a group H acting on a space X, on the left. (Right actions work just as well.) Form $G = H \times X$, and regard the pair (h, x) as an edge from x to hx. Define $(h, h'x)(h', x) = (hh', x)$, $i_x = (e, x)$ and $(h, x)^{-1} = (h^{-1}, hx)$. This groupoid is sometimes written instead as $\{(y, h, x) : x, y \in X, h \in H, \text{ and } y = hx\}$.

This kind of groupoid, called *a transformation group groupoid*, and related ones, account for many of the groupoids that arise in analysis. The group acts freely iff the groupoid is principal. Note also that the subgroup of H that fixes a point x of X is naturally isomorphic to $\{\gamma : r(\gamma) = s(\gamma) = x\}$, which explains the name of the stabilizer subgroupoid.

There is a way to make new groupoids from given ones that is not permitted for transformation groups themselves, called *restriction*. If G is a groupoid on X and $Y \subseteq X$, then $G|Y = \{\gamma \in G : s(\gamma) \in Y \text{ and } r(\gamma) \in Y\}$ is a groupoid on Y in a natural way. Interesting and useful examples of groupoids can be made in this way from transformation group groupoids. For example, let the group $\mathbb{R}^n$ act on itself by translation. giving a groupoid G. If Y is an orthant in $\mathbb{R}^n$, or some other cone, then $G|Y$ can be used in analysing Weiner-Hopf operators [M-R].

It is important to note that (r, s) can be one-to-one without being a homeomorphism onto its range, relative to the product topology on $X \times X$. For example, transformation groups often have dense orbits that are not homeomorphic to the coset spaces that they correspond to in the algebraic sense. One specific example of this is given by the integers acting on a circle by an irrational rotation, and a related one called the *Kronecker flow* is as follows:

Let $X = \mathbb{T}^2$, thought of as $\mathbb{R}^2/\mathbb{Z}^2$, take a to be a positive irrational number, let $\mathbb{R}$ act on $\mathbb{R}^2$ by $t \cdot (x, y) = (x + t, y + at)$, and take the quotient action of $\mathbb{R}$ on $\mathbb{T}^2$. This can be done because the action of $\mathbb{R}$ commutes with that of $\mathbb{Z}^2$. In this case $\mathbb{R} \times \mathbb{T}^2$ is locally compact, but the equivalence relation as a subset of $\mathbb{T}^2 \times \mathbb{T}^2$ is not Polish, nor are the orbits homeomorphic to $\mathbb{R}$ as subsets of $\mathbb{T}^2$.

Definition. A groupoid G on X is called *Polish* iff both G and X are equipped with Polish topologies, r and s are continuous and open, and multiplication and inversion are continuous.

A transformation group groupoid is Polish if both the group and the space are Polish. A result of Becker and Kechris [B-K] shows that if X is a standard Borel space and H is a Polish group acting in a Borel way on X, then X can be given a Polish topology for which the action is continuous and the Borel sets are the original ones.

Questions: Can this Becker-Kechris result be extended to more general groupoids? What hypotheses are needed? What about a Polish groupoid acting on a standard Borel space? It seems likely that the case of equivalence relations is central, being quite different from that of transformation groups. Perhaps there is some way to give the equivalence relation a local product structure of the kind seen in foliations and then exploit that.

Remark: If X is Polish and the equivalence relation E on X has countable equivalence classes, and is a Borel set in $X \times X$, then E can be endowed with a Polish topology as a groupoid in such a way as to preserve the Borel structure on E inherited from $X \times X$. To do this we use a result of Lusin [Ke, p. 123; Sr, p. 205] according to which a countable-to-one Borel image of a Borel set has a Borel selection. As pointed out by Feldman and Moore [F-M], this allows us to express E as a countable union of Borel sets on each of which both r and s are one-to-one. Any Borel set $S \subseteq X \times X$ such that $r|S$ and $s|S$ are both one-to-one is the graph of a Borel bijection between Borel sets in X. Using Lemma A, we see that there is a larger Polish topology on X so that all the domains and ranges of these mappings are open and closed and all the mappings are homeomorphisms.

The hypothesis of open-ness of r and s is needed for one basic fact that plays a vital role in the development of Polish groupoids. In topological groups, it is easy to show that the set of products of an open set with any set is open, being a union of open sets. In the case of groupoids, more is needed. The proof goes as follows [R2, p. 361].

If A and B are sets in a groupoid, we define $AB = \{\gamma\gamma' : \gamma \in A, \gamma' \in B,$ and $s(\gamma) = r(\gamma')\}$. If G is Polish and A and B are both open, then $s(A)$ is an open set in X, so $r^{-1}(s(A))$ is open in G. This can be viewed as continuity of the possibility to form products (continuity of composability). Now any γ_0 in AB is a product of an element γ_1 of A and an element γ_2 of B. Because r is continuous, there is an open set W containing γ_0 such that $r(W)$ is contained in $r(A)$. Because r is open, the set $r(W)$ is open. Let U_2 be an open neighborhood of γ_2 contained in B. By continuity of multiplication and inversion, there exist an open neighborhood of γ_1, say U, contained in A, and an open neighborhood of γ_0, say V_1, contained in W, so that if $\gamma^{-1}\gamma'$ is defined for $\gamma \in U$ and $\gamma' \in V_1$, then $\gamma^{-1}\gamma' \in U_2$. Since r is open and continuous there is an open neighborhood V of γ_0 contained in V_1 such that $r(\gamma') \in r(U)$ if $\gamma' \in V$. Thus $V \subseteq AB$, proving what we claimed.

3 Foliations

In this section, we illustrate the possibility of using groupoids as a tool in thinking about manifolds in general and foliations of manifolds in particular. We also point out how the dichotomy of 'good quotient' vs. 'very bad quotient' applies to foliations.

Roughly speaking, we can think of a foliation of a manifold as a way to partition the manifold smoothly with (connected) manifolds of some fixed lower dimension. These smaller manifolds may be only immersed rather than imbedded, and are called *leaves*. There is also a local product structure that provides the proper kind of smoothness.

Perhaps the simplest example is the case of a product of two manifolds, $M \times N$ in which the leaves are the submanifolds $\{x\} \times N$. Only slightly more complicated is the case in which one manifold is a bundle over another and the fibers are taken as leaves. One of the best known foliations is that of a two-torus, $\mathbb{T}^2$, foliated by the orbits of the Kronecker flow described in Section 2. This foliation illustrates at least some of the possible complications. It also is an example of the fact that a Lie group acting smoothly on a manifold leads to a foliation by the orbits of the action as long as the orbits all have the same dimension. See [C-N, p. 29].

Every foliation of a manifold M has associated with it at least three closely related groupoids. The one that contains the most information about the foliation is called the *graph of the foliation* or *the holonomy groupoid of the foliation* [W, C, Rei, C-N, P]. One thing this groupoid tells is the manner in which leaves of the foliation 'spiral' in toward other leaves or even themselves. It is locally Euclidean, but not necessarily Hausdorff. Sometimes it is said to unwind the foliation.

Another groupoid associated with a foliation of a manifold M is the equivalence relation of belonging to the same leaf. This equivalence relation is also the canonical equivalence relation associated with the holonomy groupoid. We are going to look more closely at the equivalence relation induced on the disjoint union, T, of local transversals. As far as the orbit spaces are concerned these two are equivalent, and the second one is easier to think about.

This equivalence relation, R, on T is Borel as a subset of $T \times T$, and the equivalence classes are all countable. Hence we can modify topologies to make R a Polish groupoid as described in Section 2. In terms of the dichotomies into 'nice' versus 'bad' orbit spaces, [R2] shows that the Borel versions of being nice are equivalent to topological versions. Thus changing the topology cannot change the orbit space from T_0 to non-T_0.

Our context for foliations is that of C^∞ manifolds, and we refer to C^∞ objects of all kinds as 'smooth'. The standard definition of 'manifold' begins with a Hausdorff space, M, and introduces the notion of a coordinate chart, that being an open set together with a homeomorphism of it onto a open set in some $\mathbb{R}^n$ that is homeomorphic to an open ball. If (U, φ) and (V, ψ) are coordinate charts, they are called *compatible* provided that $\psi \circ \varphi^{-1}$ is smooth on its domain of definition, namely $\varphi(U \cap V)$. For M to be a smooth manifold we require the existence of a covering of M by compatible coordinate charts.

It is possible to describe smooth manifolds alternatively as parametrized spaces in a certain sense, namely as well-behaved quotient spaces of more concrete spaces. Since groupoids are so convenient for thinking about quotient spaces, this adds a groupoid perspective to the subject of abstract manifolds. The idea is to extend the way we think of curves and surfaces in $\mathbb{R}^n$ to the general case. We ordinarily study curves as parametrized, and it is common enough for surfaces. It is necessary to allow the parameter space not to be connected in the general case.

To construct a manifold, start with a discrete set I, usually countable, form $\mathbb{R}^n \times I$, and let X be an open set in $\mathbb{R}^n \times I$ such that for every $i \in I$ the set $X \cap (\mathbb{R}^n \times I)$, denoted by X_i, is homeomorphic to an open ball in $\mathbb{R}^n$. The space X will parametrize a manifold if we take a suitable quotient. Suppose that E is an equivalence relation on X implemented by diffeomorphisms, in the sense that whenever $((x, i), (y, j)) \in E$ there is a diffeomorphism f_{ji} from a neighborhood D_i of x to a neighborhood D_j of y such that $z \in D_i$ implies $(f_{ji}(z), z) \in E$. Suppose also that the E equivalence classes in X are closed, so that X/E is Hausdorff, and that on each X_i the equivalence relation is trivial. Denote the quotient map by π. Then there is a manifold structure on X/E so that the coordinate neighborhoods are sets of the form $\pi(X_i)$, and the coordinate map for $\pi(X_i)$ is $(\pi|X_i)^{-1}$ followed by the projection into $\mathbb{R}^n$.

If a manifold is described as we just did, to get a foliation we arrange that every X_i is a product of the form $T_i \times L_i \times \{i\}$, where the sets T_i are all cubes of the same dimension and centered at 0, the sets L_i are all cubes of the same dimension centered at 0, and we require that E respects the product structure. The latter means that whenever $\pi(X_i) \cap \pi(X_j) \neq \emptyset$, there exists a diffeomorphism $g = g_{ji}$ from part of T_i to part of T_j and a smooth map $h = h_{ji}$ from part of $T_i \times L_i$ to L_j so that on its natural domain, the diffeomorphism $(\pi|X_j)^{-1} \circ (\pi|X_i)$ takes a point (x, y, i) to the point $(g(x), h(x, y), j)$. When the coordinate charts have this relationship, they are called foliation charts. The foliation has *dimension* equal to that of the cubes L_i and *codimension* equal to that of the cubes T_i.

A single set of the form $\pi(\{t\} \times L_i \times \{i\})$ is called a *plaque*, and a leaf is a maximal connected union of plaques. A set of the form $\pi(T_i \times \{y\} \times \{i\})$ is called *a local transversal* of the foliation, and we write T for the union of the sets $T_i \times \{i\}$.

If c is a curve in a leaf, there will be a finite sequence of plaques in the leaf that cover it. The plaques come from a finite sequence of foliation charts. Choose (x, y, k) so that $c(0) = \pi(x, y, k)$, and (w, z, l) so that $c(1) = \pi(w, z, l)$, and suppose that $\pi(X_k)$ is the chart in the finite sequence that contains $c(0)$ and $\pi(X_l)$ is the one that contains $c(1)$. As described beginning on page 62 of [C-N], this can all be done so that the foliation structure gives us a smooth map f_c of an open set U_k in T_k to an open set U_l in T_l, and for $u \in U_k$ we have $\pi(\{u\} \times L_k \times \{k\})$ and $\pi(\{f_c(u)\} \times L_l \times \{l\})$ connected by plaques, each of which intersects the next one. This map f_c is called the *holonomy* of c, and depends only on the class of c under homotopies that fix the endpoints and deform c only among paths in the same leaf. The graph of the foliation can be described in terms of the triples (x, f_c, y) for which c is a curve from x to y in the leaf containing the two points.

4 Local Group Actions

Just as manifolds are spaces covered by open sets homeomorphic to open balls in Euclidean spaces, and hence called locally Euclidean, a topological groupoid G on X will be said to be *locally a group action* provided X is covered by open sets U such that $G|U$ is isomorphic to a transformation group groupoid, and when two such U's intersect the two actions on $U \cap U'$ are connected by an automorphism of the group in question. (Perhaps the automorphism should be allowed to vary from orbit to orbit.) El Sayed Sallam studied the 'equivariant cohomology' of such groupoids in the case that the group is a circle for his Ph.D. work at the University of Colorado, completed in 1998 [S].

Just as with manifolds, if a groupoid G is locally a group action, we can think of G as the quotient of the disjoint union of the $G|U$ which are transformation group groupoids. That disjoint union is itself a transformation group groupoid.

Sometimes, a quotient of a transformation group groupoid is also a transformation group groupoid, but not always. Suppose that H and K are groups acting on a space Y and the actions commute. This is equivalent to having an action of $H \times K$ on Y. In this case, there is a well-defined action of H on the space, X, of K-orbits in Y. That is the quotient of the groupoid $H \times Y$ by the action of K on Y, and the result is a transformation group groupoid, $H \times X$.

For a different example, let Y be $\mathbb{T}^2$, let $H = \mathbb{T}$ acting on Y by $h \cdot (z, w) = (z, hw)$. Define an action of $K = \mathbb{Z}_2$ on Y by having the non-trivial element a take (z, w) to $(-z, \bar{w})$. Then Y/K is the Klein bottle. If A is an open arc in $\mathbb{T}$ of less than π, the quotient map is one-one on $A \times \mathbb{T}$, and the action of the circle passes to the quotient. The quotient of the groupoid $H \times Y$ is not given by a group action but it is locally a group action; it is naturally an equivalence relation on the Klein bottle.

Dr. Sallam found that for local circle actions this is somewhat typical. There is always a (connected when X is connected) double covering of X on which $\mathbb{T}$ acts, and the given groupoid is the quotient of that transformation group groupoid. What makes this possible is the fact that the automorphism group of the circle has only two elements.

This example shows another way that non-transformation group groupoids can be of interest.

5   Dichotomies for Polish Groupoids

There are techniques of descriptive set theory itself that have been used to establish dichotomies for equivalence relations on Polish spaces when the relation forms a Borel set in the cartesian product of the space with itself [H-K-L]. These techniques do not require that the equivalence relation derive from a group action, so they are of maximal generality.

For Polish groupoids, techniques are available that are closer to those used for transformation groups. They rely on the metric space structure explicitly and are thus more elementary. It is hard not to speculate on the possibility of using somewhat more elementary methods of descriptive set theory than those used in [H-K-L], though reducing the sophistication to elementary metric space ideas seems unlikely. Having a locally compact groupoid does not seem to make the argument easier than it is in the Polish case, except that the necessary hypothesis that the equivalence relation is an F_σ set is automaticially satisfied.

In most dichotomy results there is one particular step in the chain that is more difficult. In the Polish groupoid setting, it is the step of showing that if the space of orbits is not T_0, i.e., if there are two orbits each of which contains the other in its closure, then there is a non-trivial ergodic measure. The basic idea of [E, K-W] is to find a set in X that is homeomorphic to the Cantor set, a countable product of 2-point spaces, and on which the equivalence relation restricts to the one in which two points are equivalent if they differ only in finitely many coordinates (called tail equivalence).

For group actions, a group element that moves a point x_0 of X to another point in some small neighborhood of x_0 automatically carries some small neighborhood of x_0 to a disjoint open set contained in the original small neighborhood of x_0. In a groupoid, we must work instead with open sets in G, and we arrange for decreasing sequences of open sets whose intersections give special G-sets in G (sets on which r and s are both one-to-one and hence define transformations of a part of X to another part of X). The construction

[R2] produces a Cantor set C contained in X, i.e., a homeomorphic image of $\{0,1\}^{\mathbb{N}}$. For any element $a \in \{0,1\}^n$, let C_a denote the image in C of $\{x \in \{0,1\}^{\mathbb{N}} : (x_1, x_2, \ldots, x_n) = a\}$. If a and b are in $\{0,1\}^n$, the process provides a G-set to transform C_a to C_b. Indeed, we get G-sets that make up an inverse semigroup of partial homeomorphisms of C as described in [P].

References

[B-K] H. Becker and A. S. Kechris, *The Descriptive Set Theory of Polish Group Actions,* London Mathematical Society Lecture Note Series 232, Cambridge University Press, Cambridge, 1996.

[Bou] N. Bourbaki, *Topologie Générale - Ch. 9: Utilisation des Nombres Réels,* Hermann, Paris, 1958.

[C-N] C. Camacho and A. L. Neto, *Geometric Theory of Foliations,* Birkhäuser, Boston, 1985.

[C] A. Connes, Sur la Theorie Non Commutative de l'Integration, in: A. Dold and B. Eckmann, eds., *Algèbres d'Operateurs,* Lecture Notes in Mathematics 725, Springer-Verlag, Berlin, 1978, 19-143.

[E1] E. G. Effros, Transformation groups and C^*-algebras, *Ann. of Math.* **81** (1965), 36-55.

[E2] E. G. Effros, Polish transformation groups and classification problems, in: Rao and McAuley, eds., *General Topology and Modern Analysis,* Academic Press, New York, 1981, 217-227.

[F-M] J. Feldman and C. C. Moore, Ergodic equivalence relations, cohomology and von Neumann algebras, I and II, *Trans. Amer. Math. Soc.* **234** (1977), 289-359.

[G] J. Glimm, Locally compact transformation groups, *Trans. Amer. Math. Soc.* **101** (1961), 124-138.

[K-W] Y. Katznelson and B. Weiss, The construction of quasi-invariant measures, *Israel J. Math.* **12** (1971), 1-4.

[Ke] Alexander S. Kechris, *Classical Descriptive Set Theory,* Springer-Verlag, New York, 1995.

[Ku] K. Kuratowski, *Topology,* Vol. I, Academic Press, New York, 1966.

[M] K. C. H. Mackenzie, *Lie Groupoids and Lie Algebroids in Differential Geometry,* London Mathematical Society Lecture Note Series, vol. 124, Cambridge University Press, Cambridge, 1987.

[Ma1] G. W. Mackey, *The Theory of Unitary Group Representations,* University of Chicago Press, Chicago, 1976.

[Ma2] G. W. Mackey, Ergodic theory and virtual groups, *Math. Ann.,* **166** (1966), 187-207.

[M-R] P. Muhly and J. Renault, C^*-algebras of multivariable Wiener-Hopf operators, *Trans. Amer. Math. Soc.* **274**(1982), 1-44.

[P] Alan L. T. Paterson, *Groupoids, Inverse Semigroups, and their Operator Algebras,* Birkhäuser, Boston, 1999.

[R1] A. Ramsay, Virtual groups and group actions, *Adv. in Math.,* **6** (1971), 253-322.

[R2] A. Ramsay, The Mackey-Glimm dichotomy for foliations and other Polish groupoids, *J. Funct. Analysis,* **94** (1990), 358-374.

[Rei] B. L. Reinhart, *Differential Geometry of Foliations,* Ergebnisse der Mathematik und ihrer Grenzgebiete, Vol. 99, Springer-Verlag, Berlin, 1983.

[S] El Sayed Sallam, *Cohomology of Groupoid Structure,* Ph. D. Thesis, University of Colorado, Boulder, 1998.

[Sr] S. M. Srivastava, *A Course on Borel Sets,* Springer-Verlag, New York, 1998.

[W] H. E. Winkelnkemper, The graph of a foliation, *Ann. Global Anal. Geom.,* **1** (1983), 51-75.

A Survey of Generic Dynamics

Benjamin Weiss
Institute of Mathematics
Hebrew University
Jerusalem, Israel

0   Introduction

Topological dynamics studies the action of a group G acting as homeomorphisms of a topological space X, usually taken to be compact. In ergodic theory, a measure structure $(\mathcal{B}, \mu)$ is added and one studies properties modulo μ-null sets. In *generic dynamics* we study properties modulo the topologically negligible sets — the sets of first category or meager sets. Compactness is not such a property, so the natural spaces for our investigations will be Polish spaces — the complete separable metric spaces. Continuous actions of G on such Polish spaces always have compactifications, so we could stay in the class of compact spaces — but this restricts the framework unnecessarily.

Our survey will deal mainly with three topics which we introduce now by describing their analogues in ergodic theory and topological dynamics. The first concerns ergodicity, weak mixing and the basic results of P. Halmos and J. von Neumann that characterize weak mixing as the absence of pure point spectrum, or as the ergodicity of cartesian products. As we will see, there is a very nice reduction of general systems to ergodic ones in generic dynamics and there are also interesting characterizations of weak mixing in the sense of multipliers. For classical measure preserving systems this characterization is as follows: an ergodic measure preserving system $(X, \mathcal{B}, \mu, T)$ is weakly mixing if and only if for any other ergodic system $(Y, \mathcal{C}, \nu, S)$ the product system $(X \times Y, \mathcal{B} \times \mathcal{C}, \mu \times \nu, T \times S)$ is also ergodic.

The cleanest result that we have in generic dynamics gives a characterization of those systems whose product with every weakly mixing system is ergodic in terms of their recurrence properties. We do not know if weak mixing itself is equivalent to some natural multiplier property and in particular the connection between point spectrum and weak mixing remains to be established.

The second topic deals with orbit equivalence. Here the classical theory started with the celebrated theorem of H. Dye according to which any two free ergodic actions of $\mathbb{Z}$ are orbit equivalent. An early surprise in generic dynamics was the discovery that all ergodic actions are generically orbit equivalent. This is in marked contrast to the ergodic situation where there is a very sharp dichotomy between amenable groups G and non amenable groups G. It turns out that all ergodic amenable actions are orbit equivalent to $\mathbb{Z}$-actions but no non amenable group has a free action that is orbit equivalent to a $\mathbb{Z}$-action. Moreover the same group, say, F_k, the free group on k generators, $k \geq 2$, has ergodic actions that are not orbit equivalent. Rather recently, D. Gaboriau has shown that no free ergodic action of F_k can be orbit equivalent to one of F_ℓ for $k \neq \ell$.

The third topic concerns Bernoulli systems and their isomorphisms. In ergodic theory this theory is intimately connected with entropy and the basic result of A. Kolmogorov, Y. Sinai and D. Ornstein that says that Bernoulli systems are classified by the single invariant – entropy. In generic dynamics, M. Keane proved some years ago that all Bernoulli systems were generically isomorphic. His proof has remained unpublished and I will present here another proof of this result. This is another illustration of the fact that the category of generic dynamics is much looser than measure dynamics.

We include also an open problem concerning bounded sequences of Baire measurable functions that arose in our investigations with D. Sullivan into hyperfiniteness in generic dynamics. Before taking up the main topics we give a brief discussion of how generic dynamics is related to single orbit dynamics. There is in fact an exact correspondence between pointed ergodic systems on Polish spaces and certain metrics on the group G. Here is the correspondence in detail for $G = \mathbb{Z}$.

Let us begin with an ergodic homeomorphism T of a Polish space X, and fix a point $x_0 \in X$ with a dense orbit.

Fixing a metric ρ on X, and a point x_0 with a dense orbit one can define a metric $\hat{\rho}$ on $\mathbb{Z}$ by $\hat{\rho}(n, m) = \rho(T^n x_0, T^m x_0)$. This metric has the property that translation is continuous, but not in general uniformly continuous. There is a remarkable theorem of Lavrentiev that enables one to reverse this deduction. Namely given such a metric $\hat{\rho}$ on $\mathbb{Z}$, it enables one to construct an ergodic system (X, T). This gives an alternate view of generic dynamics as the study of certain metrics on $\mathbb{Z}$. This point of view will be important for us later on and we proceed to explain it in detail. Here is Lavrentiev's theorem (see [Ku, p. 429]):

Theorem. *If φ is a homeomorphism between subsets $X_0 \subset X, Y_0 \subset Y$ of Polish spaces X and Y, then there are G_δ subsets $X_0 \subset \hat{X} \subset X$, $Y_0 \subset \hat{Y} \subset Y$ and a homeomorphism $\hat{\varphi}$ between $\hat{X}$ and $\hat{Y}$ that extends φ.*

It is not hard to deduce from this the following proposition.

Proposition. *If $X_0 \subset X$ is a subset of a Polish space X invariant under a countable group G of homeomorphisms from X_0 onto X_0, there is a G_δ subset $X_0 \subset \hat{X} \subset X$ and an action of G by homeomorphisms of $\hat{X}$ that extends the given action of G on X_0.*

Returning to $\mathbb{Z}$ equipped with a metric $\hat{\rho}$ such that translation is continuous we can form the completion of $\mathbb{Z}$ to obtain a Polish space X with $X_0 = \mathbb{Z}$ a dense subset. Applying the proposition will give a G_δ, $\hat{X} \supset X_0$ and a homeomorphism $T \colon \hat{X} \to \hat{X}$ that extends the shift by one on $\mathbb{Z}$. As a G_δ subset of a Polish space, $\hat{X}$ is itself a Polish space and we have thus constructed a system $(\hat{X}, T)$ in our category out of the initial data $(\mathbb{Z}, \hat{\rho})$. This is the point of view that underlies most of the proofs that we will encounter below. It is with pleasure that I thank H. Furstenberg and E. Glasner for many discussions about this work.

1 Ergodicity and Weak Mixing

The natural notion of ergodicity for a generic dynamical system (X, G) is the existence of a dense orbit. This is equivalent to the assertion that any G-invariant open set is dense. Letting $\{U_n\}$ be a countable basis for the topology one sees that the points with a dense orbit coincide with the G_δ set:

$$X_0 = \bigcap_n \bigcup_{g \in G} gU_n,$$

and restricting to X_0 we are now in the situation where *all* orbits are dense. Thus in this category, "ergodicity" and "minimality" coincide. It is worth pointing out that, unlike the situation in topological dynamics any system decomposes into ergodic subsystems. Indeed one can define an equivalence relation on X by setting xRy if and only if

$$\overline{\{gx : g \in G\}} = \overline{Gx} = \overline{Gy}.$$

In $X \times X$ one easily sees that this is a G_δ subset and so we have a nice decomposition of X into ergodic sets.

For a dynamical system to be interesting it should exhibit some recurrence properties. The simplest system, in which one puts the discrete topology on G, leads to systems with no recurrence possible, with the exception of periodic points, and we shall usually exclude them from our considerations. In analogy with the situation in ergodic theory, we say that a system (X, G) is *weakly mixing* if the product system $(X \times X, G)$, where G acts on pairs (x_1, x_2) by $g(x_1, x_2) = (gx_1, gx_2)$, is ergodic. In analogy with ergodic theory we can formulate the following question:

Question. *If (X, G) is weakly mixing and (Y, G) is ergodic (with no isolated points) is $(X \times Y, G)$ ergodic?*

The following simple lemma due to H. Furstenberg [F] gives some evidence for a positive answer in case G is commutative.

Lemma. *If (X, G) is weakly mixing and G is commutative, then $(X^{(k)}, G)$ the k-fold product of the system (X, G), is ergodic for all $k \geq 2$.*

Proof. The proof will be by induction on k. With $k = 2$ being the assumption, we may suppose that the result is true for some $k \geq 2$ and establish the result for $k + 1$. Given non empty open sets $U_i, V_i, 0 \leq i \leq k$, we must find some $g \in G$ such that

$$U_i \cap gV_i \neq \emptyset \quad \text{for all} \quad 0 \leq i \leq k.$$

This is sufficient since sets of the form $\prod_0^k U_i$ form a basis for the topology of $X^{(k+1)}$. Applying the assumption to $(U_0 \times V_0)$ and $(U_1 \times V_1)$ in $X \times X$ we find some g_0 so that

$$U_0 \cap g_0 U_1 \neq \emptyset \quad \text{and} \quad V_0 \cap g_0 V_1 \neq \emptyset.$$

Apply now the induction hypothesis to the sets

$$U_0 \cap g_0 U_1, U_2, U_3, \cdots U_k; V_0 \cap g_0 V_1, V_2, V_3, \cdots V_k,$$

we find a g such that

$$(U_0 \cap g_0 U_1) \cap g(V_0 \cap g_0 V_1) \neq \emptyset, \quad U_i \cap g V_i \neq \emptyset, \quad 2 \leq i \leq k.$$

Since $gg_0 = g_0 g$, the first relation implies that both $U_0 \cap g V_0 \neq \emptyset$ and $U_1 \cap g V_1 \neq \emptyset$ as required. ∎

For non commutative groups this lemma may fail in a rather extreme way. A simple case where $(X^{(3)}, G)$ is ergodic but $(X^{(4)}, G)$ fails is the following example which was pointed out to me by S. Mozes. Let G be a countable dense subgroup in $SL_2(\mathbb{R})$, and let X be $\mathbb{R}$ with the action being that of fractional linear transformations. Since $SL_2(\mathbb{R})$ is 3-transitive on $\mathbb{R}$, $(X^{(3)}, G)$ is ergodic, but the classical cross ratio is a continuous function on distinct 4-tuples which is invariant under G, and hence G is not ergodic on $X^{(4)}$. Here is an example where $(X^{(2)}, G)$ is ergodic but $(X^{(3)}, G)$ is not. Let X be the unit circle S^1 and G be the group generated by $\rho = $ rotation by an irrational angle, and $\psi = $ a continuous orientation preserving self map of S^1 with two fixed points, say at ± 1, with one being attracting and the other repelling, e.g.,

$$\rho(e^{i\pi t}) = e^{i\pi(t+\sqrt{2})}, \quad 0 \leq |t| \leq 1.$$

$$\psi(e^{i\pi t}) = \begin{cases} e^{i\pi t^2} & , 0 \leq t \leq 1, \\ e^{-i\pi t^2} & , -1 \leq t \leq 0. \end{cases}$$

Since G consists of orientation preserving maps it cannot be ergodic on $X^{(3)}$, but it is not hard to verify that it does act ergodically on $X^{(2)}$.

Returning to commutative groups G, the existence of invariant measures with global support allows for a positive answer, as the following proposition (drawn from a revised version of [GW]) shows:

Proposition. *Suppose that G is commutative, (X, G) weakly mixing and (Y, G) an ergodic system that admits an invariant measure μ with global support. Then $(X \times Y, G)$ is ergodic.*

Proof. Let $A, B \subset X$ and $U, V \subset Y$ be non empty open sets. Since (Y, G) is ergodic, the set $W = U \cap \bigcup_{g \in G} gV$ is non empty. We set $\mu(\bigcup_{g \in G} gV) = v > 0, \mu(W) = w > 0$. Let $F \subset G$ be a finite set such that

$$\mu(\bigcup_{g \in F} gV) \geq v - \frac{1}{2}w.$$

From the lemma applied to (X, G), we find a $g_0 \in G$ such that $A \cap g_0 g B \neq \emptyset$ for all $g \in F$. Observe that $\mu(\bigcup_{g \in F} g_0 g V) = \mu(\bigcup_{g \in F} g V) \geq v - \frac{1}{2} w$ and thus U must still intersect $\bigcup_{g \in F} g_0 g V$, hence for some $f \in F$, $U \cap g_0 f V \neq \emptyset$ and thus we have shown that $X \times Y$ is ergodic. ∎

The proof just given will establish that for any group G, if we assume $(X^{(k)}, G)$ ergodic for all k, then $(X \times Y, G)$ is ergodic for any ergodic (Y, G) admitting an invariant measure with global support. After these preliminary results, we turn to a more detailed investigation of the case for $\mathbb{Z}$-actions. We will begin by giving an example showing that there are two weakly-mixing $\mathbb{Z}$-actions whose product is not even ergodic – in marked contrast to the situation in ergodic theory where the product of weakly mixing systems is always weakly mixing. After this negative result we shall give a fairly precise characterization of those systems (X, T) which have the property that for any weakly mixing (Y, S) the product $(X \times Y, T \times S)$ is ergodic.

Our examples of weakly mixing systems will be constructed as subshifts of $\Omega = \{0, 1\}^{\mathbb{Z}}$. They will be constructed by describing a single point $\omega \in \Omega$ and then taking for X the orbit closure of ω under the shift mapping $\sigma \colon \Omega \to \Omega$ defined by $(\sigma\omega)(n) = \omega(n+1)$ for all $n \in \mathbb{Z}$. This automatically ensures that (X, σ) is ergodic, since it has a dense orbit. In all of our constructions ω will also be a recurrent point, so that X will be without isolated points. To ensure that X is weakly mixing, we shall construct together with ω, another point ω' such that the orbit of (ω, ω') will be dense in $X \times X$. We also recall some terminology from topological dynamics – namely a subset $D \subset \mathbb{Z}$ is said to be *syndetic* if it has bounded gaps, and the complement of a non syndetic set is called *thick*. More explicitly, a set $E \subset \mathbb{Z}$ is *thick* if it contains arbitrarily large intervals.

Proposition. *Let $E \subset \mathbb{Z}$ be a symmetric thick set. Then there is a weakly mixing system (X, T) without isolated points, and a non empty open set $U \subset X$ such that*

$$\{n \neq 0 \colon T^n U \cap U \neq \emptyset\} \subset E.$$

Before proving the proposition let us note that as an immediate consequence we have that there are pairs of weakly mixing systems without isolated points whose products are not ergodic. Indeed it is easy to find disjoint symmetric thick sets.

Proof of Proposition. (1) The points ω, ω' will be constructed in stages. For both of them we will set $\omega(i) = \omega(i) = 0$ for all $i < 0$ and $\omega(0) = \omega'(0) = 1$. In general $w(i)$ and $\omega'(i)$ will be zero – unless we specify that they equal 1. The set U will be the set of points in X whose 0-th coordinate is 1. Our job is then to make sure that ω is a recurrent point, and that the orbit of (ω, ω') will be dense in $X \times X$ while making certain that distances between the 1's that occur in ω and ω' all belong to the given set E. To keep track of things we shall treat the successive stages as points in Ω. Thus we have defined so far

$$\omega_0(i) = \omega'_0(i) = \begin{cases} 0 & , i < 0, \\ 1 & , i = 0, \\ 0 & , i > 0. \end{cases}$$

(2) Choose now some $i_1 \in E$ and begin by setting:

$$\omega_1(i) = \begin{cases} \omega_0(i) & \text{,for } i < i_1, \\ 1 & \text{,for } i = i_1. \end{cases}$$

This is to provide recurrence for ω. To get the fact that the orbit of (ω, ω') will be dense we now have to insert more copies of the U in ω and ω' in various relative shifts. Denote by $I_n = [-n, +n]$, and choose successively ℓ_j's in E so as to satisfy:

(i) $\ell_1, \ell_1 - i_1 \in E$
(ii) $\ell_2 + I_{2\ell_2} \subset E$
(iii) $\ell_3 + I_{2\ell_2} \subset E$
 ⋮

$(2i_1 + 1)\ell_{2i_1+1} + I_{2\ell_{2i_1}} \subset E$

If we put now for ℓ_j, $\omega_1(\ell_j) = 1$ and $\omega'_1(\ell_j + j - 1 - i_1) = 1$ and 0 otherwise, we will have guaranteed that any two occurrences of a 1 in either ω or ω' are separated by a distance that belongs to E, while we have in (ω, ω') all relative shifts of U – between $[-i_1, +i_1]$.

(3) For the next stage we repeat the same thing, however we work with v_r, v'_r which are the symmetric central blocks of ω_1, ω'_1 that contain all non zero elements. A new feature at this stage is that we insert into ω_2 both v_r and v'_r (one at a time) so as to ensure that ω' will be in the orbit closure of ω. The point is that all of this can be done while maintaining our restriction that $T^n U \cap U \neq \emptyset$ only for $n \in E$. When we finish constructing ω_2, ω'_2 we will have that $(v_1, \sigma^k v_1)$ will occur in (ω_2, ω'_2) for all $|k| \leq i_2$, where i_2 is the maximal i for which $\omega_1(i) \neq 0$.

Continuing this procedure we will construct ω, ω' that satisfy:

(a) ω' is in the orbit closure of ω.
(b) for all $k \in \mathbb{Z}$, (ω, σ^k) belongs to the orbit closure of (ω, ω').
(c) for any $i \neq j$ with $\omega(i) = \omega(j) = 1$, $j - i \in E$ and the same for ω'.
These three properties clearly yield the assertions of the proposition. ∎

We turn now to our positive result.

Definition 1. A system (Y, S) is *s-recurrent* if for any non empty open set U in Y the set $N(U, U) = \{n \in \mathbb{Z} : U \cap T^n U \neq \emptyset\}$ is syndetic. It is called *s-ergodic* if in addition it is ergodic.

The next theorem characterizes the multipliers of weakly mixing systems to ergodic ones:

Theorem. *A system (Y, S) satisfies:*
(*) $(X \times Y, T \times S)$ is ergodic for all weakly mixing (X, T) if and only if (Y, S) is s-ergodic.

Proof. I. Sufficiency. Using the fact that if (X, T) is weakly mixing, the sysytem $(X^{(k)}, T^{(k)})$ is ergodic for all k, it is easy to see that for any nonempty open sets U, V

$$N(U, V) = \{n : U \cap T^n V \neq \emptyset\}$$

contains arbitrarily long intervals of consecutive integers. Again, it is straightforward to see that for an s-ergodic system (Y, S) and non empty open sets A, B,

$$N_Y(A, B) = \{n : A \cap S^n B \neq \emptyset\}$$

is syndetic. It follows that there will be some n belonging to both $N(U, V)$ and $N_Y(A, B)$ and thus $(X \times Y, T \times S)$ is ergodic.

II. In the other direction, if (Y, S) fails to be s-ergodic then either it is non ergodic in which case $X \times Y$ is certainly not, or it is not s-recurrent in which case for some non empty open $A \subset Y$, $N_Y(A, A)$ contains a symmetric thick set. The previous proposition then yields a weakly mixing system (X, T) for which $(X \times Y, T \times S)$ is not ergodic.   ∎

There is another, rather interesting characterization of these s-recurrent systems which shows their significance for compact dynamical systems.

Definition 2. A compact system without isolated points (Y, S) is *contracting* if there is a non empty open set U and a sequence $n_i \to \infty$ such that $\lim_{i \to \infty} S^{n_i}(\overline{U}) = y_0$, where the limit is in the natural topology on the closed subsets of Y. Systems that are not contracting are called NC systems.

The first observation relating s-recurrence to contracting systems is:

Lemma 1. *Every compact non s-recurrent system has a nontrivial contracting factor.*

Proof. Let A be an open set in the non s-recurrent system (Y, S) such that $N(A, A)$ is not syndetic. Let ρ be a continuous function from Y to $[0, 1]$ such that for some $y_0 \in A$, $\rho(y_0) = 1$ but $\rho(y) = 0$ for all $y \notin A$. Let $\Psi : \to [0, 1]^{\mathbb{Z}}$ be defined by $\Psi(y)(n) = \rho(S^n y)$, and set $\hat{Y} = \Psi(Y)$ with $(\hat{Y}, \sigma)$ the shift restricted to $\hat{Y}$. Clearly this is a factor of (Y, S). Set $U = \{\hat{y} \in \hat{Y} : \hat{y}(0) > 0\}$, and use the fact that $N(A, A)$ is not syndetic to find a sequence $n_i \to \infty$ such that for all $|j| \leq i$, $n_i + j \notin N(A, A)$. It is easy to see that if $\hat{y} \in U$, then

$$\hat{y}(n_i + j) = 0 \quad \text{for} \quad |j| \leq i$$

and thus diam$(\sigma^{n_i}(U)$ tends to zero as i tends to infinity showing that $(\hat{Y}, \sigma)$ is contracting. ∎

On the other hand s-recurrence and contraction are not compatible. This is the content of:

Lemma 2. *Any compact s-recurrent system with no isolated points is in NC.*

Proof. Suppose to the contrary that (Y, S) is both s-recurrent and contracting. Let A be a non empty open set such that $\lim_{i \to \infty}$ diam$S^{n_i} A = 0$. Note that since Y is compact we needn't specify a metric. By passing to a subsequence we can arrange that $\lim_{i \to \infty} S^{n_i} A = y_0 \in Y$.

Take some open set $B \subset A$ such that $\overline{B} \subset A$, and use the fact that $N(B, B)$ is syndetic to find some L such that for all i, and some $0 \leq \ell_i < L$, we have $n_i + \ell_i \in N(B, B)$. Passing to a subsequence we may assume that all the $\ell_i = \ell$, and we conclude that $S^\ell y_0 \in \overline{B} \subset A$. From this we get that there is some M with $S^M A \subset A$ and $A \setminus S^M A$ contains an open set U. It follows that for all integers n, $S^{Mn} U \cap U = \emptyset$.

However this means that U cannot contain any recurrent points – which violates the s-recurrence. ∎

Putting together these two lemmas with the observation that a factor of an s-recurrent system is again s-recurrent we have:

Proposition 3. *A compact system (Y, S) is s-recurrent if and only if all factors are NC.*

2 Generic Isomorphisms of Bernoulli Shifts

The Bernoulli shifts are one of the basic examples in ergodic theory and topological dynamics and their study has given rise to a not insubstantial portion of the abstract theory. About 15 years ago, Mike Keane showed that the 2-shift and 3-shift were generically isomorphic — in stark contrast to the measure theoretical and topological situation. The main goal of this section is to present a generalization of his result to arbitrary groups and alphabets. As will be evident from the proof a much more general theorem is valid. As special cases of the more general theorem any mixing shift of finite type will be seen to be generically isomorphic to the 2-shift. Here we refer to $\mathbb{Z}$-actions, already for $\mathbb{Z}^2$, the mixing condition for a system of finite type is rather weak and is consistent with rather subtle long range correlations.

For simplicity we shall first describe the proof for $\mathbb{Z}^d$, and

$$(X = \{0,1\}^{\mathbb{Z}^d}, T_a), \ (Y = \{0,1,2\}^{\mathbb{Z}^d}, S_a)$$

the 2-shift and 3-shift respectively over $\mathbb{Z}^d$. For the proof we shall construct two points, $\bar{x} \in X, \bar{y} \in Y$ each with dense orbits under the shift action, and we will define a map ϕ between their orbits by $\phi(T_a\bar{x}) = S_a\bar{y}, a \in \mathbb{Z}^d$.

Let $S_k = \{a \in \mathbb{Z}^d : |a_i| \le k, \ 1 \le i \le d\}$ be the centered square of size $(2k+1)^d$ and for some N_k that tends to infinity (and will be specified below) define

$$U_k = \{x \in X : \exists \bar{a} \in S_k \text{ such that } \forall b \in S_{N_k} \ \ x(b) = \bar{x}(\bar{a}+b)\}.$$

This is an open set and for all m, $\bigcup_{k=m}^{\infty} U_k$ is an open dense set in X if the orbit of $\bar{x}$ is dense. The dense G_δ on which ϕ will extend to a continuous mapping will be

$$X_0 = \bigcap_{m=1}^{\infty} \bigcup_{k=m}^{\infty} U_k.$$

For Y, Y_0 will be defined in an analogous fashion:

$$V_k = \{y \in Y : \exists \bar{a} \in S_k \text{ such that } y(b) = y(\bar{a}+b) \text{ for all } b \in S_{N_k}\}$$

$$Y_0 = \bigcap_{m=1}^{\infty} \bigcup_{k=m}^{\infty} V_k.$$

Step 1. Let N_1 be sufficiently large so that we can place in $S_{N_1}, 3^{|S_1|}$ disjoint translates of S_1, each separated from the other by a distance of at least 10, and also separated from the boundary of S_{N_1} by a distance at least 10. Define then $\bar{x}|_{S_{N_1}}$ by putting $\bar{x}(a)$ equal to 0 at all $a \in S_{N_1}$ that are not in the translates of S_1 mentioned above, and not within distance 5 of the boundary of S_{N_1}. For $a \in S_{N_1}$ within distance 5 from the boundary put $\bar{x}(a) = 1$, and for $3^{|S_1|}$ translates of S_1 define $\bar{x}(a)$ in such a way that each of the elements of $\{0,1\}^{S_1}$ occurs at least once. There are two important properties of $\bar{x}|_{S_{N_1}}$:

I. Every element of $\{0,1\}^{S_1}$ occurs at least once.

II. The only way $\bar{x}|_{S_{N_1}}$ can overlap a translate of itself is along the 5-boundary, i.e., if in some $x \in X$ there are two positions b and c where one sees $\bar{x}|_{S_{N_1}}$, then the 5-interiors are disjoint. Here we are using the natural language "seeing $\bar{x}|_{S_{N_1}}$ at b in x" for the equalities

$$\bar{x}(a) = x(b+a), \quad \text{all } a \in S_{N_1}.$$

For $\bar{y}|_{S_{N_1}}$ we use the same pattern of translates of S_1, the same "background" of 0's, and 1's along the boundary, and fill in the translates of S_1 with all elements of $\{0,1,2\}^{S_1}$.

Step k. For the kth step we will use S_{N_k}, with V_k chosen to be sufficiently large so that it can accommodate $2 \cdot 3^{|S_k|}$ disjoint translates of S_k disjoint from the central $S_{N_{k-1}}$ and separated by a distance of at least $10 \cdot N_{k-1}$. The boundary of S_{N_k} to a depth of $5 \cdot N_{k-1}$ will be filled with 1's, and the background, between the central $S_{N_{k-1}}$, and all the translates of S_k will be filled with 0's. Then for $\bar{x}|_{S_{N_k} \backslash S_{N_{k-1}}}$ we will fill in half of the translates of the S_k's so that each element of $\{0,1\}^{S_k}$ appears at least once.

For the definition of $\bar{y}|_{S_{N_k} \backslash S_{N_{k-1}}}$ a little bit of care is required. We must now make certain that whenever the pattern corresponding to some $\bar{x}|_{S_{N_j}}$, for $j < k$ appears in an element of $\{0,1\}^{S_k}$, then in the corresponding place in $\bar{y}|_{S_{N_k}}$ will be found $\bar{y}|_{S_{N_j}}$. The fact that these patterns occur disjointly for the same j (or possibly entirely in the interior for $j' \neq j$) ensures that this is possible. We also take care that no $\bar{y}|_{S_{N_j}}$ is filled in that doesn't correspond to a $\bar{x}|_{S_{N_j}}$.

To ensure that $\bar{y}$ has a dense orbit, in the second half of the translates of the S_k we proceed to first fill in one copy of each element of $\{0,1,2\}^{S_k}$, and then in the corresponding positions in $\bar{x}|_{S_{N_k}}$ we make sure to put $\bar{x}|_{S_{N_j}}$ in a place corresponding to $\bar{y}|_{S_{N_j}}$, and only there.

As a result of the above we have defined $\bar{x}|_{S_{N_k}}$, $\bar{y}|_{S_{N_k}}$ in such a way that each one contains a copy of all of the S_k-blocks in its system, and we also have guaranteed that the previous correspondences of $\bar{x}|_{S_{N_j}}$, $j < k$ with $\bar{y}|_{S_{N_j}}$ have been maintained in the following sense:

(i) If for some $c \in S_{N_k}$, $j < k$, $S_{N_j} + c \subset S_{N_k}$ and $\bar{x}(a + c) = \bar{x}(a)$, all $a \in S_{N_j}$, then we also have $\bar{y}(a + c) = \bar{y}(a)$ for all $a \in S_{N_j}$;

(ii) Conversely, if for some $c \in S_{N_k}, j < k$, $S_{N_j} + c \subset S_{N_k}$ and $\bar{y}(a + c) = \bar{y}(a)$ for all $a \in S_{N_j}$ then we also have $\bar{x}(a + c) = \bar{x}(a)$, for all $a \in S_{N_j}$.

We claim now that the correspondence ϕ between the translates of $\bar{x}$ and $\bar{y}$ defines a homeomorphism between X_0 and Y_0 that is equivariant with respect to the respective shifts. This claim is a direct consequence of the definitions of X_0, Y_0 and the properties of $\bar{x}$ and $\bar{y}$ that have been spelled out.

Indeed the property (i) above guarantees that if *any* translate of $\bar{x}$ lies in U_j, say $T_a \bar{x} \in U_j$, then one also has $S_a \bar{y} \in V_j$. This shows that ϕ is continuous from X_0 to Y_0, while (ii) shows in analogous fashion that ϕ^{-1} is well defined and continuous from Y_0 to X_0. The result of this discussion is the following:

Theorem 1. *The 2-shift and 3-shift over $\mathbb{Z}^d$ are generically isomorphic.*

It is an easy exercise to generalize this construction to the Bernoulli shifts over arbitrary alphabets. This includes also the case of any 0-dimensional space as an alphabet — here as the size of the window (the N_k of S_{N_k}) increases one also increases the resolution. This allows the above construction to be carried out in a finite fashion at each step while preseving the same features that enabled the proof of Theorem 1 to work.

The only property of the full shift that is really being used is the specification property — which allows us to specify arbitrary allowable blocks in certain positions which have a minimum separation. Without entering into a detailed discussion the interested reader can easily generalize the above, for example, to mixing Markov shifts over $\mathbb{Z}$.

Finally, we will give a brief discussion of the case of a general countable group G. In case G is finitely generated we can look at the Cayley graph of G and define $S_k(G)$ to be the ball of radius k around the identity in G. In this case the discussion above can be copied verbatim, bearing in mind that the distance is now being measured in terms of the Cayley graph. If G is not finitely generated, then we simply describe an increasing sequence of finite sets $S_1 \subset S_2 \subset \ldots$ with $\bigcup S_k = G$ and choose S_{N_k} large enough so that it contains enough translates of S_k that are well separated compared to $S_{N_{k-1}}$. The mere fact that G is an infinite group and each S_k is finite is enough to ensure that such a choice is possible. This enables one to establish for example:

Theorem 2. *If G is a countable group and A, B are two 0-dimensional spaces of cardinality at least 2, then (A^G, shift) and (B^G, shift) are generically isomorphic.*

3 Hyperfiniteness in Generic Dynamics

Let Γ be a countable group acting continuously on a Polish space X. An open set $U \subset X$ will be called an *A-set*, $A \subset \Gamma$, if for any two distinct elements a_1, a_2 of A the sets $a_1 U$ and $a_2 U$ are disjoint. As usual, R_Γ denotes the equivalence relation defined by Γ, so that $x R_\Gamma y$ means $\gamma x = y$ for some $\gamma \in \Gamma$. A finite subequivalence relation $K \subset R_\Gamma$ will be called *regular* if for some finite collection of finite subsets of Γ, $A_1, \ldots, A_k$ there are sets U_i which are A_i-sets satisfying:

(i) $A_i U_i \cap A_j U_j = \emptyset$, for $i \neq j$.

(ii) $\bigcup\limits_{i=1}^{k} A_i U_i$ is dense in X.

(iii) The equivalence classes of R restricted to $\bigcup\limits_{i=1}^{k} A_i U_i$ are precisely the fibers $\{A_i x_i\}$ for $x_i \in U_i$, $1 \leq i \leq k$.

If there is an increasing sequence of regular finite sub-equivalence relations $R_1 \subset R_2 \subset \cdots \subset R_\Gamma$ such that $\bigcup_1^\infty R_n$ exhausts R_Γ on a dense G_δ in X we say that R_Γ is *generically hyperfinite*. It is straightforward to see that if R_Γ is generically hyperfinite, then on a dense G_δ set $X_0 \subset X$ one can define a single continuous mapping T so that on X_0, $R_\Gamma = R_T$, where as usual $x R_T y$

means $T^m x = y$ for some $m \in \mathbb{Z}$.

Theorem. *If Γ acts continuously on X, and there is an $x_0 \in X$ with $\gamma x_0 \neq x_0$ for $\gamma \in \Gamma \setminus \{e\}$ and Γx_0 dense in X, then R_Γ is generically hyperfinite.*

Proof. (1) Construct in Γ an increasing sequence of finite subsets $e \in A_1 \subset A_2 \subset \ldots$ such that if $A_n^k \subset A_n$ is defined for $k < n$ by

$$A_n^k = \{a \in A_n : A_k a \subset A_n\},$$

i.e. the "interior" of A_n relative to A_k, then for all $k = 1, 2, \ldots$

$$\bigcup_{n > k} A_n^k = \Gamma.$$

This is easily done for example by defining inductively

$$A_n = \bigcup_{j=0}^{n} A_{n-1} \gamma_j$$

where $e = \gamma_0, \gamma_1, \ldots$ is some enumeration of the elements of Γ and A_0 is understood to be $\{\gamma_0\}$.

(2) Since $\gamma x_0 \neq x_0$ for all $\gamma \in \Gamma$, the fact that Γ acts continuously implies that for any finite set $A \subset \Gamma$ there is a nbhd of x_0 that is an A-set. Begin by letting U_1 be a nbhd of x_0 that is an A_1-set and in addition satisfies

$(*)_1$ $\hspace{3cm} A_1(\partial U_1) \cap \Gamma x_0 = \emptyset.$

Define R_1 to be the regular finite subequivalence relation whose equivalence classes are:

(i) $\{A_1 x\}$, for $x \in U_1$,
(ii) singletons for $x \notin A_1 U_1$.

(3) Define $B_2 \supset A_2$ minimally so that $B_2 x_0$ consists of complete R_1-equivalence classes. Explicitly whenever for some $a \in A_2$, $\hat{a} x_0 \in \hat{A}_1 U_1$, say $a x_0 \in h U_1$, add to A_2 all elements of the form $g h^{-1} a$, $g \in A_1$, and in this way form B_2 which is clearly finite. Because of $(*)$ there is a nbhd V of x_0 so that for all $b \in B_2$, bV either lies entirely in some $h U_1, h \in A_1$, or $bV \cap \overline{A_1 U_1} = \emptyset$. Find now a B_2-set, $U_2 \subset V$ such that

$(*)_2$ $\hspace{3cm} B_2(\partial U_2) \cap \Gamma x_0 = \emptyset$

and define the finite subequivalence relation R_2 to consist of the following equivalence classes:

(i) $\{B_2 x\}$ for $x \in U_2$,
(ii) $\{A_1 x\}$ for $x \in U_1 \setminus B_2 U_2$,

(iii) singletons for $x \notin A_1 U_1 \cup A_2 U_2$.

(4) In general we will define finite sets $B_n \supset A_n \cup B_{n-1}$ $(B_1 = A_1)$ and B_n-sets, U_n, that contain x_0 and satisfy:

(i) for $b_i \in B_i$, $b_j \in B_j$, $i < j \leq n$, either $b_i U_i \cap b_j U_j = \emptyset$ or $b_i U_i \supset b_j U_j$,

(ii) if $i < j \leq n$ and $b_j \in B_j$, $b_j x_0 \in b_i U_i$, then $B_i b_i^{-1} b_j \subset B_j$,

(iii) $\partial U_i \cap \Gamma x_0$, $i \leq n$.

The equivalence relations R_n will be defined inductively by $x R_n y$ iff for some $i \leq n$, $\exists u_i \in U_i$ $a_i, b_i \in B_i$ with $a_i u_i = 2, b + i u_i = y$, and

(iv) with this definition R_n is a regular finite subequivalence relation whose equivalence classes are all of the form $B_i u_i$ for some $u_i \in U_i$, $i \leq n$.

We have shown how to construct R_1 and B_2 and R_2 and it is clear that they satisfy the above. Suppose now we have (i)–(iv) up to n, we have to see how to continue.

(5) Define B_{n+1} by

$$B_{n+1} = A_{n+1} \cup \{B_i b_i^{-1} a_{n+1} : i \leq n, a_{n+1} x_0 \in b_i U_i, a_{n+1} \in A_{n+1}\}.$$

Let us check that (ii)$_{n+1}$ is satisfied. Suppose then that for some $b_{n+1} \in B_{n+1}, b_{n+1} x_0 \in b_j U_j$, with $j \leq n$. The general form of b_{n+1} is $c_i b_i^{-1} a_{n+1}$, where $a_{n+1} \in A_{n+1}, a_{n+1} x_0 \in b_i U_i$ and $b_i, c_i \in B_i$, $i \leq n$. We distinguish three cases.

(1) $i = j$. Here, since U_i is a B_i set, we must have $c_i = b_j$, and then the result follows immediately from the definition of B_{n+1}.

(2) $i < j$. We have $c_i b_i^{-1} a_{n+1} x_0 \in b_j U_j$, thus $a_{n+1} x_0 \in b_i c_i^{-1} b_j U_j$. Since (i) holds we have $b_j U_j \subset c_i U_i$ and in particular $b_j x_0 \in c_i U_i$. Now by (ii)$_j$, $b_i c_i^{-1} b_j \in B_j$, and thus by the definition of B_{n+1}, $B_j (b_i c_i^{-1} b_j)^{-1} a_{n+1} \subset B_{n+1}$ or $B_j b_j^{-1} (c_i b_i^{-1} a_{n+1}) \subset B_{n+1}$ as required.

(3) $i > j$. Now we get $c_i U_i \subset b_j U_j$, whence $c_i x_0 \in b_j U_j$ and thus $B_j b_j^{-1} c_i \subset B_i$, and since $B_i b_i^{-1} a_{n+1} \subset B_{n+1}$ by construction, we get $(B_j b_j^{-1} c_i)(b_i^{-1} a_{n+1}) \subset B_{n+1}$ or $B_j b_j^{-1} (c_i b_i^{-1} a_{n+1}) \subset B_{n+1}$, as required.

(6) Once B_{n+1} has been defined satisfying (ii)$_{n+1}$, we use (iii)$_n$ to find a nbhd of x_0, U_{n+1}, that is a B_{n+1}-set, and satisfies (i)$_{n+1}$ and (iii)$_{n+1}$. This is all an immediate consequence of the continuity of the action of Γ and of the fact that $\gamma x_0 \neq x_0$ for all $\gamma \in \Gamma \setminus \{e\}$.

Next we define R_{n+1} by the recipe preceding (iv) and all that is left to check is that (iv)$_{n+1}$ is satisfied. It suffices to consider the following situation: For some $u_{n+1} \in U_{n+1}$, $b, c \in B_{n+1}$ we have

$$x = b u_{n+1}, \quad y = c u_{n+1}$$

and for some $u_i \in U_i$, $i \leq n$, $g, h \in B_i$, we have

$$y = g u_i, \quad z = h u_i.$$

Our task is to find a $d \in B_{n+1}$ such that $z = du_{n+1}$, which will prove the transitivity of the equivalence relation. Now since $cu_{n+1} = gu_i$ we have by $(i)_{n+1}$ that $cU_{n+1} \subset gU_i$, and in particular $cx_0 \in gU_i$. Thus by $(ii)_{n+1}$ we have

$$hg^{-1}c \in B_{n+1}, \text{ and } hg^{-1}cu_{n+1} = hg^{-1}y = hg^{-1}gu_i = hu_i = z,$$

hence setting $d = hg^{-1}c$ we have what is required.

We have thus shown how to construct B_n, U_n and R_n for all n satisfying (i)–(iv).

(7) Define now

$$Y_k = \bigcup_{n>k} A_n^k U_n$$

and set $Y = \bigcap_1^\infty Y_k$. By (1), and the fact that $x_0 \in U_n$ for all n, we see that $Y_k \supset \Gamma x_0$ and thus is a dense open set. To complete the proof, we must simply show that on Y, $\bigcup_1^\infty R_n = R_\Gamma$. It follows from the construction that for any $x \in Y_k$, and $a_k \in A_k$, $xR_j a_k x$ for some $j > k$. Indeed, suppose $x \in A_j^k U_j$, then $x = a_j u_j$ for some $a_j \in A_j^k, u_j \in U_j$, but then $a_k a_j \in A_j$ by the definition of A_j^k, and $a_k x R_j x$ by the definition of R_j. Hence on $Y = \bigcap Y_k$ we have that for any $a \in \Gamma = \bigcup_1^\infty A_k$, $xR_j ax$ for some j, thus on Y, $\bigcup_1^\infty R_n = R_\Gamma$ which completes the proof of the theorem.

4   Equivariant Projections

As is well known, generic properties and typical properties in the sense of measure, do not always go hand in hand. The relation between the two is very nicely described in the monograph by J. Oxtoby [O]. For a simple example of how they can differ, consider the law of large numbers for Bernoulli random variables X_i that take values $0, 1$ with equal probability. The sample space can be identified with the unit interval equipped with Lebesgue measure, and the X_i becomes the ith digit in the binary expansion of a number $t \in [0, 1]$. E. Borel proved that those t's for which the limiting frequency is equal to $1/2$ form a set of Lebesgue measure 1. On the other hand, the set of those points t with any limiting frequency at all is a set of first category, as is easily seen. In other words it is a generic property that no limiting frequency exists at all!

This section is devoted to an interesting open problem concerning the Baire measurable functions, that arose in the attempt made by D. Sullivan and me to answer Szpilrajn's problem on the uniqueness of Lebesgue measure on S^2 as a finitely additive rotation invariant measure on the Borel mod meager sets. That problem was solved a few years ago by R. Dougherty and M. Foreman [DF] by a direct construction. Our program was rather indirect and the hyperfiniteness result in generic dynamics was originally proved as a big step in our program. Without going into details, the idea was to fix a countable free group of rotations acting on S^2 and then define a certain bounded cohomology of the resulting equivalence relation, using Baire measurable functions. The existence of an invariant measure would give a projection to constants and thus we could conclude that the cohomology was non trivial since the free group has a non trivial bounded cohomology. On the other hand, the hyperfiniteness of the equivalence relation would prove that the bounded cohomology is trivial – already at the level of the Baire measurable functions. The snag developed in establishing an equivariant projection of sequences of Baire measurable functions — and this is the open problem that I would like to explain now.

For comparison, I shall begin with the story for bounded Lebesgue measurable functions. Let S denote the space $\ell_\infty(L_\infty([0,1], \text{Lebesgue}))$, i.e. an element of S is a sequence $f = \{f_n\}_1^\infty$ of L_∞ functions on $[0,1]$ with $\|f\| = \sup_n \|f_n\|_\infty < +\infty$. A linear projection $\pi \colon S \to L_\infty$ will be called *natural* if it satisfies the following:

(i) $\pi(\{f_n\}_1^\infty) = f_1$, if $f_1 = f_2 = f_3 = \cdots$,
(ii) if all $f_n \geq 0$, then also $\pi(\{f_n\}) \geq 0$,
(iii) $\pi(\{f_{n+1}\}_1^\infty) = \pi(\{f_n\}_1^\infty)$,
(iv) $\varliminf f_n \leq \pi(\{f_n\}) \leq \varlimsup_n f_n$.

Even though (iv) follows from the others, we have indicated it separately to emphasize its importance.

One way to achieve all of this is to take any Banach limit LIM defined on bounded sequences and apply it pointwise to $\{f_n(t)\}$, i.e. form: $\mathop{\text{LIM}}\limits_{n \to \infty} f_n(t)$. Needless to say there is no guarantee that such an expression is Lebesgue measurable. It was established by Mokobodski that there are Banach limits LIM with exactly the required property. He called such limits *medial* ([MO]).

This method of constructing a natural projection gives an additional property – namely:

(v) π is equivariant with respect to the action of any countable group Γ that preserves the Lebesgue measure.

The meaning of this condition is that if a new sequence $\{f_n \circ \gamma\}$ is formed by composing with $\gamma \in \Gamma$, then $\pi(\{f_n \circ \gamma\}) = (\pi\{f_n\}) \circ \gamma$. It is this equivariance that was for us the key property.

Turning to the corresponding question for Baire measurable functions, it was shown by Sierpinski that for *no* pointwise Banach limit is it the case that $\text{LIM} f_n(t)$ will always be a Baire measurable function. In fact, if $f_n(t)$ gives the n-th binary digit in the binary expansion of t then in essence such a pointwise limit would give a finitely additive measure in the subsets of $\mathbb{N}$ – identified as points in $\{0,1\}^{\mathbb{N}}$ which vanishes on finite sets and is Baire measurable and he showed that such measures cannot exist.

To see this let $\varphi \colon \{0,1\}^{\mathbb{N}} \to [0,1]$ be such a measure, and let $\Gamma = \sum\limits_{1}^{\infty} \mathbb{Z}/\mathbb{Z}_2$ be the countable sum of two element groups that acts in a natural way on $\{0,1\}^{\mathbb{N}}$ by coordinatewise addition. The fact that the measure vanishes on finite sets means that φ is invariant under Γ, thus each

$$A_i = \varphi^{-1}\left(\frac{i-1}{5}, \frac{i}{5}\right), \quad 1 \leq i \leq 5$$

is Γ-invariant and has the Baire property. Thus exactly one of them, say A, must differ from a dense open set in $\{0,1\}^{\mathbb{N}}$ by a meager set. Denote by θ the flip – i.e. $\theta(x)_i = \bar{x}_i$, with $\bar{0} = 1, \bar{1} = 0$. Since θ is continuous and A is residual, $A \cap \theta(A)$ is also residual and thus A must equal A_3.

Consider now the mapping $(x,y) \mapsto x \cdot y$ defined by coordinatewise multiplication. It is continuous and open and thus

$$\{(x,y) \colon x \in A, x \cdot y \in A\}$$

is residual in $X \times X$, and the same for the mapping $x - x \cdot y$, thus we find $x \in A$ and $y \in X$ with both $x \cdot y \in A$ and $x - x \cdot y \in A$. Setting $z = x \cdot y$ and $w = x - x \cdot y$, we get three elements, x, z, w all in A with z and w disjoint and $x = z + w$, whence $\theta(x) = \theta(z) + \theta(w)$, which clearly cannot happen for three elements in A_3.

This simple argument shows that medial limits do not exist for Baire measurable functions. Nonetheless, a natural projection does exist on bounded sequences of bounded Baire functions that is natural in the sense of (i) – (iv) above. One can construct such projections by imitating Banach's original proof of the existence of a Banach limit on bounded real sequences.

Here is an outline of a proof using the continuum hypothesis. Denote by B_∞ bounded sequences of Borel mod meager measurable functions and consider it is a vector space over $\mathbb{Q}$. Suppose that V_0 is a countable subspace of B_∞ for which we already have defined a natural projection π_0 onto B – the bounded Borel mod meager functions. If we wish to extend π_0 to $V_0 + \xi, \xi = \{x_i\}^\infty$, then we must satisfy, for any $\eta \in V_0$, $r, s \in \mathbb{Q}, \eta = \{y_i\}^\infty$,

$$\pi_0(r\eta + s\xi) \leq \varlimsup_{i \to \infty} (ry_i + sx_i),$$

and for any $\zeta \in V_0, \zeta = \{z_i\}$ also

$$\varliminf(rz_i + sx_i) \leq \pi_0(r\xi + s\xi).$$

These easily reduce to the case $r = 1, s = 1$, and thus we can define $\pi_0(\xi)$ to satisfy all of these constraints if

$$\varliminf(z_i + x_1) - \pi_0(\{z_i\}) \leq \varlimsup(y_i + x_i) - \pi_0(\{y_i\}),$$

which holds, since, by assumption, on V_0

$$\pi_0(\{y_i - z_i\}) \leq \overline{\lim}\{y_i - z_i\} = \overline{\lim}\{(y_i + x_i)\}$$
$$\leq \overline{\lim}(y_i + x_i) - \underline{\lim}(z + i + x_i).$$

We can define, therefore, $\pi(\xi)$ as the infimum over all $\eta \in V_0$ of $\overline{\lim}(y_1 + x_i) - \pi_0(\eta)$ and this extends π_0 to $V_0 + \mathbb{Q}\xi$.

The shift invariance (iii), which is not explicitly present in this construction, is easy to get by applying a projection operator to the averages of a sequence $\frac{1}{n}\sum_1^n x_i$. We can now formulate our open problem:

Problem. For the space $\{0,1\}^{\mathbb{N}}$ with the action of $\Gamma = \sum_1^{\infty} \mathbb{Z}/2\mathbb{Z}$ described above, find a natural projection from bounded sequences of Borel mod meager functions that is Γ-*equivariant*.

To recapitulate: We know that no pointwise limit can be used – the Γ equivariance captures something of this locality. One last observation is that, of course, Γ is an increasing union of finite groups Γ_n in a natural way. It is easy to show that for any fixed n, there is a Γ_n-equivariant projection – but of course these projections need not converge.

References

[DF] R. Dougherty and M. Foreman, Banach-Tarski decomposition using sets with the property of Baire, *J. Amer. Math. Soc.* **7(1)** (1994), 75–124.

[F] H. Furstenberg, Disjointness in ergodic theory, minimal sets, and a problem in Diophantine approximation, *Math. Systems Theory*, **1** (1967), 1–49.

[GW] E. Glasner and B. Weiss, Sensitive dependence on initial conditions, *Nonlinearity* **6** (1993), 1067–1075.

[Ke] A. Kechris, *Classical Descriptive Set Theory*, Springer, 1995.

[Ku] K. Kuratowski, *Topology*, Vol I., Academic Press, New York 1966.

[Mo] G. Mokobodski, Limites Mediales, (Exp. de P.A. Meyer), *Sem. Probabilites, Univ. de Strasbourg, 1971–2*, LNM 321, pp. 198–204.

[O] J. Oxtoby, *Measure and Category*, Springer, 1980.

[SWW] D. Sullivan, B. Weiss and J.D. Maitland Wright, Generic dynamics and monotone complete C^*-algebras, *Trans. Amer. Math. Soc.* **295** (1986), 795–809.